Introduction to Mathematical Modeling and Chaotic Dynamics

Introduction to Mathematical Modeling and Chaotic Dynamics

Ranjit Kumar Upadhyay

Indian School of Mines
Dhanbad, India

Satteluri R. K. Iyengar

Gokaraju Rangaraju Institute of Engineering & Technology
Hyderabad, India

CRC Press
Taylor & Francis Group
Boca Raton London New York

CRC Press is an imprint of the
Taylor & Francis Group, an **informa** business

A CHAPMAN & HALL BOOK

CRC Press
Taylor & Francis Group
6000 Broken Sound Parkway NW, Suite 300
Boca Raton, FL 33487-2742

First issued in paperback 2019

© 2014 by Taylor & Francis Group, LLC
CRC Press is an imprint of Taylor & Francis Group, an Informa business

No claim to original U.S. Government works

ISBN-13: 978-1-4398-9886-4 (hbk)
ISBN-13: 978-0-367-37976-6 (pbk)

Library of Congress Cataloging-in-Publication Data

Upadhyay, Ranjit Kumar, author.
  Introduction to mathematical modeling and chaotic dynamics / Ranjit Kumar Upadhyay and Satteluri R.K. Iyengar.
      pages cm
  Includes bibliographical references and index.
  ISBN 978-1-4398-9886-4 (hardback)
  1. Differential equations--Textbooks. 2. Mathematical models--Textbooks. I. Iyengar, S. R. K. (Satteluri R. K.), 1941- author. II. Title.

QA371.U675 2014
515'.35--dc23   2013014550

Visit the Taylor & Francis Web site at
http://www.taylorandfrancis.com

and the CRC Press Web site at
http://www.crcpress.com

Contents

Preface

We attempted to write a book that can be used as a textbook for undergraduate as well as for postgraduate students. The book stresses mathematical models in natural systems, particularly ecological systems. In Chapter 1, an introduction to mathematical modeling and mathematical preliminaries including testing of stability is presented. Most of the models discussed in the book are solved by using MATLAB®. Chapter 2 deals with the modeling of systems from natural science. One- and two-dimensional continuous and discrete time models are discussed in this chapter. Chapter 3 presents an introduction to chaotic dynamics. The routes to study chaos, types of chaos, and methods of investigation for detecting chaos are discussed in this chapter. *Mathematica*® is used in the computation of Lyapunov exponents in this chapter. Chaotic dynamics in model systems from natural science is discussed in Chapter 4. Single and multiple species systems are considered in this chapter. Finally, a brief discussion on models in mechanical systems (oscillators) and electronic circuits (Chua's circuit and others) is provided in Chapter 5.

We thank all the reviewers for their valuable suggestions, which gave us an impetus in writing this book.

We thank all the experts in various fields. Their works have been cross-referred to in our book. We are grateful to the following experts who granted us permission to use some of the figures appearing in their publications: Professor Dr. Arnold Neumaier, Professor Dr. Arunas Tamasevicius, Professor Dr. Farhad Khellat, Professor Dr. J. C. Sprott, Professor Dr. Leon O. Chua, Professor Dr. M. Lakshmanan, and Professor Dr. Tomasz Kapitaniak.

We gratefully acknowledge the following publishers who have granted copyright permission to use some figures from their reputed international journals: Cambridge University Press (Cambridge, UK), EDP Sciences (Cambridge University Press, UK), Elsevier Limited, Hindawi Publishing Corporation (editor: Professor Dr. Amany Qassem), Hirzel-Verlag (Stuttgart), IOP Publishing Ltd, IOS Press (editor in chief: Professor Dr. Theodore Elias Simos), Lithuanian Association of Nonlinear Analysts (secretary: Professor Dr. Romas Baronas), New Age International (P) Ltd, Society for Industrial and Applied Mathematics (SIAM), Springer Science, the American Physical Society Publishing, and the Taylor & Francis Group.

We thank Professor Dr. Aziz-Alaoui for drawing our attention to an interesting application of logistic maps.

Our special thanks to Professor Dr. Kazuyuki Aihara, director, Collaborative Research Centre for Innovative Mathematical Modeling, and professor of Institute of Industrial Science, University of Tokyo, Japan. Dr. Kazuyuki Aihara and his colleagues have been creating innovative practical applications of bifurcation and chaos theory in various fields. We are very grateful to him for granting us permission to include the special fashion dresses created by him and his colleagues (Eri Matsui, Keiko Kimoto, and Kazuyuki). One of the dresses was presented at a famous fashion show "Tokyo Collection" held in Japan in March 2010. For further details of Dr. Kazuyuki Aihara's work, visit the web site http://www.sat.t.u-tokyo.ac.jp/center.

It is only the readers who are indicators of whether we have succeeded in our endeavor to present both the fundamentals of the subject and recent literature particularly from the field of mathematical ecology in a simple and lucid manner.

We profusely thank Taylor & Francis for accepting to publish the book and bringing it to a presentable form.

We shall be extremely happy to receive suggestions and comments to improve the quality of the book.

MATLAB® and Simulink® are registered trademarks of The MathWorks, Inc. For product information, please contact:

The MathWorks, Inc.
3 Apple Hill Drive
Natick, MA 01760-2098 USA
Tel: 508 647 7000
Fax: 508-647-7001
E-mail: info@mathworks.com
Web: www.mathworks.com

Authors

Dr. Ranjit Kumar Upadhyay is a professor in the Department of Applied Mathematics, Indian School of Mines (ISM), Dhanbad, India. He earned his PhD under the supervision of Professor Satteluri R. K. Iyengar, Department of Mathematics, Indian Institute of Technology (IIT), New Delhi, India. He has been teaching applied mathematics and mathematical modeling courses for more than 16 years. His research areas are interdisciplinary in nature and include chaotic dynamics of real-world situations, population dynamics for marine and terrestrial ecosystems, disease dynamics, reaction–diffusion modeling, environmental modeling, differential equations, and dynamical systems theory. He has published many research papers in international journals of high repute and also with international collaborators. He has been on the editorial board for many reputed journals. He is a guest editor of a special issue titled "Nonlinear Phenomena in Biology and Medicine" being brought out by *Computational and Mathematical Methods in Medicine.* He has written an invited review article for the *International Journal of Bifurcation and Chaos.* He was a visiting research fellow under the Indo-Hungarian Educational exchange program in Eötvös University, Budapest, Hungary. He is a member of the International Society of Computational Ecology, Hong Kong and the American Mathematical Society.

Dr. Satteluri R. K. Iyengar was a professor and former head of the Department of Mathematics, Indian Institute of Technology (IIT), New Delhi, India. He worked as a professor for more than 22 years. His areas of research work are numerical analysis and mathematical modeling. He is a joint author of a number of books on numerical analysis and advanced engineering mathematics. He has many research publications in international journals of repute. He was a postdoctoral fellow at Oxford University Computing Laboratory, Oxford, United Kingdom, and also at the University of Saskatchewan, Canada. He was awarded the Distinguished Service Award by the Indian Institute of Technology, New

Delhi, India during the golden jubilee year (2010–2011). He was also awarded the Distinguished Indian Award in 2007 by the Pentagram Research Center (P) Limited, Hyderabad, India for his contributions. After retiring from IIT Delhi, he now works as dean (academic affairs) and professor of mathematics at Gokaraju Rangaraju Institute of Engineering & Technology, Hyderabad, India.

1

Introduction to Mathematical Modeling

Introduction

Over the last several decades, mathematical modeling has been playing a major role in understanding and solving many real-life problems, under certain conditions. Most mathematical models have been like individual works of art that reflected the scientific views and personal characteristics of the modeler. However, many attempts are being made to unify the mathematical models in order to provide a standardized and reliable method of investigation accessible for every scientist. Modeling is a multistep process involving the following:

   i. Identifying the problem

   ii. Constructing or selecting the appropriate model

   iii. Figuring out what data need to be collected

   iv. Deciding the number of variables and predictors to be chosen for greater accuracy

   v. Analytically or numerically computing the solution and testing the validity of models

   vi. Implementing the models in real-world situations

Usually, modeling is an iterative process in which we start from a crude model and gradually refine it until it is suitable for solving the problem, and modeling enables us to gain insight into the original situation. The purpose of the model is to understand the underlying phenomenon and perhaps to make predictions about future behavior. If the predictions do not compare well with reality, we need to refine our model or formulate a new model and start the cycle again.

What Is Mathematical Modeling?

Mathematical modeling is a discipline that attempts to describe real-world phenomena in mathematical terms and then solves them. Mathematical models were constructed and analyzed by different people for different reasons. The foremost reason is that direct experimental evidence cannot be obtained by simulating a real-life situation. For example, the spread of a communicable disease or the outcome of a space program (e.g., trajectory planning, flight simulation, and shuttle re-entry) cannot be decided based on experiments or field studies alone. There may be situations where certain conclusions have to be drawn before the problem itself can be completely formulated.

A real-world problem is modeled using mathematical equations to describe the process with the help of a suitable number of variables, parameters, and so on. The methodology consists of carefully formulating the definitions of the concepts to be discussed and explicitly stating the assumptions that shall be the basis for the reasoning employed. The actual problem is formulated by using these definitions and assumptions, whereas the conclusions are drawn by employing rigorous logic, based upon observations derived from the mathematical analysis. Thus, mathematical modeling is a process involving transformation of a physical situation into mathematical analogies with appropriate conditions and the solving and study of the problem in almost every aspect of its development. Modeling primarily deals with the study of the characteristics governing the observable and operating features. A model is supposed to be a prototype of the system under investigation. In mathematical modeling, we neither perform any practical activity nor interact with the actual situation directly. For example, we do not take any sample of blood from the body to know about the physiology. The rapid development of high-speed computers with the increasing desire to find some answers to everyday life problems has led to the enhanced demand for modeling in almost every area.

It is now a well-accepted fact that the flow of knowledge in science and technology depends to a great extent on the development of advanced mathematical tools. New mathematical techniques are being proposed and are being successfully applied to explore a variety of interesting topics in many application areas. A few of these topics and some application areas include (see Neumaier [21] for a comprehensive list): computer science (image processing and computer graphics), engineering (microchip analysis, power supply network optimization, planning of production units, stability of high-rise buildings, bridges and airplanes, and crash simulation), medicine (radiation therapy planning, blood circulation models, and computer-aided tomography), meteorology (weather prediction and climate prediction), biology (protein folding and human genome project), psychology (formalizing diaries of therapy sessions), chemistry (chemical reaction dynamics and molecular modeling), physics (laser dynamics and quantum field theory predictions), economics (data analysis), finance (risk analysis and value estimation of options), social sciences (election voting patterns), and so on.

Most real-world problems are highly nonlinear and a large number of them can be modeled in the form of a system of nonlinear ordinary or partial differential equations. Computer simulations of such mathematical models are being used extensively to solve such problems.

The process of mathematical modeling can be divided into three major steps:

i. Obtain a clear idea of the various types of laws governing the problem.
ii. Idealize or simplify the problem by introducing certain assumptions and to convert the problem into mathematical equations.
iii. Solve the mathematical equations and interpret the results; this requires knowledge of analytical, numerical, and graphical tools.

A mathematical model can be defined as a group of logical connections, formalized dependences, and formulas that enables the study of real-world problems without its experimental analysis. As mentioned earlier, it is preparing a mathematical model and then testing it is an iterative process. We start with a simple model and then gradually refine it by looking back at the assumptions and simplifications so that the results match the real-world situation or experiment. Mathematical modeling is like diving into a mathematical ocean and emerging with a solution for the real life problem. If the solution is not acceptable, we make modifications to the model and may also make the solution procedure more rigorous. We repeat the procedure until a satisfactory solution is obtained. A mathematical model is an idealization and never gives a completely accurate representation of a physical situation. The purpose of the model is to understand the phenomenon and perhaps make predictions about future behavior. A good model simplifies reality so as to permit mathematical calculations but is accurate enough to provide valuable conclusions. It is important to realize the limitations of a model.

Examples of the interesting real-life problems for which mathematical models are available or are being constructed include:

i. Mathematical models to show how diseases (such as swine flu and anthrax) might spread in humans in the event of an epidemic/bioterrorism.
ii. Mathematical modeling of social insects such as honeybees, termites, and so on to find out how they use local information to generate a complex and functional pattern of communication.
iii. Mathematical modeling of urban city planning.
iv. Mathematical modeling of the traffic flow on highways or the stock market options.
v. Mathematical models to understand the working of heart, brain, lungs, kidneys, and the endocrine system.

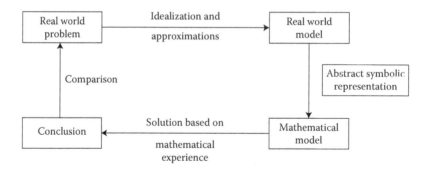

FIGURE 1.1

Symbolic representation of a mathematical model. (From Kapur, J. N. 1994. *Mathematical Modeling*, Wiley Eastern Limited. (Presently) New Age International (P) Ltd., New Delhi. Copyright 1994, New Age International (P) Ltd. Reprinted with permission.)

vi. Mathematical models to study the spread of forest fire/mines fire depending on the types of tree/coal, weather, and nature of the ground surface.

vii. Mathematical models to understand the fluid flow in drains, lakes, rivers, spillways, and so on.

viii. Mathematical models for global warming.

ix. Mathematical models to demonstrate the action of medicine in the human system.

x. Mathematical models for developing video games.

Symbolically, we may define mathematical modeling as follows [15] (Figure 1.1).

Characteristics of Mathematical Models

A good model should be as simple as possible. A model helps one to go beyond the surface of a phenomenon to an understanding of the mechanisms and relationships.

The characteristics and requirements of mathematical models include [15]:

i. *Realism of models*: A mathematical model should represent the reality as closely as possible, but at the same time it should be mathematically tractable.

ii. *Robustness of models*: A mathematical model is said to be robust if small changes in the parameters lead to small changes in the behavior of the model. Any dynamical behavioral is considered to be robust when it is detected in a dense set formed by key parameters of model systems. The decision is made by using sensitivity analysis of the models.

iii. *Nonuniqueness of models*: A particular situation need not have only one mathematical model. We can search for better and different models.

iv. *Hierarchy of models*: For every situation, we can obtain a hierarchy of models. Models obtained at each stage should be more realistic and better than the previous one.

v. *Self-consistency of models*: A mathematical model that involves equations and inequations should be self-consistent.

vi. *Estimation of parameters*: Every mathematical model contains some parameters which control the dynamics of the model system, and these need to be estimated. Optimal control theory is one of the methods for estimating these vital parameters. For example, it is desirable to estimate the proportion of a population of voters who will vote for a particular candidate. That proportion is the unobservable parameter; the estimate is based on a small random sample of voters.

vii. *Generality and applicability of models*: Some models are used in a large number of situations and there are others which are applicable to some specific situations only. For example, logistic maps and the Laplace equation are used in a wide variety of situations.

viii. *Imperfections of models and the cost of modeling*: No model is complete in all respects. There is always a chance for improvement. However, there is an inherent cost involved for each improvement.

ix. *Transferability of mathematical models*: If a single model is applicable to many fields, then the model is very useful and this is known as the transferability of the model. For example, the Laplace equation model is used in many fields such as irrotational flows, electrostatic potential, gravitational potential, and so on.

x. *Criteria for successful models*: A successful model (a) gives good agreement between observations and predictions, and (b) has simplicity in the model equations.

Classification of Mathematical Models

A mathematical model can be classified either as per the mathematical technique used, such as differential equations and/or stochastic differential equations, statistical techniques, numerical data analysis, graphs, and so on, or as per the subject matter, such as physics, medicine, anthropology, engineering, and so on.

Classification as per the Nature of Basic Equations

a. The basic equations may be a system of linear or nonlinear ordinary or partial differential equations or difference equations.

Examples based on linear equations:

   i. $X_{n+1} = 2X_n$ is a linear recursion relation that generates X_{n+1} from X_n by doubling it.

   ii. Linear and damped harmonic oscillators:

   Force: $F = -kx$ and $F = -kx - b(dx/dt)$.

   Equation of motion: $m(d^2x/dt^2) = -kx$ and $m(d^2x/dt^2) = -kx - b (dx/dt)$.

Examples based on nonlinear equations:

   i. Cubic anharmonic oscillator:

   Force: $F = -kx - \mu x^3$.

   Equation of motion: $m(d^2x/dt^2) = -kx - \mu x^3$.
   In Kepler's problem,

   Force: $F = -k\mathbf{r}/r^3$.

   Equation of motion: $m(d^2r/dt^2) = -d/dr\,((-k/r) + (l^2/2mr^2))$.

   ii. Bernoulli shift $X_{n+1} = 2X_n$ (mod 1) maps the unit interval [0, 1] onto itself. The mapping can be described as follows—choose any input number X_0 between 0 and 1, multiply it by 2 (stretching), discard the integer part of the resulting number and retain the fractional part as the output X_1 (folding back). Repeat the operation n times and obtain the sequence $X_0, X_1, X_2, \ldots, X_n$. The stretching out doubles the initial separation, whereas the folding back confines them by re-injection into the unit interval and, in doing so, sends them out to different destinations inside the unit square, amplifying the separation catastrophically. We observe here the snowballing of the initial error. The mapping is highly nonlinear. Altogether, the stretching and folding back operations produce the sensitivity to initial conditions that is essential for the generation of chaos.

b. The basic equations may be *static* or *dynamic* as the time variations in the model system are or are not taken into consideration. Consider the relation

$$F = f\,(p(x, y, z, t)),\qquad\qquad (1.1)$$

where F is the predicted parameter, p is the generalized argument, and x, y, and z are the spatial coordinates and t is time. When the parameter p and hence F depend on both the spatial coordinates and time, the models are called time-dependent, nonautonomous or dynamic. When the parameter p and hence F are independent of time t, the models are called stationary, autonomous, or static [14]. They can be represented by the relation $F = f(p(x, y, z))$.

Examples for autonomous systems:

i. An exponential oscillator: $\ddot{x} + e^x = 0$.

ii. Toda oscillator: $\ddot{x} + b\dot{x} + a(e^x - 1) = 0$.

iii. Brusselator equations (a model chemical reaction of two species):

$$\dot{x}_1 = a_1 - x_1 - a_2 x_1 + x_1^2 x_2, \quad \dot{x}_2 = a_2 x_1 - x_1^2 x_2.$$

Examples for nonautonomous systems [17]:

i. Driven Morse (exponential) oscillator: $\ddot{x} + \alpha\dot{x} + \beta e^{-x}(1 - e^{-x}) = f \cos \omega t$.

ii. Duffing oscillator (driven nonlinear oscillator): $\ddot{x} + \alpha\dot{x} + \omega_0^2 x + \beta x^3 = f \sin \omega t$.

iii. Driven van der Pol oscillator: $\ddot{x} - \alpha(1 - x^2)\dot{x} + \omega_0^2 x = f \sin \omega t$.

iv. Hill's equation: $\ddot{x} + (\alpha + p(t))x = 0$, where $p(t)$ is a periodic function. If $p(t) = \beta \cos t$, then it is called a Mathieu equation.

c. The basic equations may be *stochastic* or *deterministic* as the chance factors are or are not taken into account. If the generalized argument p in Equation 1.1 has a single meaning (with some error of calculation), but is not estimated in terms of statistical distributions, we can define an exact value of the predicted parameter F. These models are called deterministic [18,19]. However, measurements of nature have no precise relationship among variables, but include some statistical component. When the generalized argument forms a distribution of the possible values, characterized by statistical indexes such as mean, dispersion, and standard deviation, a model is called stochastic [5,20]. The predicted value in this case does not have a single meaning, but it is presented by a spectrum of possible values. Models which include randomness are called stochastic. The following are some examples. (i) We may try to calculate the trajectory of a missile launched in a known direction with known velocity. If exact calculation was possible, such a model would be entirely *deterministic*. However, factors such as variable wind velocity that may throw a missile slightly off course are to be taken into consideration. (ii) Monthly sale of a newspaper in a city, in which there are many unknown factors, is a time-dependent phenomenon. It is not possible to write a deterministic model that allows for exact calculation of future behavior of the phenomenon. It may be possible to derive a *stochastic model* or *probability model* that can be used to predict future sales of the newspaper. Models for time series that are needed, for example, to achieve optimal forecasting and control are stochastic models. Forecasting is important in industry, stock market, economics, and so on.

d. The basic equations may be *discrete* or *continuous* as the variables involved are discrete or continuous. Continuous models represent continuous changes of a variable with time. This type of model allows us to define the generalized argument p and the predicted parameter F in Equation 1.1 at every point in the time interval $[t_0, t_n]$. For a discrete model, we use discrete time steps $t_0 < t_1 < \cdots < t_i < \cdots < t_n$ to describe changes in the object of modeling during the time interval $[t_0, t_n]$. The time step $\Delta t = t_i - t_{i-1}$ can be uniform or nonuniform [8]. For example, logistic and Hénon maps are discrete, whereas the Lorenz equations and the Duffing oscillator are continuous. Many animal populations grow in discrete time, due to having well-defined breeding seasons, whereas human populations grow continuously in time.

Some Simple Examples of Optimization Problems

Many real-life problems can be solved using optimization techniques. A businessman wants to minimize costs and maximize profits. A tourist or traveler wants to minimize transportation time and cost. Fermat's principle in optics states that light follows the path that takes the least amount of time. In solving such practical problems, the greatest challenge is often to convert the word problem into a mathematical optimization problem by setting up the function that is to be maximized or minimized. Depending on the physical structure of the optimization problem, we choose linear, nonlinear, neural network, fuzzy system techniques, and so on to find an optimal solution [25]. Methods of calculus are used to find the unconstrained maxima or minima of the function of several variables. Problems with equality constraints may be solved by using *Lagrange's method* and problems with inequality constraints may be solved by *Kuhn–Tucker's optimality conditions*. A well-known technique is the corner point method, which suggests that the functional may attain maxima/minima at one of the corner points of the feasible region. Economists study marginal demand, marginal revenue, and marginal profit, which are derivatives of the cost, demand, revenue, and profit functions.

In the following example, we derive the conditions under which profit can be maximized.

Example 1.1

Suppose that $C(x)$ is the total cost that a company incurs in producing x units of a certain commodity. The function C is called a *cost function*. The *average cost function* $c(x) = C(x)/x$ represents the cost per unit. The marginal cost is the rate of change of C with respect to x. That is, the *marginal cost function* $= C'(x)$. Now, consider the marketing of the produced items. Let $p(x)$ be the price per unit that the company wants to charge if it sells x units. The function $p(x)$ is called the *price function* (or *demand function*) and we would expect it to be a decreasing function of x. If x units are sold and the price per unit is $p(x)$, then the total revenue = *revenue function* (or *sales function*) $= R(x) = x\, p(x)$. The rate of change of the revenue function with respect to the number of units sold = *marginal revenue function* $= R'(x)$.

If x units are sold, then the total profit = *profit function* $= P(x) = R(x) - C(x)$.

The derivative of the profit function = *marginal profit function* $= P'(x)$.

To maximize the profit, we look for the critical values of P, that is, the values where the marginal profit is zero. If $P'(x) = R'(x) - C'(x) = 0$, then $R'(x) = C'(x)$. Therefore, if the profit is a maximum, then marginal revenue = marginal cost.

Note that $P''(x) = R''(x) - C''(x) < 0$. This condition implies that the rate of increase of marginal revenue is less than the rate of increase of the marginal cost. Thus, the profit will be a maximum when $R'(x) = C'(x)$ and $R''(x) < C''(x)$.

In the following example, we illustrate a situation in which the cost can be minimized.

Example 1.2

A sealed can, cylindrical in shape, is to hold 10 L of edible oil. Find the dimensions that will minimize the cost of the material for manufacturing the can.

SOLUTION

To minimize the cost of the material, we minimize the total surface area of the sealed cylinder (top, bottom, and lateral sides). Let the dimensions of the cylinder be radius $= r$ cm and height $= h$ cm. The total surface area is $S = 2\pi r^2 + 2\pi rh$.

But the volume $= 10$ L $= 10{,}000$ cm$^3 = \pi r^2 h$.

Hence,

$$S = 2\pi r^2 + 2\pi r \left(\frac{10{,}000}{\pi r^2} \right) = 2\pi r^2 + \frac{20{,}000}{r}.$$

The stationary points are the solutions of $S'(r) = 4\pi r - (20{,}000/r^2) = 0$. We obtain $r = 10(5/\pi)^{1/3}$. Hence,

$$h = \frac{10{,}000}{\pi r^2} = \frac{10{,}000}{100\pi} \left(\frac{\pi}{5} \right)^{2/3} = 20 \left(\frac{5}{\pi} \right)^{1/3} = 2r.$$

Since $S''(r) = 4\pi + (40{,}000/r^3) > 0$ at the critical point $r = 10(5/\pi)^{1/3}$, the extreme value is a minimum. Thus, to minimize the cost of the can, the dimensions of the can should be

$$h = 2r = 20(5/\pi)^{1/3} \approx 23.35 \text{ cm.}$$

Example 1.3

Find the production level that will maximize the profit of a toy manufacturing company with the given cost and demand functions

$$C(x) = 1650 + 36x - 0.01x^2 + 0.0012x^3 \quad \text{and} \quad p(x) = 72 - 0.01x.$$

SOLUTION

To maximize the profit, we consider the revenue function as

$$R(x) = x\, p(x) = 72x - 0.01x^2.$$

The marginal revenue function is given by $R'(x) = 72 - 0.02x$.
The marginal cost function is given by $C'(x) = 36 - 0.02x + 0.0036x^2$.
For the profit to be maximum, the marginal revenue = marginal cost. Therefore,

$$72 - 0.02x = 36 - 0.02x + 0.0036x^2.$$

Solving, we obtain $x = 100$, and $R''(x) = -0.02$, $C''(x) = -0.02 + 0.0072x$. Now, $R''(x) - C''(x) < 0$ for all $x > 0$. Therefore, the production level of 100 units will maximize the profit.

Limitations Associated with Mathematical Modeling

Some of the limitations/problems associated with mathematical modeling include:

i. Sometimes, the model may not address the situation you want to describe. The model may be too simple to mirror adequately or too complex to aid in the understanding of the situation.

ii. A model may be very sensitive to the initial conditions or to the parameter values.

iii. Successful guidelines may not be available for choosing the number of parameters and for estimating the values of these parameters.

iv. The model may create a mathematical problem that does not lend itself to a mathematical solution.

v. The results may be too technical in nature or may not be in a form that can be implemented.

vi. Funds may be inadequate for implementing a suggested solution.

Modeling Approaches

We present different approaches for modeling real-world situations, each approach providing an insight into a different aspect. A combination of seven approaches may provide the best understanding of the situation: (i) empirical; (ii) theoretical; (iii) stochastic or probabilistic; (iv) deterministic; (v) statistical; (vi) simulation; (vii) discrete and continuous.

Empirical Approach

Empirical models are based on an experimental hypothesis. When the problem-solving process is data driven, we call the approach an empirical approach. Empirical models lead to *laws of nature,* which represent the fundamental characteristics of nature. Many empirical models were formulated by Newton, Einstein, and others. When data are collected from an experiment or field, empirical relations between the data may be obtained through various procedures. One such method is the least-squares approximation. We fit a least-squares curve for the data and then use this curve to predict the outcomes where there are no data. However, the empirical approach is least useful. The disadvantage of this approach is that we cannot be confident that the fitted curve can be used outside the range of data considered. This approach is often used to create a continuous dataset to be used as the input for another model.

We present some simple problems to illustrate the use of least-squares approximation.

Example 1.4

The average carbon dioxide (CO_2) level in the atmosphere measured in parts per million (ppm) in a metropolitan city is given in Table 1.1. Using these data, (i) fit a straight line passing through the endpoints; and (ii) fit a linear least-squares model. Find the maximum error in magnitude in the first case. Predict the year in which the CO_2 level may exceed 300 ppm.

SOLUTION

i. The slope of the line joining the endpoints = $(279.8 - 252.6)/(2006 - 1990) = 1.7$.

TABLE 1.1

CO_2 Levels in ppm

Year (t)	1990	1992	1994	1996	1998	2000	2002	2004	2006
CO_2 Level = y	252.6	257.3	262.2	264.1	267.5	271.5	275.1	276.0	279.8

The equation of the line: $y = 1.7t - 3130.4$. The maximum error in magnitude = 2.8.
The year in which the CO_2 level may exceed 300 ppm:

$$1.7t - 3130.4 > 300 \text{ gives } t = 2018 \text{ (approx.).}$$

ii. The linear least-squares model is given by $y = 1.650833333t - 3031.020556$.
The year in which the CO_2 level may exceed 300 ppm:

$$1.650833333t - 3031.020556 > 300 \text{ gives } t = 2018 \text{ (approx.).}$$

Example 1.5

A physicist studying a decaying process decides to fit a model of the form $f(t) = ae^{-bt}$ to the data given in Table 1.2. Determine the values of a, b by the method of least squares. Compute the least-squares error.

SOLUTION

Taking logarithms, we write the given model in the form

$$\ln f(t) = \ln a - bt, \quad \text{or} \quad y = A - bt, \quad \text{where } A = \ln a \text{ and } y = \ln f(t).$$

The nonlinear model is now transformed into a linear model.
The method of least squares gives the values of A, a, b as

$$A = 0.372304413, \quad a = e^A = 1.45107467, \quad b = 0.91393925.$$

The least-squares model is given by $f(t) = 1.45107467e^{-0.91393925t}$. The least-squares error is 0.0004196.

Theoretical Approach

Theoretical models are inspired by empirical models. We use the basic laws of nature to construct the mathematical models. The modeling process is based more on theory.

TABLE 1.2

Data of a Decaying Process

Time (t)	0.4	0.6	0.8	1.0	1.2	1.4	1.6	1.8	2.0
$f(t)$	0.99	0.84	0.70	0.59	0.49	0.40	0.34	0.28	0.23

We give below an example of a model describing the oscillations of a simple pendulum.

Example 1.6 [Simple Pendulum]

We derive the equations that govern the oscillations of a simple pendulum. The assumptions are that (i) the string has negligible mass, and (ii) the air resistance is negligible. The only forces acting on the system are (i) tension T in the string, and (ii) gravitational force mg, which acts vertically downward (Figure 1.2). This law gives the equations

$$m\frac{d^2x}{dt^2} = -T\sin\theta, \quad m\frac{d^2y}{dt^2} = T\cos\theta - mg.$$

If l is the length of the string, then we have $x = l\sin\theta$, $y = l - l\cos\theta$. Substituting in the above equations, we get

$$ml\cos\theta\frac{d^2\theta}{dt^2} + \left[T - ml\left(\frac{d\theta}{dt}\right)^2\right]\sin\theta = 0,$$

$$ml\sin\theta\frac{d^2\theta}{dt^2} - \left[T - ml\left(\frac{d\theta}{dt}\right)^2\right]\cos\theta = -mg.$$

Eliminating the second terms and then the first terms in these equations, we get the equations

$$l\frac{d^2\theta}{dt^2} + g\sin\theta = 0, \quad T = ml\left(\frac{d\theta}{dt}\right)^2 + mg\cos\theta.$$

Since it is difficult to solve these nonlinear equations, we can solve them under various assumptions.

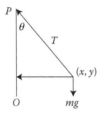

FIGURE 1.2
Simple pendulum.

Consider the case of small oscillations. Then $\sin\theta \approx \theta$, and $\cos\theta \approx 1$. Under these assumptions, the above equations simplify as

$$\frac{d^2\theta}{dt^2} + \frac{g}{l}\theta = 0, \quad \text{and} \quad T = ml\left(\frac{d\theta}{dt}\right)^2 + mg.$$

Since θ is small, we can also neglect $(d\theta/dt)^2$. From the second equation, we obtain $T = mg$. Therefore, in the ideal case, tension T is constant even for small swings of the pendulum. To solve the first equation, let us choose the initial conditions as the following:

At $t = 0$: initial angle of the swing: $\theta = \theta_0$, initial velocity: $d\theta/dt = 0$.
The solution is given by $\theta = \theta_0 \cos(\sqrt{g/l}\, t)$.
Using this value for θ, we can find the position of the bob by computing $x = l\sin\theta$, $y = l - l\cos\theta$.

Stochastic or Probabilistic Approach

Using the stochatic or probabilistic approach, we try to estimate the probability of a certain outcome based on the available data. This approach should ideally be used when there is a high degree of variability in the problem. Examples include (i) modeling a small population when reproduction rates need to be predicted over a time interval; (ii) modeling the economic fluctuations or economic growths; (iii) modeling the insurance, telecommunications, and traffic theory problems; and (iv) biological models.

Deterministic Approach

In the deteministic approach, the chance factors are not taken into account. We ignore the random variations and formulate the mathematical equations that describe the relationship between the variables of the problem. This approach is widely used and can produce accurate results providing valuable insights into the problem. An example is the modeling of (predicting) satellite orbits. In a population model, we aim to obtain an equation relating birth rates and death rates which themselves are related to the population size at any given time. The exponential and logistic models of population growth can be formulated in terms of differential equations as

$$dP/dt = (b - d)P \quad \text{and} \quad dP/dt = rP(1 - P/P_{max}), \text{ respectively.}$$

Statistical Approach

The statistical approach concerns testing of the hypothesis, that is, to find out from which category the data have been obtained. The datasets are assumed to have some particular type of distribution (with the associated means, variances, or standard deviations). This distribution can be used to predict the outcome of further trials. The statistical approach is widely used in psychology, paleontology, and biological sciences.

Simulation Approach

The simulation approach is used when the problem under investigation cannot be easily modeled analytically or the relevant data required for modeling the system cannot be collected. In this approach, a computer program simulates (produces) a set of data that mimics a real outcome. It provides a useful means by which datasets can be generated. The computer program can be run many times and the necessary information gained in the process. The simulation approach provides realistic but not the best models. The best models are usually those which are simple, yet they provide results which are useful. For example, in designing an aircraft, the design engineer investigates the air flow around the aircraft. This can be done by using the theory of fluid mechanics and dynamical system theory. It is easier to simulate the study by building a scale model of the aircraft and by investigating the behavior of the model in a wind tunnel. Another example of simulation modeling is a hospital scenario, where the hospital administration would like to know the optimal number of doctors required to manage the patients in various shifts. A simulation of the model can be carried out by using suitable data on a computer.

Discrete and Continuous Approaches

In the discrete approach, the time variable is treated as a discrete variable rather than as a continuous variable. This may mean, for example, that it is sufficient or meaningful to measure certain physical variables after finite intervals of time, say an hour, a week, a month, and so on rather than on a continuous basis. Examples of a discrete approach are the following: population of an insect species in a forest, radioactive decay, rainfall and temperature of a metropolitan city, and so forth. These systems are represented by difference equations/recursion relations/iterated maps.

In the continuous approach, time is always treated as a continuous (flow) variable and the models appear to be easier to handle than the discrete approach due to the development of calculus and differential equations. The continuous models are simpler only when analytical solutions are available, otherwise we have to approximate it so that we can handle it numerically. Some examples of the continuous approach are the Rössler model, Lorenz model, Helmholtz oscillator, Duffing oscillator, and others.

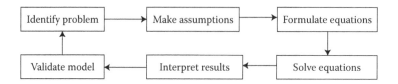

FIGURE 1.3
Schematic diagram of a modeling cycle. (From Barnes, B., Fulford, G. R. 2002. *Mathematical Modeling with Case Studies*. Taylor & Francis, Boca Raton, FL. With permission.)

Modeling/Cyclic Processes

A mathematical model is a representation of a complex real-world problem. Some mathematical equations describe the processes involved in the problem. A model is an attempt at mimicking the real-world situation and cannot describe it completely. The purpose of the model is to understand the phenomenon and maybe to make predictions about future behavior. If the predictions do not compare well with reality, we need to refine our model or formulate a new model and start the cycle again.

The important stages of construction of a mathematical model as given by Barnes and Fulford [4] are presented in Figure 1.3. If the model does not produce results consistent with observations, then we return to step (ii) (assumptions) and modify the assumptions. Berry and Houston [6] have also given a schematic diagram of a modeling cycle.

A Modeling Diagram

Here, we present the modeling diagram as suggested by Arnold Neumaier (see Figure 1.4) [21]. The nodes of the diagram represent information to be collected, sorted, evaluated, and organized. The edges of the diagram represent activities of two-way communication between the nodes and the corresponding source of information.

  S: Problem statement (arising out of real-world situations).

  M: Mathematical model (uses concepts/variables, relations, restrictions, defines a goal and decides priorities/quality assignments).

  T: Theory (of application, of mathematics, literature search).

  N: Numerical methods (software).

  P: Programs (flow diagrams, implementation, user interface, and documentation).

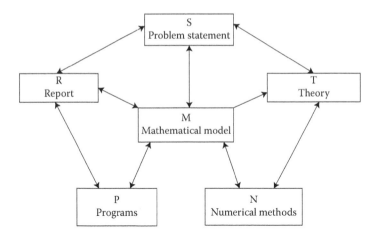

FIGURE 1.4
Modeling diagram as suggested by Arnold Neumaier. (From Neumaier, A. 2004. Mathematical model building, Chapter 3. In: *Modeling Language in Mathematical Optimization*, J. Kallrath, ed. Kluwer, Boston. Reprinted with permission from Professor Dr. Arnold Neumaier.)

R: Report (analysis of results, validation, visualization, limitations, and recommendations).

The modeling diagram has 6 nodes and 10 edges; that is, it suggests 16 different processes. Every stage of the modeling process is important. Modeling is complete only if the contribution along all edges becomes insignificant. It is possible that the formulation of the problem itself may change due to the insight gained by the modeling process.

Compartment Models

The compartment model framework is a natural and valuable means of modeling the processes, which have inputs and/or outputs over time. Many of the real-world processes may be considered as compartmental models; that is, the processes have inputs and outputs from a "compartment" over time [4].

For example, we may consider a problem in atmospheric sciences with the compartments taken as ocean, land, town, and so forth.

The following are some compartment models.

i. *The amount of CO_2 in the Earth's atmosphere* [4]:

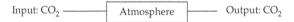

We are interested in determining the amount of CO_2 retained in the compartment at any given time to know whether over long periods of time, the levels are increasing or decreasing. For modeling the amount of CO_2 occupying the compartment, the rate of change can be considered as the "rate in" minus the "rate out," that is,

{net rate of change of CO_2} = {rate in} – {rate out}.

ii. *Assimilation of drug/medicine in the bloodstream*: The human body processes medicines in either a linear or exponential manner for most substances. The Michaelis–Menten equation describes the linear absorption phenomenon very closely. The method assumes that the human body is processing a particular medicine with a rate of change which is linear. Consider the following example. Assume that a drug is provided continuously for a period of 24 h and stopped. We are interested in finding the amount of medicine in the body at any given time. We consider this problem as a compartment model, where the rate of change is inflow minus outflow. If M is the amount of administered medicine in milligrams, $D(t)$ is the rate at which the medicine is administered and $P(M)$ is the processing rate, then the model equation can be written as

$$\frac{dM}{dt} = D(t) - P(M).$$

Let us assume the following data:

$D(t) = 10$ mg/h for the first 24 h and then zero afterwards, and $P(M) = 0.1\,M$. Let $M(0) = 0$, that is, there is no medicine in the body at $t = 0$. The model simplifies to

$$\frac{dM}{dt} + 0.1M = 10, \quad 0 < t < 24.$$

The solution under the given initial condition is $M(t) = 100$ $(1 - e^{-0.1t})$. After 24 h, the body has $M(24) = 90.93$ mg of medicine. For $t > 24$, the system is governed by the equation $dM/dt = -0.1M$. Taking the condition $M(24) = 90.93$ mg, the solution of this equation is obtained as $M(t) = 1002.34e^{-0.1t}$. We find that the medicine has been completely absorbed or has gone out of the system in about 140 h.

iii. *The decay processes of radioactive materials*: The decay process is governed by the law of natural decay. The law states that the rate of change of a substance (population) $X(t)$ is proportional to the amount of the substance (population) available at that time. We can consider this problem as a compartment model where there is no input but continuous output of nuclei over a period of time. The law states that $(dX/dt) \propto X(t)$. Since the substance decays, we have the governing differential equation as $(dX/dt) = -kX(t)$, $k > 0$. If initially, the amount of the substance is N, that is, $X(0) = N$, then the solution is given by $X(t) = Ne^{-kt}$, which agrees well with the experimental results.

When a substance is cooling, the above law is also called *Newton's law of cooling*. The exact form of Newton's law of cooling is $T - T_s = (T_0 - T_s)e^{-kt}$, where T is the temperature at time t, T_s is the surrounding temperature, and T_0 is the initial temperature.

Consider the following examples.

Example 1.7

If 10% of a radioactive material disintegrates in 100 years, how long does it take for 90% of the material to disintegrate?

SOLUTION

Let $X(t)$ be the amount of material available at any instant of time and N be the initial amount of radioactive material.

From the above discussion, the solution is given by $X(t) = Ne^{-kt}$.

In 100 years, 10% of the material disintegrates; that is, 90% or $0.9N$ is available. Therefore,

$$0.9N = Ne^{-100\,k}, \quad \text{or} \quad k = -\ln(0.9)/100.$$

When 90% of the material disintegrates, that is, $0.1N$ is available, we get

$$0.1N = Ne^{-kt}, \quad \text{or} \quad t = -\frac{\ln(0.1)}{k} = \frac{100\ln(0.1)}{\ln(0.9)} = 2185 \text{ years}$$

(approximately).

Example 1.8

A dead man's body was discovered in a room with a constant temperature of 65°F. The forensic science laboratory was requested to estimate the time of his death. Assume that the body temperature at the time of his death was 98.8°F. The body temperature was 94.8°F when the first reading was taken by the forensic laboratory.

SOLUTION

The body was shifted to a cooled room. After death, the body will radiate heat into the cooler room, causing the body's temperature to decrease. The laboratory will try to estimate the time of his death by observing the body's current temperature and calculating how long it would have taken to lose heat to reach this point.

Let $T(t)$ be the body temperature at time t. By Newton's law of cooling, the body will radiate heat energy into the room as

$$\frac{dT}{dt} = -k[T(t) - 65], \quad T(\text{reference time } t = 0) = 94.8\,^\circ\text{F}.$$

Separating the variables, the solution is obtained as $T(t) = 65 + Ae^{-kt}$.

Using the initial condition, we obtain $T(t) = 65 + 29.8e^{-kt}$.

The body temperature is again measured after, say, 90 min and the temperature was found to be 90.0°F. Using these conditions, we obtain $k = -(1/90) \ln(25/29.8) = 0.00195$.

The temperature of the body satisfies the equation $T(t) = 65 + 29.8e^{-0.00195t}$.

At the time of death, the body temperature was 98.8°F. Inserting into the above equation, we obtain

$$t = -\frac{1}{0.00195} \ln\left(\frac{98.8 - 65}{29.8}\right) = -64.6 \text{ min (approximately).}$$

Hence, the man died about 64.6 min before the first measurement of the body temperature (94.8°F) was made.

Mathematical Preliminaries

Some of the theoretical results presented in this section are based on the following source material: Ahmad and Rao [1], Jordan and Smith [13], Perko [23], Strogatz [28], and Wiggins [31].

Consider an autonomous system of the form

$$\frac{dx}{dt} = x' = f(x), \quad \text{where } f \in C[R^n, R^n] \tag{1.2}$$

in the variables $x_1, x_2, \ldots, x_n$. We assume that $f(x)$ is sufficiently smooth and the solution of system (1.2) exists and is unique. Let $f(0) = 0$ and $f(x) \neq 0$ for x in the neighborhood of the origin. Then, system (1.2) admits the zero solution, $x \equiv 0$, which is an isolated critical point of the system.

For our discussion, we shall use the Euclidean norm

$$\|x\|^2 = x_1^2 + x_2^2 + \cdots + x_n^2. \tag{1.3}$$

Let $V(x)$ be a real-valued scalar continuous function in the variables $x_1, x_2, \ldots, x_n$. Let Γ be an open set containing the origin in R^n. Then, $V(x)$ is said to be

i. *Positive definite* on Γ if and only if $V(0) = 0$ and $V(x) > 0$ at all points $x \neq 0$ on Γ.

ii. *Positive semidefinite* on Γ if and only if $V(0) = 0$, $V(x) = 0$ at a few points $x \neq 0$ in Γ, and $V(x) > 0$ at all of the remaining points $x \neq 0$ in Γ.

iii. *Negative definite* on Γ if and only if $-V(x)$ is positive definite on Γ.

iv. *Negative semidefinite* on Γ if and only if $-V(x)$ is positive semidefinite on Γ.

Let $V(t, x)$ be a real-valued time-varying scalar continuous function in the variables $t, x_1, x_2, \ldots, x_n$. Then, $V(t, x)$ is said to be

i. *Positive definite* if there exists a positive-definite function $V(x)$ such that $V(t, 0) \equiv 0$ and $V(t, x) \geq V(x)$ at all points $x \neq 0$.

ii. *Positive semidefinite* if $V(t, 0) \equiv 0$, $V(t, x) = 0$ at a few points $x \neq 0$, and $V(t, x) > 0$ at all the remaining points $x \neq 0$.

iii. *Negative definite* if and only if $-V(t, x)$ is positive definite.

iv. *Negative semidefinite* if and only if $-V(t, x)$ is positive semidefinite.

K-class of functions: A function $\varphi(r)$ is called a *K*-class function if $\varphi \in C[(0, h), R^+]$, $\varphi(0) = 0$ and φ is strictly monotonically increasing with r. It is also called a *wedge function*.

Since $V(x)$ is continuous

$$V(x) \leq \max_{\|z\| \leq r} V(z), \quad \text{and} \quad V(x) \geq \min_{r \leq \|z\| \leq d} V(z)$$

for sufficiently small r, $0 < c \leq r \leq d$, where $\|x\| = r$.

Therefore, there exist strictly monotonic increasing functions $\alpha(r)$, $\beta(r)$, that is, α, β are *K*-class functions, with $\alpha(0) = 0$, $\beta(0) = 0$, such that

$$\alpha(r)\|x\| \leq V(x) \leq \beta(r)\|x\|.$$

From this inequality, an alternative definition to positive definiteness of $V(x)$ can be written as

$$V(0) = 0, \quad V(x) \geq \alpha\|x\|, \quad \text{for all } x \in \Gamma. \tag{1.4}$$

Example 1.9

Test whether the following functions are positive definite

    i. $V(x_1, x_2) = x_1^2 + x_2^2$.
    ii. $V(x_1, x_2, x_3) = x_1^2 + x_3^2$.
    iii. $V(x_1, x_2, x_3) = (x_1 + x_2)^2 + x_3^2$.
    iv. $V(x_1, x_2) = x_1^4 + x_2^4$.

SOLUTION

    i. We have $V(0, 0) = 0$ and $V(x_1, x_2) > 0$ at all points $x = (x_1, x_2) \neq 0$ in the neighborhood of $(0, 0)$. Hence, V is positive definite on R^2.
    ii. We have $V(0, 0, 0) = 0$ and $V(x_1, x_2, x_3) = 0$ on the x_2-axis. Also, $V(x_1, x_2, x_3) > 0$ at all points $x = (x_1, x_2, x_3) \neq 0$ in the neighborhood of $(0, 0, 0)$. Hence, V is positive semidefinite on R^3.
    iii. We have $V(0, 0, 0) = 0$ and $V(x_1, x_2, x_3) = 0$ on the line $x_3 = 0$, $x_1 = -x_2$. Also, $V(x_1, x_2, x_3) > 0$ at all other points $x \neq 0$ in the neighborhood of $(0, 0, 0)$. Hence, V is positive semidefinite on R^3.
    iv. We have $V(0, 0) = 0$. Also, from the inequality $(x_1^2 - x_2^2)^2 \geq 0$, we obtain

$$x_1^4 + x_2^4 - 2x_1^2 x_2^2 \geq 0, \quad \text{or} \quad 2(x_1^4 + x_2^4) - (x_1^2 + x_2^2)^2 \geq 0,$$

$$\text{or} \quad (x_1^4 + x_2^4) \geq \left(\frac{1}{2}\right)(x_1^2 + x_2^2)^2, \quad \text{or} \quad V = (x_1^4 + x_2^4) \geq \left(\frac{r^4}{2}\right),$$

where $r = \|x\| = (x_1^2 + x_2^2)^{1/2}$.
Hence, by definition (1.4), V is positive definite on R^2.

Example 1.10

Test whether the following time-varying functions are positive definite:

    i. $V(t, x_1, x_2) = (e^t + e^{-t})(x_1^2 + x_2^2)$.
    ii. $V(t, x_1, x_2) = t(x_1^2 + x_2^2) + 2x_1 x_2 \cos t$.
    iii. $V(t, x_1, x_2) = (2 + e^{-t})((x_1^2/(1 + x_1^4)) + x_2^2)$.

SOLUTION

    i. $V(t, 0) = 0$, and $V(t, x_1, x_2) \geq V(x_1, x_2)$ where $V(x_1, x_2) = x_1^2 + x_2^2$ is positive definite. Hence, $V(t, x_1, x_2)$ is positive definite.
    ii. $V(t, x_1, x_2) = t(x_1^2 + x_2^2) + 2x_1 x_2 \cos t \geq 0$. There does not exist a positive-definite function $V(x_1, x_2)$ such that $V(t, x_1, x_2) \geq V(x_1, x_2)$. Hence, V is positive semidefinite.
    iii. $V(t, 0) = 0$, and $V(t, x_1, x_2) \geq V(x_1, x_2)$ where $V(x_1, x_2) = ((x_1^2/(1 + x_1^4)) + x_2^2)$ is positive definite. Hence, $V(t, x_1, x_2)$ is positive definite.

To discuss the positive definiteness of more general forms of V than the forms considered in Example 1.9, we consider the quadratic form

$$Q = V(x) = x^T A x = \sum_{i,j=1}^{n} a_{ij} x_i x_j, \qquad (1.5)$$

where A is an $n \times n$ matrix $A = (a_{ij})$.

The quadratic form $V(x)$ is positive definite if A is positive definite. A necessary and sufficient condition is that the leading minors of A are positive.

If $x = x(t)$ is any solution of the autonomous system (1.2), then

$$\frac{d}{dt} V(x(t)) = \frac{\partial V}{\partial x_1} \frac{dx_1}{dt} + \frac{\partial V}{\partial x_2} \frac{dx_2}{dt} + \cdots + \frac{\partial V}{\partial x_n} \frac{dx_n}{dt}$$

$$= \frac{\partial V}{\partial x_1} f_1(x(t)) + \frac{\partial V}{\partial x_2} f_2(x(t)) + \cdots + \frac{\partial V}{\partial x_n} f_n(x(t))$$

$$= V^*(x(t)). \qquad (1.6)$$

Note that Equation 1.6 is the derivative of V with respect to t of the system (1.2).

Denote S_r as the set $S_r = \{x \in R^n : \|x\| < r\}$.

We state some results regarding the stability of the zero solution of the system (1.2).

Theorem 1.1 [1]

If there exists a positive-definite scalar function $V(x)$ such that $V^*(x) \leq 0$ on S_r, then the zero solution of the system (1.2) is stable.

Theorem 1.2 [1]

If there exists a positive-definite scalar function $V(x)$ such that $V^*(x)$ is negative definite on S_r, then the zero solution of the system (1.2) is asymptotically stable.

Theorem 1.3 [1]

If there exists a scalar function $V(x)$ such that $V(0) = 0$, $V^*(x)$ is positive definite on S_r, and if in every neighborhood N of the origin, $N \subset S_r$, there is a point x_0 where $V(x_0) > 0$, then the zero solution of the system (1.2) is unstable. This result was due to Chetayev.

Remark 1.1

Note that the above theorems give us a procedure for testing the stability or instability of the zero solution of Equation 1.2 without explicitly solving Equation 1.2. However, it does not provide a method for constructing a *Lyapunov function V(x)*.

Remark 1.2

Lyapunov functions satisfying the conditions of Theorem 1.1 or Theorem 1.2 which determine the stability or asymptotic stability of the zero solution are called weak and strong Lyapunov functions, respectively.

We consider a few examples by constructing the Lyapunov function $V(x)$ arbitrarily.

Example 1.11

Test the stability of the zero solution in the following problem:

$$x_1' = -3x_2 - \left(\frac{1}{2}\right)x_1 x_2^2, \quad x_2' = 2x_1 - \left(\frac{1}{3}\right)x_2.$$

SOLUTION

Let us choose $V(x) = ax_1^2 + bx_2^2$. Then,

$$V^*(x) = 2ax_1 f_1 + 2bx_2 f_2$$

$$= 2ax_1[-3x_2 - (1/2)x_1 x_2^2] + 2bx_2[2x_1 - (1/3)x_2]$$

$$= 2x_1 x_2[-3a + 2b] - ax_1^2 x_2^2 - (2/3)bx_2^2.$$

Set $-3a + 2b = 0$. We may choose $a = 2$, $b = 3$. We get $V(x) = 2x_1^2 + 3x_2^2$. Hence,

$$V^*(x) = -2x_2^2(x_1^2 + 1).$$

Now, $V^*(x)$ is negative semidefinite as it vanishes on $x_2 = 0$, that is, on the x_1-axis. By Theorem 1.1, the zero solution $(0, 0)$ is stable.

Example 1.12

Test the stability of the zero solution in the following problem:

$$x_1' = -2x_1 + 3x_2^2, \quad x_2' = -3x_2 - 6x_1 x_2.$$

SOLUTION

Let us choose $V(x) = ax_1^2 + bx_2^2$. Then,

$$V^*(x) = 2ax_1 f_1 + 2bx_2 f_2$$

$$= 2ax_1[-2x_1 + 3x_2^2] + 2bx_2[-3x_2 - 6x_1x_2]$$

$$= -4ax_1^2 - 6bx_2^2 + (6a - 12b)x_1x_2^2.$$

Set $6a - 12b = 0$. We may choose $a = 2$, $b = 1$. We get $V(x) = 2x_1^2 + x_2^2$. Hence,

$$V^*(x) = -(8x_1^2 + 6x_2^2).$$

Now, $V^*(x)$ is negative definite. By Theorem 1.2, the zero solution $(0, 0)$ is asymptotically stable.

Remark 1.3

In Example 1.12, $(dV/dt) < 0$, except at the origin $(0, 0)$. This implies that every solution path P crosses the ellipse $V(x) = 2x_1^2 + x_2^2 = k$ with decreasing major and minor axes, in an inward direction. As $t \to \infty$, the points on the path P move to the equilibrium point at the origin. However, it is possible that in some systems, the progress of P to the origin $(0, 0)$ may be blocked by a limit cycle.

Example 1.13

Test the stability of the zero solution in the following problems:

    i. $x_1' = x_2 + rx_1$, $\quad x_2' = -x_1 + rx_2$, $\quad r = (x_1^2 + x_2^2)^{1/2}$.

    ii. $x_1' = -x_1 + x_2 - 6x_3 + x_1x_2^2$, $\quad x_2' = -x_1 - x_2 + 3x_3 - x_1^2x_2 + x_3^2$,

        $x_3' = 6x_1 - 3x_2 - x_3 - x_2x_3$.

    iii. $x_1' = x_2 - \sin^5 x_1$, $\quad x_2' = -3x_1 - \sin^5 x_2$, $\quad$ for $\sin x_1 > 0$, $\sin x_2 > 0$.

SOLUTION

    i. Choose $V(x) = x_1^2 + x_2^2$. Then,

$$V^*(x) = 2x_1 f_1 + 2x_2 f_2$$

$$= 2x_1[x_2 + rx_1] + 2x_2[-x_1 + rx_2]$$

$$= 2r(x_1^2 + x_2^2) = 2r^3.$$

Now, $V^*(x)$ is positive definite. By Theorem 1.3, the zero solution $(0, 0)$ is unstable.

ii. Choose $V(x) = ax_1^2 + bx_2^2 + cx_3^2, a > 0, b > 0, c > 0$. Then,

$$V^*(x) = 2ax_1 f_1 + 2bx_2 f_2 + 2cx_3 f_3$$

$$= 2ax_1[-x_1 + x_2 - 6x_3 + x_1 x_2^2] + 2bx_2[-x_1 - x_2 + 3x_3$$

$$- x_1^2 x_2 + x_3^2] + 2cx_3[6x_1 - 3x_2 - x_3 - x_2 x_3]$$

$$= -2ax_1^2 + 2x_1 x_2(a - b) - 12x_1 x_3(a - c) + 6x_2 x_3(b - c)$$

$$- 2bx_2^2 - 2cx_3^2 + 2x_1^2 x_2^2(a - b) + 2x_2 x_3^2(b - c).$$

Setting $a - b = 0$ and $b - c = 0$, we get $a = b = c$. Hence,

$$V(x) = a(x_1^2 + x_2^2 + x_3^2), \quad a > 0.$$

$$V^*(x) = -2a(x_1^2 + x_2^2 + x_3^2) = -2V.$$

$V^*(x)$ is negative definite. By Theorem 1.2, the zero solution is asymptotically stable.

Now, integrating the equation $V^*(x) = -2V$, and evaluating the integration constant using the initial condition, we obtain

$$V(x_1(t), x_2(t), x_3(t)) = V(x_1(0), x_2(0), x_3(0))e^{-2t}.$$

Therefore, the zero solution is also exponentially asymptotically stable.

iii. Let us choose $V(x) = ax_1^2 + bx_2^2, a > 0, b > 0$. Then,

$$V^*(x) = 2ax_1 f_1 + 2bx_2 f_2$$

$$= 2ax_1[x_2 - \sin^5 x_1] + 2bx_2[-3x_1 - \sin^5 x_2]$$

$$= -2(ax_1 \sin^5 x_1 + bx_2 \sin^5 x_2) + 2(a - 3b)x_1 x_2.$$

Set $a - 3b = 0$. We may choose, $a = 3, b = 1$. We get

$$V(x) = 3x_1^2 + x_2^2, \quad \text{and} \quad V^*(x) = -2(3x_1 \sin^5 x_1 + x_2 \sin^5 x_2).$$

Now, $V^*(x)$ is negative definite. By Theorem 1.2, the zero solution $(0, 0)$ is asymptotically stable.

Example 1.14

Show that $x(t) = 0, \ t \geq t_0$, is an unstable solution of the equation $\ddot{x} + \dot{x} \sin 2x - x = 0$.

SOLUTION

The equivalent system is $\dot{x}_1 = x_2, \dot{x}_2 = x_1 - x_2 \sin 2x_1$, where $x = x_1$. Choose, $V(x_1, x_2) = x_1 x_2$. In every neighborhood of the origin, $V(x_1, x_2)$ takes both positive and negative values. We obtain

$$V^*(x) = x_1^2 + x_2^2 - x_1 x_2 \sin(2x_1).$$

In any neighborhood of the origin, $|x_1|$, $|x_2|$ are small and $\sin(2x_1) \approx 2x_1$. The third term on the right-hand side is of order of one higher than the first two terms. It is seen that $V^*(x) > 0$ and therefore positive definite. By Theorem 1.3, the zero solution is unstable.

Construction of the Lyapunov Function and Testing of Stability

We have noted above that there is no simple procedure for constructing the Lyapunov function $V(x)$. However, for some linear and nonlinear autonomous systems, *Krasovskii's method* [16], gives a procedure for constructing a Lyapunov function.

Krasovskii's Method

We assume that $f_i(x)$ in Equation 1.2 contains x_i. Determine the Jacobian matrix $J(x)$, where

$$J(x) = \frac{\partial f}{\partial x} = \begin{bmatrix} \partial f_1/\partial x_1 & \partial f_1/\partial x_2 & \cdots & \partial f_1/\partial x_n \\ \partial f_2/\partial x_1 & \partial f_2/\partial x_2 & \cdots & \partial f_2/\partial x_n \\ \cdots & \cdots & \cdots & \cdots \\ \partial f_n/\partial x_1 & \partial f_n/\partial x_2 & \cdots & \partial f_n/\partial x_n \end{bmatrix}.$$

Define a matrix $M(x) = J^T(x) + J(x)$, where $J^T(x)$ is the transpose of J. The Lyapunov function is taken as

$$V(x) = f^T(x)f(x) = f_1^2(x) + f_2^2(x) + \cdots + f_n^2(x).$$

Obviously, $V(x)$ is positive definite. We have

$$\frac{df}{dt} = \frac{\partial f}{\partial x}\frac{dx}{dt} = J(x)f(x).$$

Then,

$$V^*(x) = f^T(x)f'(x) + f'^T(x)f(x)$$

$$= f^T(x)J(x)f(x) + [J(x)f(x)]^T f(x)$$

$$= f^T(x)J(x)f(x) + f^T(x)J^T(x)f(x)$$

$$= f^T(x)[J(x) + J^T(x)]f(x) = f^T(x)M(x)f(x),$$

where $M(x) = J(x) + J^T(x)$.

If $M(x)$ is negative definite in the neighborhood of the origin, then the quadratic form $V^*(x) = f^T(x)M(x)f(x)$ is also negative definite. Therefore, if $M(x)$ is negative definite in the neighborhood of the origin, then the zero solution of Equation 1.2 is asymptotically stable. However, $V(x)$ may not be a simple function.

Remark 1.4

The method works if $M(x)$ is negative definite (a sufficient condition) in the neighborhood of the origin but if $M(x)$ is not negative definite, we cannot say anything about the stability.

We cannot apply the method on Examples 1.11 and 1.13. Let us test the method on Example 1.12.

Example 1.15

Test the stability of the zero solution in the following problem (Example 1.12) using Krasovskii's method.

$$x'_1 = -2x_1 + 3x_2^2, \quad x'_2 = -3x_2 - 6x_1x_2.$$

SOLUTION
We have

$$V(x_1, x_2) = f_1^2(x_1, x_2) + f_2^2(x_1, x_2)$$

$$= (-2x_1 + 3x_2^2)^2 + (-3x_2 - 6x_1x_2)^2.$$

Note that $V(0, 0) = 0$ and $V(x_1, x_2)$ is positive definite. We have

$$J(x) = \begin{bmatrix} -2 & 6x_2 \\ -6x_2 & -3 - 6x_1 \end{bmatrix}, \quad J^T(x) = \begin{bmatrix} -2 & -6x_2 \\ 6x_2 & -3 - 6x_1 \end{bmatrix},$$

and

$$M(x_1, x_2) = J(x) + J^T(x) = \begin{bmatrix} -4 & 0 \\ 0 & -6(1 + 2x_1) \end{bmatrix}.$$

Now, $M(x_1, x_2)$ is negative definite and hence $V^*(x)$ is negative definite in the neighborhood of the origin $(0, 0)$. Hence, the zero solution is asymptotically stable.

Lyapunov Function for Linear Systems with Constant Coefficients

Consider the following system of n linear first-order differential equations with constant coefficients:

$$x' = Ax, \tag{1.7}$$

where A is a real $n \times n$ constant matrix. We consider the case when the matrix A has distinct eigenvalues (real or complex roots). Suppose that A has r real eigenvalues $\lambda_1, \lambda_2, \ldots, \lambda_r$. Denote the complex pairs of eigenvalues as $(\lambda_i, \lambda_{i+1}) = p_i \pm q_i$, $i = r+1, r+3, \ldots, n-1$. The complex pairs of roots are counted as two consecutive eigenvalues.

Let the diagonal matrix $D = diag(\lambda_1, \lambda_2, \ldots, \lambda_n)$. Using the transformation $x = Py$, (P is a nonsingular matrix) in Equation 1.7, we obtain

$$Py' = APy, \quad \text{or} \quad y' = P^{-1}APy = Dy. \tag{1.8}$$

Since we have applied a similarity transformation, the eigenvalues of D are the same as the eigenvalues of A. *For this system to be asymptotically stable, we require that all the real eigenvalues are negative and the complex pairs of eigenvalues have negative real parts.* (This implies that A is a stable matrix.) The system is unstable if one or more real eigenvalues are positive or some of the complex eigenvalues have positive real parts.

Now, we select a Lyapunov function for Equation 1.8 as the inner product

$$V(y) = (y, By), \tag{1.9}$$

where the $n \times n$ real, constant, symmetric matrix B is to be determined. Differentiating, we obtain

$$V^*(y) = (y', By) + (y, By') = (Dy, By) + (y, BDy)$$

$$= (y, (\bar{D}^T B + BD)y) = (y, (\bar{D}B + BD)y) \tag{1.10}$$

since D is a diagonal matrix.

In order that V^* be negative definite, we require that

$$V^*(y) = -(y, y) = -(y_1^2 + y_2^2 + \cdots + y_n^2) \tag{1.11}$$

when all the eigenvalues are real and negative, or as

$$V^*(y) = -(y, y) = -(y_1^2 + y_2^2 + \cdots + |y_{r+1}|^2 + |y_{r+2}|^2 + \cdots + |y_n|^2)$$

when r eigenvalues are real and negative and the remaining complex pairs of eigenvalues have negative real parts.

Comparing Equations 1.10 and 1.11, we require that B satisfies the equation

$$\bar{D}B + BD = -I. \tag{1.12}$$

A simple calculation gives

$$b_{i,j} = 0, i \neq j; \quad b_{i,i} = -\left[\frac{1}{2\lambda_i}\right], \quad i = 1, 2, \dots, r;$$

$$(b_{i,i}, b_{i+1,i+1}) = -\left[\frac{1}{2p_i}\right], \quad i = r+1, r+3, \dots, n-1.$$

Therefore,

$$B = diag\left[-\frac{1}{2\lambda_1}, -\frac{1}{2\lambda_2}, \dots, -\frac{1}{2\lambda_r}, -\frac{1}{2p_{r+1}}, -\frac{1}{2p_{r+1}}, \dots, -\frac{1}{2p_{n-1}}\right].$$

From Equation 1.9, we get

$$V(y) = -\frac{1}{2\lambda_1}y_1^2 - \frac{1}{2\lambda_2}y_2^2 - \dots - \frac{1}{2\lambda_r}y_r^2 - \frac{1}{2p_{r+1}}(|y_{r+1}|^2 + |y_{r+2}|^2) - \dots.$$

If the eigenvalues of A are all real and distinct, then $\bar{D} = D$ and Equation 1.12 becomes $DB + BD = -I$. Then, B is given by

$$B = diag\left[-\frac{1}{2\lambda_1}, -\frac{1}{2\lambda_2}, \dots, -\frac{1}{2\lambda_n}\right]$$

and

$$V(y) = -\frac{1}{2\lambda_1}y_1^2 - \frac{1}{2\lambda_2}y_2^2 - \dots - \frac{1}{2\lambda_n}y_n^2.$$

Remark 1.5

If a suitable Lyapunov function can be determined following the above procedure, then the stability of the zero solution is assured. However, if no such function can be determined, we cannot provide any conclusion about the stability. From the above discussion, $\lambda = 0$ is not an eigenvalue of A.

Remark 1.6

The above procedure cannot deal with the critical case (i.e., when the real parts of the eigenvalues of A are nonpositive and the real part of at least one eigenvalue is zero).

Remark 1.7

The solution trajectories of the system (1.7) tend (converge) to zero as $t \to \infty$ if and only if the real eigenvalues of A are negative and the complex eigenvalues have negative real parts. In such a case, the origin of the system is called an *attractor*.

Example 1.16

Reduce the differential equation $z''' + 6z'' + 11z' + 6z = 0$, to a three-dimensional system of the form $x' = Ax$. Hence, construct a Lyapunov function, if possible, for the system. Is the origin $(0, 0, 0)$ asymptotically stable?

SOLUTION

Set $z = x_1$. Then,

$$x_1' = x_2, \quad x_2' = x_3, \quad x_3' = -6x_3 - 11x_2 - 6x_1.$$

We obtain the system of equations as $x' = Ax$, where

$$A = \begin{bmatrix} 0 & 1 & 0 \\ 0 & 0 & 1 \\ -6 & -11 & -6 \end{bmatrix}.$$

The characteristic equation of A is $\lambda^3 + 6\lambda^2 + 11\lambda + 6 = 0$. The eigenvalues are $\lambda = -1, -2, -3$. The matrix of eigenvectors (modal matrix) is obtained as

$$P = \begin{bmatrix} 1 & 1 & 1 \\ -1 & -2 & -3 \\ 1 & 4 & 9 \end{bmatrix} \quad \text{and} \quad P^{-1}AP = D = \begin{bmatrix} -1 & 0 & 0 \\ 0 & -2 & 0 \\ 0 & 0 & -3 \end{bmatrix}.$$

The transformation $x = Py$ gives $y' = Dy$.

To construct a Lyapunov function, we find the matrix B such that $DB + BD = -I$. We obtain

$$B = diag(1/2, 1/4, 1/6).$$

The Lyapunov function is given by

$$V(y) = (y, By) = \frac{1}{2}y_1^2 + \frac{1}{4}y_2^2 + \frac{1}{6}y_3^2.$$

We conclude that the origin $(0, 0, 0)$ is asymptotically stable.

We can write the Lyapunov function in terms of the original variables. From $x = Py$, we get $y = P^{-1}x$. We obtain

$$y = \frac{1}{2}\begin{bmatrix} 6x_1 + 5x_2 + x_3 \\ -6x_1 - 8x_2 - 2x_3 \\ 2x_1 + 3x_2 + x_3 \end{bmatrix}.$$

$$V(x) = \frac{1}{8}(6x_1 + 5x_2 + x_3)^2 + \frac{1}{16}(6x_1 + 8x_2 + 2x_3)^2 + \frac{1}{24}(2x_1 + 3x_2 + x_3)^2.$$

With this expression for the Lyapunov function $V(x)$, it can also be shown that $V^*(x)$ is negative definite.

The Routh–Hurwitz Criterion for Stability

In the previous section, we have noted that the system $x' = Ax$ is asymptotically stable if all the real eigenvalues of A are negative and the complex pairs of eigenvalues have negative real parts, which implies that A is a stable matrix.

An alternative method to show that A is a stable matrix is by testing the Routh–Hurwitz criterion.

We have the following theorem.

Theorem 1.4 (Routh–Hurwitz Criterion)

Let

$$P(\lambda) = a_0\lambda^n + a_1\lambda^{n-1} + a_2\lambda^{n-2} + \cdots + a_n.$$

Define

$$D = \begin{vmatrix} a_1 & a_3 & a_5 & \cdots & \cdots & a_{2n-1} \\ a_0 & a_2 & a_4 & \cdots & \cdots & a_{2n-2} \\ 0 & a_1 & a_3 & \cdots & \cdots & a_{2n-3} \\ 0 & a_0 & a_2 & \cdots & \cdots & a_{2n-4} \\ \cdots & \cdots & \cdots & \cdots & \cdots & \cdots \\ 0 & 0 & 0 & \cdots & \cdots & a_n \end{vmatrix},$$

where $a_k = 0$ for $k < 0$ or $k > n$. Then, the roots of the equation $P(\lambda) = 0$ are negative or have negative real parts if and only if $a_i > 0$ for all i and the leading principal minors of D are positive. That is,

$$D_1 = a_1 > 0, \quad D_2 = \begin{vmatrix} a_1 & a_3 \\ a_0 & a_2 \end{vmatrix} > 0, \quad D_3 = \begin{vmatrix} a_1 & a_3 & a_5 \\ a_0 & a_2 & a_4 \\ 0 & a_1 & a_3 \end{vmatrix} > 0, \quad \text{etc.}$$

If these conditions are satisfied, then the matrix D whose characteristic equation is $P(\lambda) = 0$ is said to be *Hurwitz stable*.

The theorem implies the following:

i. If the characteristic equation is a quadratic, that is, $P(\lambda) = a_0\lambda^2 + a_1\lambda + a_2$, then $a_0 > 0$, $a_1 > 0$, and $a_2 > 0$ are the necessary and sufficient conditions that the roots of the equation $P(\lambda) = 0$ are negative or have negative real parts.

ii. If the characteristic equation is a cubic, that is, $P(\lambda) = a_0\lambda^3 + a_1\lambda^2 + a_2\lambda + a_3$, then $a_0 > 0$, $a_1 > 0$, $a_2 > 0$, $a_3 > 0$, and $a_1a_2 - a_0a_3 > 0$ are the necessary and sufficient conditions that the roots of the equation $P(\lambda) = 0$ are negative or have negative real parts.

Example 1.17

Test whether the roots of the following characteristic equations are negative or have negative real parts.

    i. $P(\lambda) = \lambda^3 + 7\lambda^2 + 14\lambda + 8$.
   ii. $P(\lambda) = \lambda^3 + 5\lambda^2 + 17\lambda + 13$.

SOLUTION

i. We have $a_0 = 1, a_1 = 7, a_2 = 14, a_3 = 8$ and $a_1a_2 - a_0a_3 = 90$. Therefore, the roots of $P(\lambda) = 0$ are negative or have negative real parts. The actual roots are -1, -2, and -4.

ii. We have $a_0 = 1$, $a_1 = 5$, $a_2 = 17$, $a_3 = 13$, and $a_1a_2 - a_0a_3 = 72$. Therefore, the roots of $P(\lambda) = 0$ are negative or have negative real parts. The actual roots are -1, $-2 \pm 3i$.

Stability Discussion Based on the Linearization Procedure

Consider the system $x' = f(x)$, or in component form

$$x_i' = f_i(x_1, x_2, \ldots, x_n), \quad i = 1, 2, \ldots, n$$

$$f_i(0, 0, \ldots, 0) = 0$$

(1.13)

defined in the open set Γ of R^n containing the origin.

We assume that (i) f_i are sufficiently smooth and (ii) partial derivatives exist and are small.

An *equilibrium solution* of Equation 1.13 is a point $x_0 \in R^n$ such that $f(x_0) = 0$, that is, a solution which is constant in time. It is also called an equilibrium point, fixed point, stationary point, critical point, or steady state.

To carry out local stability analysis, we linearize system (1.13) about x_0 as

$$x_i' = \sum_{j=1}^{n} a_{ij} x_j + R_i(x_1, x_2, \ldots, x_n), \tag{1.14}$$

where the coefficients a_{ij} are the first partial derivatives evaluated at x_0 and are constants. That is,

$$a_{ij} = \left(\frac{\partial f_i}{\partial x_j} \right)_{x_0}.$$

$R_i, i = 1, 2, \ldots, n$ are the remainder terms and are of higher order than the first. In matrix notation, we write Equation 1.14 as

$$x' = Ax + R(x), \tag{1.15}$$

where A is the coefficient matrix

$$A = \begin{bmatrix} a_{11} & a_{12} & \cdots & a_{1n} \\ a_{21} & a_{22} & \cdots & a_{2n} \\ \cdots & \cdots & \cdots & \cdots \\ a_{n1} & a_{n2} & \cdots & a_{nn} \end{bmatrix}.$$

The smallness of the higher-order partial derivatives is defined by the inequality

$$\|R(x)\| \leq N\|x\|^{1+2\alpha} \tag{1.16}$$

in the neighborhood of x_0, N and α are positive constants and Euclidean norm (1.3) is used.

Consider the linearized system as

$$x' = Ax. \tag{1.17}$$

An equilibrium point x_0 is called a

i. *Stable node or sink:* If all of the eigenvalues of matrix A have negative real parts.

ii. *Unstable node or source:* If all the eigenvalues of A have positive real parts.

iii. *Hyperbolic equilibrium point:* If all the eigenvalues of matrix A have nonzero real parts.

iv. A *saddle point:* If it is a hyperbolic equilibrium point and A has at least one eigenvalue with a positive real part and at least one eigenvalue with a negative real part.

If the eigenvalues of A are purely imaginary and nonzero, the nonhyperbolic fixed point is called a *center*.

The set of mappings $e^{At}: R^n \to R^n$ is called the *flow of the linear system* (1.17) describing the motion of the point $x_0 \in R^n$ along the solution trajectory of the system (1.17). If all the eigenvalues of A have nonzero real parts, then the flow $e^{At}: R^n \to R^n$ is called a *hyperbolic flow* and the system (1.17) is called a *hyperbolic linear system*.

A subspace $E \subset R^n$ is said to be invariant with respect to the flow $e^{At}: R^n \to R^n$ if $e^{At}E \subset E$ for all $t \in R$.

A simple physical example of a nonlinear sink is given by a simple pendulum moving in the vertical plane (see Example 1.6 and Figure 1.2). The assumptions are that (i) the string has negligible mass, and (ii) air resistance is negligible. The only forces acting on the system are (i) tension T in the string, and (ii) gravitational force mg, which acts vertically downward.

The length of the string is l. The bob of the pendulum moves along the circumference of a circle of radius l. The angular velocity of the bob is $d\theta/dt$ and the velocity is $l\, d\theta/dt$. The total force acting along the tangent to the circle at time t is

$$F = -\left(kl\frac{d\theta}{dt} + m\sin\theta \right),$$

where k is the coefficient of friction.

The acceleration of the bob, tangent to the circle is $a = l(d^2\theta/dt^2)$. From Newton's law, we have

$$ml\frac{d^2\theta}{dt^2} = -\left(kl\frac{d\theta}{dt} + m\sin\theta \right).$$

Setting $\omega = \theta'$, we obtain the equivalent first-order system

$$\theta' = \omega, \quad \text{and} \quad \omega' = -(1/l)\sin\theta - (k/m)\omega.$$

This nonlinear autonomous system in R^2 has equilibrium points at $(\theta, \omega) = (n\pi, 0)$, $n = 0, \pm1, \pm2, \ldots$.

Now, consider the equilibrium point at $(0, 0)$. The vector field defining the above system is $f(\theta, \omega) = (\omega, -(1/l)\sin\theta - (k/m)\omega)$. The Jacobian of $f(\theta, \omega)$ is given by

$$J[f(\theta,\omega)] = \begin{bmatrix} 0 & 1 \\ -(\cos\theta/l) & -k/m \end{bmatrix},$$

and

$$J[f(0,0)] = \begin{bmatrix} 0 & 1 \\ -1/l & -k/m \end{bmatrix}.$$

The eigenvalues of J are $\lambda_{1,2} = [-p \pm \sqrt{p^2 - (4/l)}]/2$, where $p = k/m$. The real parts of the eigenvalues are negative. Therefore, the equilibrium point $(\theta, \omega) = (0, 0)$ is a sink. The above mathematical model is only an approximation to reality [11].

Linear stability (local stability) is defined in the following theorems.

Theorem 1.5

Let R satisfy inequality (1.16). Then, if A is a stable matrix, the zero solution $x(t) = 0$ of Equation 1.15 is asymptotically stable.

Theorem 1.6

Let R satisfy inequality (1.16). Then, if A has at least one eigenvalue whose real part is positive, then the zero solution $x(t) = 0$ of Equation 1.15 is unstable.

Remark 1.8

From Theorems 1.5 and 1.6, we conclude that if A does not have any zero eigenvalue and the eigenvalues have nonzero real parts, then the stability properties of the original system are same as the stability properties of the linearized system. However, a linearly stable solution may be nonlinearly unstable.

Remark 1.9

Stability of the zero solution $x(t) = 0$ of Equation 1.15 cannot be discussed using the linearization procedure in the critical case.

Example 1.18

Test the linear stability of the zero solution $x_1(t) \equiv 0$, $x_2(t) \equiv 0$, in the following problems.

i. $x_1' = x_1 + 4x_2 - 3x_1^2$, $x_2' = 3x_1 + 2x_2 + 5x_2^4$.

ii. $x_1' = -x_2 - x_1^5$, $x_2' = 4x_1 - x_2^5$.

iii. The Lotka–Volterra population model governing the growth/decay of species

$$x_1' = ax_1 - x_1x_2, \quad x_2' = -bx_2 + x_1x_2, \quad \text{with } a = 1, b = 1.$$

SOLUTION

i. Linearizing the right-hand sides using Taylor series, we get

$$x_1' = 0 + x_1[1 - 6x_1]_0 + x_2[4] = x_1 + 4x_2,$$

$$x_2' = 0 + x_1[3] + x_2[2 + 20x_2^3]_0 = 3x_1 + 2x_2.$$

We obtain the system as $x' = Ax + R(x)$, where $A = \begin{bmatrix} 1 & 4 \\ 3 & 2 \end{bmatrix}$.

The nonlinear terms R_1, R_2 satisfy an inequality of form (1.16).

The characteristic equation of A is $\lambda^2 - 3\lambda - 10 = 0$, and the eigenvalues are $\lambda = 5$, $\lambda = -2$. By Theorem 1.6, the zero solution is unstable.

ii. Linearizing the right-hand sides using Taylor series, we get

$$x_1' = 0 + x_1[-5x_1^4]_0 + x_2[-1] = -x_2,$$

$$x_2' = 0 + x_1[4] + x_2[-5x_2^4]_0 = 4x_1.$$

We obtain the system as $x' = Ax + R(x)$, where $A = \begin{bmatrix} 0 & -1 \\ 4 & 0 \end{bmatrix}$.

The nonlinear terms $R(x)$ satisfy an inequality of the form (1.16). The characteristic equation of A is $\lambda^2 + 4 = 0$, and the eigenvalues are $\lambda = \pm 2i$. This is a critical case where the real parts are zero. The test based on linearization cannot be applied.

However, we can study the stability by constructing a Lyapunov function. Consider the positive-definite function $V(x_1, x_2) = 4x_1^2 + x_2^2$.

We obtain

$$V^*(x_1, x_2) = 8x_1(-x_2 - x_1^5) + 2x_2(4x_1 - x_2^5) = -2(4x_1^6 + x_2^6).$$

$V^*(x_1, x_2)$ is negative definite. Therefore, the zero solution is asymptotically stable.

iii. Setting $x_1 - x_1 x_2 = 0$, $-x_2 + x_1 x_2 = 0$, we obtain the critical points as $(0, 0)$ and $(1, 1)$.

First, consider the critical point $(0, 0)$. The linearized system about $(0, 0)$ is given by

$$x' = Ax, \quad \text{where } A = \begin{bmatrix} 1 & 0 \\ 0 & -1 \end{bmatrix}.$$

The eigenvalues are $\lambda = \pm 1$. Therefore, $(0, 0)$ is a saddle point and hence it is unstable. Now, consider the critical point $(1, 1)$. The linearized system about $(1, 1)$ is given by

$$y' = Ay, \quad \text{where } A = \begin{bmatrix} 0 & -1 \\ 1 & 0 \end{bmatrix},$$

where $y_1 = x_1 - 1$, $y_2 = x_2 - 1$.

The eigenvalues are $\lambda = \pm i$. It is a critical case. Stability cannot be studied by the linearization method.

We try to construct a Lyapunov function after shifting the origin to the point $(1, 1)$. Setting $y_1 = x_1 - 1$, $y_2 = x_2 - 1$, we get the system as

$$y_1' = -y_2 - y_1 y_2, \quad y_2' = y_1 + y_1 y_2. \tag{1.18}$$

Choose a positive-definite function $V(y_1, y_2)$ as

$$V(y_1, y_2) = y_1 + y_2 + \log \left[1/\{(1 + y_1)(1 + y_2)\} \right].$$

We obtain,

$$V^*(y_1, y_2) = \frac{\partial V}{\partial y_1} f_1 + \frac{\partial V}{\partial y_2} f_2 = -y_1 y_2 + y_1 y_2 = 0.$$

By Theorem 1.1, the critical point $(0, 0)$ for the derived system (1.18) is stable but not asymptotically stable. This implies that the critical point $(1, 1)$ of the given system is a center of the system.

Example 1.19

Discuss the stability of the zero solution of the Lorenz equations using the linearization method

$$x_1' = a(x_2 - x_1), \quad x_2' = bx_1 - x_2 - x_1 x_3, \quad x_3' = x_1 x_2 - cx_3, a > 0, b > 0, c > 0.$$

How many equilibrium points exist for the cases $b \le 1$, $b > 1$? For $b < 1$, discuss the global stability of the origin.

SOLUTION

Setting $a(x_2 - x_1) = 0$, $bx_1 - x_2 - x_1x_3 = 0$, $x_1x_2 - cx_3 = 0$, we get the equilibrium points as

$$E_0(0,\ 0,\ 0),\quad E_1 = (\sqrt{c(b-1)}, \sqrt{c(b-1)}, b-1)$$

and

$$E_2 = (-\sqrt{c(b-1)}, -\sqrt{c(b-1)}, b-1).$$

For $b \le 1$, $E_0(0, 0, 0)$ is the only equilibrium point. For $b > 1$, the system has three equilibrium point E_0, E_1, E_2. Now, we discuss the linear stability of the origin.

The linearization about the origin gives

$$x_1' = a(x_2 - x_1),\quad x_2' = bx_1 - x_2,\quad x_3' = -cx_3.$$

The equation for x_3 is decoupled. The solution of this equation is $x_3 = Ae^{-ct} \to 0$ as $t \to \infty$.

The other two equations are given by

$$\mathbf{x}' = A\mathbf{x}, \quad \text{where } \mathbf{x} = \begin{bmatrix} x_1 \\ x_2 \end{bmatrix}, \text{ and } A = \begin{bmatrix} -a & a \\ b & -1 \end{bmatrix}.$$

The characteristic equation of A is $\lambda^2 + (a+1)\lambda + a(1-b) = 0$, whose roots are given by

$$\lambda = [-(a+1) \pm \sqrt{(a-1)^2 + 4ab}]/2.$$

Now, trace $(A) = \tau = -(a+1) < 0$, and determinant $(A) = a(1-b)$.

If $b > 1$, determinant $(A) < 0$ and the origin is a saddle point. Also, for $b > 1$,

$$(a-1)^2 + 4ab > (a-1)^2 + 4a = (a+1)^2.$$

Hence, one of the characteristic roots is negative and the other is positive. Therefore, the saddle point has one outgoing and two incoming directions including the decaying x_3 direction.

If $b < 1$, $(a-1)^2 + 4ab < (a-1)^2 + 4a = (a+1)^2$. Hence, both the characteristic roots are negative. Therefore, all the directions are incoming and the origin is a sink (stable node).

Now, we discuss the global stability of the origin.

For $b < 1$, we can show that every trajectory approaches the origin as $t \to \infty$, that is, the origin is globally stable. Consider a Lyapunov function as

$$V(x_1, x_2, x_3) = \frac{1}{a}x_1^2 + x_2^2 + x_3^2.$$

The surfaces of constant V are concentric ellipsoids about the origin. Now,

$$\frac{1}{2}V^*(x) = \frac{1}{a}x_1[a(x_2 - x_1)] + x_2(bx_1 - x_2 - x_1 x_3) + x_3(x_1 x_2 - cx_3)$$

$$= -[x_1^2 + x_2^2 - (b + 1)x_1 x_2 + cx_3^2]$$

$$= -\left[\left(x_1 - \frac{b+1}{2}x_2\right)^2 + \left(1 - \left(\frac{b+1}{2}\right)^2\right)x_2^2 + cx_3^2\right].$$

If $b < 1$ and $(x_1, x_2, x_3) \neq (0, 0, 0)$, then $V^*(x) < 0$ along all trajectories. The trajectories keep moving to the lower part of V and penetrate the smaller and smaller ellipsoids as $t \to \infty$. Since V is bounded below by zero, $V(x(t)) \to 0$ and, hence $x(t) \to 0$. On the right-hand side of the expression for $V^*(x)$, the coefficient of x_2^2 is positive for $b < 1$. Now, $V^*(x) = 0$ implies that $(x_1, x_2, x_3) = (0, 0, 0)$. Hence, the origin is globally stable for $b < 1$.

Global Asymptotic Stability

First, we define the region of attraction of the origin. The *region of attraction* of the origin of the autonomous system (1.13) is defined as the set S of points $x_0 \in R^n$, such that

$$\lim_{t \to \infty} x(t, x_0) = 0. \tag{1.19}$$

The region of attraction is also called the *region of asymptotic stability* of the zero solution.

If the region of attraction of the origin of the autonomous system (1.13) is the entire space R^n, then we say that the origin is globally asymptotically stable or completely stable. That is, every solution of the system irrespective of the initial value x_0 converges to zero as $t \to \infty$.

We state the following theorem regarding the global stability of an equilibrium point.

Theorem 1.7 [1]

Assume that there exists a scalar function $V(x)$ such that

  i. V is positive definite on R^n, and $V(x) \to \infty$ as $\| x \| \to \infty$,
  ii. $V^*(x) \le 0$ on R^n.

Then, all the solutions are bounded as $t \to \infty$.
This type of stability is also known as *Lagrange stability*.

Remark 1.10 [1]

If condition (ii) of Theorem 1.7 is replaced by the condition that V^* is negative definite at all points $x \in R^n$, then the zero solution of Equation 1.13 is globally asymptotically stable.

Remark 1.11 [1]

If in addition to the conditions of Theorem 1.7, if the origin is the only invariant subset of E, then the zero solution of Equation 1.13 is globally asymptotically stable.

Remark 1.12 [1]

If a system has more than one critical point, then none of the critical points are globally asymptotically stable.

Example 1.20

Show that the zero solution in the following systems is globally asymptotically stable.

  i. $x_1' = -x_2 - (1/2)x_1 x_2^2, \quad x_2' = 3x_1 - (1/3)x_2$.
  ii. $x_1' = x_2, \quad x_2' = -x_1 + \alpha x_1^2 x_2. \quad \alpha \le 0$.

SOLUTION

  i. The origin is the only critical point of the system. Let us choose that $V(x) = a\, x_1^2 + b\, x_2^2$. Then,

$$V^*(x) = 2a\, x_1 f_1 + 2bx_2 f_2$$

$$= 2a\, x_1[-x_2 - (1/2)x_1 x_2^2] + 2bx_2[3x_1 - (1/3)x_2]$$

$$= 2x_1 x_2[-a + 3b] - ax_1^2 x_2^2 - (2/3)bx_2^2.$$

Set $3b - a = 0$. We may choose $a = 3$, $b = 1$. We get $V(x) = 3x_1^2 + x_2^2$ and $V^*(x) = -x_2^2(3x_1^2 + 2)$.

Now, $V^*(x) \leq 0$ at all points $(x_1, x_2) \in R^2$ and $V(x_1, x_2) \to \infty$ as $x_1^2 + x_2^2 \to \infty$. Hence, all the solutions are bounded as $t \to \infty$. Also, $V^*(x) = 0$ when $x_2 = 0$, that is, at all points on the x_1 axis. E is the set of all points on the x_1 axis. Clearly, $(0, 0)$ is the only invariant subset of E. By Remark 1.11, zero solution $(0, 0)$ is globally asymptotically stable.

ii. The origin is the only critical point of the system. Let us choose $V(x) = a\, x_1^2 + b\, x_2^2$. Then,

$$V^*(x) = 2a\, x_1 f_1 + 2bx_2 f_2$$

$$= 2a\, x_1[x_2] + 2bx_2[-x_1 + \alpha\, x_1^2 x_2]$$

$$= 2x_1 x_2[a - b] + 2b\alpha\, x_1^2 x_2^2.$$

Set $a - b = 0$. We may choose $a = b = 1$. We obtain

$$V(x) = x_1^2 + x_2^2, \quad \text{and} \quad V^*(x) = 2\alpha\, x_1^2 x_2^2.$$

Now, $V^*(x) \leq 0$ at all points $(x_1, x_2) \in R^2$ and $V(x_1, x_2) \to \infty$ as $x_1^2 + x_2^2 \to \infty$. Hence, all the solutions are bounded as $t \to \infty$. $V^*(x)$ is negative definite for $\alpha < 0$, $x_1 \neq 0$, $x_2 \neq 0$. Also, $V^*(x) = 0$ when $x_2 = 0$, that is, at all points on the x_1 axis, and when $x_1 = 0$, that is, at all points on the x_2-axis. E is the set of all points on the x_1- and x_2-axis. Clearly, $(0, 0)$ is the only invariant subset of E. By Remark 1.11, the zero solution is globally asymptotically stable for $\alpha < 0$. Also, $V^*(x) = 0$ for $\alpha = 0$. The zero solution is stable for $\alpha = 0$.

Limit Cycles

First, we define *limit sets*. Consider the autonomous system $x' = f(x)$. Assume that a solution of the system exists and is unique. Let $f(0) = 0$ and $f(x) \neq 0$ for $x \neq 0$ in the neighborhood of the origin. Let $x(t, x_0)$ denote any solution of the system, where x_0 is an initial value of x such that $x(0, x_0) = x_0$. The solution curve $x(t, x_0)$ is called an *orbit* through x_0. Denote this orbit by $C(x_0)$. That is, $C(x_0) = \{x \in R^n : x = x(t, x_0), t \in R\}$.

If $t \in R^+$, then we call the orbit a *positive semiorbit* and denote it as $C^+(x_0)$.

If $t \in R^-$, then we call the orbit a *negative semiorbit* and denote it as $C^-(x_0)$.

The *positive limit set* or ω-limit set $L(C^+)$ is the set of all limit points of C^+.

The *negative limit set* or α-limit set $L(C^-)$ is the set of all limit points of C^-.

Limit cycle: An isolated closed path is called a limit cycle. Limit cycles can only occur in nonlinear systems. A linear system has no limit cycles. A linear system $x' = Ax$ can have closed orbits, but they would not be isolated, that is, if $x(t)$ is a periodic solution, then so is $cx(t)$ for any constant $c \neq 0$. Hence, $x(t)$ is surrounded by a one-parameter family of closed orbits. Normally, it is difficult to tell whether a given system has a limit cycle or any closed orbits from the governing equations alone. A periodic solution C of the nonlinear system (1.13) is called a limit cycle if it is a limit set of another solution of the system (1.13).

Stability of a limit cycle: Let the periodic solution C of nonlinear system (1.13) be a closed curve. We define the following:

Stable limit cycle: If the periodic solution C is the positive limit set of the solutions contained in the interior and also of the solutions contained in the exterior of C, then the limit cycle is said to be stable. That is, all the solution trajectories in the interior and also the solution trajectories in the exterior converge to C as $t \rightarrow \infty$ (see Figure 1.5a).

Unstable limit cycle: If the periodic solution C is the negative limit set of the solutions contained in the interior and also the solutions contained in the exterior of C, then the limit cycle is said to be unstable. That is, all the solution trajectories in the interior and also the solution trajectories in the exterior diverge (move away from C) as $t \rightarrow \infty$ (see Figure 1.5b).

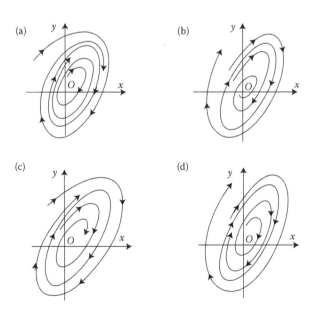

FIGURE 1.5
(a) Stable limit cycle. (b) Unstable limit cycle. (c, d) Semistable limit cycles.

Semistable limit cycle: If the periodic solution C is the negative limit set of the solutions contained in the interior but is the positive limit set of the solutions contained in the exterior of C, then the limit cycle is said to be semistable. That is, all the solution trajectories in the interior diverge from C, but the solution trajectories in the exterior converge to C as $t \to \infty$ (see Figure 1.5c). Similarly, if the periodic solution C is the positive limit set of the solutions contained in the interior but is the negative limit set of the solutions contained in the exterior of C, then the limit cycle is said to be semistable. That is, all the solution trajectories in the interior converge to C, but the solution trajectories in the exterior diverge from C as $t \to \infty$ (see Figure 1.5d).

Stable limit cycles imply self-sustained oscillations. Any small perturbation from the closed trajectory would cause the system to return to the limit cycle, making the system stick to the limit cycle.

It is easy to understand the limit cycles in two dimensions.

Consider the two-dimensional autonomous system

$$x_1' = f_1(x_1, x_2), \quad x_2' = f_2(x_1, x_2). \tag{1.20}$$

Assume that f_1, f_2 and their first partial derivatives are continuous. Sometimes, it is possible to study the nature of the limit cycle through linearization of the system. Let C be a periodic solution of period T of this system whose closed orbit is defined by $x_1 = \alpha(t)$, $x_2 = \beta(t)$. The linear system corresponding to this solution is

$$\begin{bmatrix} x_1' \\ x_2' \end{bmatrix} = \begin{bmatrix} \partial f_1/\partial x_1 & \partial f_1/\partial x_2 \\ \partial f_2/\partial x_1 & \partial f_2/\partial x_2 \end{bmatrix} \begin{bmatrix} x_1 \\ x_2 \end{bmatrix} = \begin{bmatrix} a_{11} & a_{12} \\ a_{21} & a_{22} \end{bmatrix} \begin{bmatrix} x_1 \\ x_2 \end{bmatrix}, \tag{1.21}$$

where the partial derivatives are evaluated at $(x_1 = \alpha(t), x_2 = \beta(t))$.

We know that one of the characteristic multipliers is unity (see [1], Corollary 2.2.3). That is, one of the characteristic exponents of Equation 1.21 is equal to zero. Now, trace of the matrix $= a_{11}(t) + a_{22}(t) =$ sum of the eigen values.

The other characteristic exponent is given by [1]

$$I = \frac{1}{T} \int_0^T [a_{11}(t) + a_{22}(t)] \, dt. \tag{1.22}$$

The limit cycle C is stable if I is negative and unstable if I is positive. If $I = 0$, C may or may not be semistable and we need to study the effect of the nonlinear terms.

Remark 1.13

The stability or instability of limit cycles depends on the asymptotic behavior of neighboring paths. Often, linearized systems may not provide any indication of the existence of periodic solutions or limit cycles.

An important result about periodic orbits and limit cycles is given in the following famous theorem called the *Poincaré–Bendixson* theorem [9].

Theorem 1.8 (Poincaré–Bendixson Theorem)

If C^+ is a bounded positive semiorbit of the system (1.20) and $L(C^+)$ does not contain the critical points of the system (1.20), then $L(C^+)$ is a periodic orbit. Moreover, if C^+ and $L(C^+)$ have no common regular point, then $L(C^+)$ is a limit cycle.

Example 1.21

Discuss the stability of the limit cycles in the following problems:

    i. $x_1' = -x_2 + 2x_1(x_1^2 + x_2^2 - 4), \quad x_2' = x_1 + 2x_2(x_1^2 + x_2^2 - 4).$

    ii. $x_1' = -x_2 - 2x_1(x_1^2 + x_2^2 - 4), \quad x_2' = x_1 - 2x_2(x_1^2 + x_2^2 - 4).$

    iii. $x_1' = -2x_2 + x_1(4 - x_1^2 - x_2^2)(x_1^2 + x_2^2 - 9),$

       $x_2' = 2x_1 + x_2(4 - x_1^2 - x_2^2)(x_1^2 + x_2^2 - 9).$

SOLUTION

In the problems given above, it is possible to guess the periodic closed orbits and test them for stability.

    i. Observing that the right-hand sides have the factor $(x_1^2 + x_2^2 - 4)$ and from the other factors, we find that a 2π periodic solution is given by $C: x_1 = 2\cos t, x_2 = 2\sin t$. Evaluating the first partial derivatives at $x_1 = 2\cos t, x_2 = 2\sin t$, we obtain

$$a_{11} = \frac{\partial f_1}{\partial x_1} = 2[(x_1^2 + x_2^2 - 4) + 2x_1^2] = 16\cos^2 t,$$

$$a_{22} = \frac{\partial f_2}{\partial x_2} = 2[(x_1^2 + x_2^2 - 4) + 2x_2^2] = 16\sin^2 t, \quad \text{and} \quad a_{11} + a_{22} = 16.$$

$$I = \frac{1}{T}\int_0^T [a_{11}(t) + a_{22}(t)]\, dt = \frac{1}{2\pi}\int_0^{2\pi} 16\, dt = 16 > 0.$$

    Therefore, the limit cycle C is unstable. Note that the critical point $(0, 0)$ is asymptotically stable.

ii. The problem is the same as (i) except for the negative sign in the second term on the right-hand side. We obtain from the 2π periodic solution given by $x_1 = 2\cos t$, $x_2 = 2\sin t$,

$$a_{11} = -16\cos^2 t, \quad a_{22} = -16\sin^2 t, \quad a_{11} + a_{22} = -16, \quad I = -16 < 0.$$

Therefore, the limit cycle is stable.

iii. Observing that the right-hand sides have the factors $(x_1^2 + x_2^2 - 9)$, $(4 - x_1^2 - x_2^2)$, and from the other factors, we find that there are two π periodic solutions. They are given by

$$C_1: x_1 = 2\cos 2t, \; x_2 = 2\sin 2t.$$

$$C_2: x_1 = 3\cos 2t, \; x_2 = 3\sin 2t.$$

For C_1, evaluating the first partial derivatives at $x_1 = 2\cos 2t$, $x_2 = 2\sin 2t$, we obtain

$$a_{11} = \frac{\partial f_1}{\partial x_1} = 40\cos^2 t, \quad a_{22} = \frac{\partial f_2}{\partial x_2} = 40\sin^2 t, \quad \text{and} \quad a_{11} + a_{22} = 40.$$

$$I = \frac{1}{T}\int_0^T [a_{11}(t) + a_{22}(t)]dt = \frac{1}{\pi}\int_0^\pi 40 \, dt = 40 > 0.$$

Therefore, the limit cycle C_1 is unstable.

For C_2, evaluating the first partial derivatives at $x_1 = 3\cos 2t$, $x_2 = 3\sin 2t$, we obtain

$$a_{11} = \frac{\partial f_1}{\partial x_1} = -90\cos^2 t, \quad a_{22} = \frac{\partial f_2}{\partial x_2} = -90\sin^2 t,$$

and

$$a_{11} + a_{22} = -90.$$

$$I = \frac{1}{T}\int_0^T [a_{11}(t) + a_{22}(t)] \, dt = \frac{1}{\pi}\int_0^\pi (-90) \, dt = -90 < 0.$$

Therefore, the limit cycle C_2 is stable.

Example 1.22

Using polar coordinates, discuss the stability of the limit cycle [23]

$$\dot{x} = -y + x(1 - x^2 - y^2), \quad r(0) = r_0.$$
$$\dot{y} = x + y(1 - x^2 - y^2). \quad \theta(0) = \theta_0.$$

SOLUTION

In polar coordinates, $x = r \cos \theta$, $y = r \sin \theta$, we obtain

$$\dot{r} \cos\theta - r(\sin\theta)\dot{\theta} = -r\sin\theta + r(1 - r^2)\cos\theta,$$
$$\dot{r} \sin\theta + r(\cos\theta)\dot{\theta} = r\cos\theta + r(1 - r^2)\sin\theta.$$

Solving, we obtain $\dot{r} = r(1 - r^2)$, $\dot{\theta} = 1$.

The origin ($r = 0$) is an equilibrium point of this system. The solution is given by

$$r(t, r_0) = [1 + \{(1 - r_0^2)/r_0^2\}e^{-2t}]^{-1/2}, \quad \theta(t, \theta_0) = \theta + \theta_0.$$

The flow spirals around the origin in the counter clockwise direction; it spirals outward for $0 < r < 1$ since $\dot{r} > 0$; and it spirals inward for $r > 1$ since $\dot{r} < 0$. The counter clockwise flow on the unit circle describes a stable limit cycle trajectory since $\dot{r} = 0$ on $r = 1$. The trajectory through the point $(\cos \theta_0, \sin \theta_0)$ on the unit circle is given by $x(t) = (\cos (t + \theta_0), \sin (t + \theta_0))^T$. All the initial conditions away from the origin will converge to the limit cycle.

Example 1.23

By constructing a Lyapunov function, show that the following systems have no closed orbits.

  i. $x_1' = -2x_1 + 3x_2^2$, $x_2' = -3x_2 - 6x_1x_2$.
  ii. $x_1' = -x_2 - 4x_1^3$, $x_2' = 9x_1 - 2x_2^3$.

SOLUTION

  i. Consider $V(x) = a\, x_1^2 + b\, x_2^2$. Then,

$$V^*(x) = 2a\, x_1 f_1 + 2bx_2 f_2$$

$$= 2a\, x_1[-2x_1 + 3x_2^2] + 2bx_2[-3x_2 - 6x_1x_2]$$

$$= -4a\, x_1^2 - 6bx_2^2 + (6a - 12b)x_1x_2^2.$$

Set $6a - 12b = 0$. We may choose, $a = 2$, $b = 1$. We get

$$V(x) = 2 x_1^2 + x_2^2, \quad \text{and} \quad V^*(x) = -(8x_1^2 + 6x_2^2).$$

Now, $V(x) > 0$ and $V^*(x) < 0$ for all $(x, y) \neq (0, 0)$. Hence, there are no closed orbits. In fact, all trajectories approach the origin as $t \to \infty$.

ii. Consider $V(x) = ax_1^2 + bx_2^2$. Then,

$$V^*(x) = 2a\, x_1 f_1 + 2x_2 f_2$$

$$= 2a\, x_1[-x_2 - 4x_1^3] + 2bx_2[9x_1 - 2x_2^3]$$

$$= (18b - 2a)\, x_1 x_2 - 8ax_1^4 - 4bx_2^4.$$

Set $18b - 2a = 0$. We may choose, $a = 9$, $b = 1$. We get

$$V(x) = 9 x_1^2 + x_2^2, \quad \text{and} \quad V^*(x) = -4(18x_1^2 + x_2^2).$$

Now, $V(x) > 0$ and $V^*(x) < 0$ for all $(x, y) \neq (0, 0)$. Hence, there are no closed orbits. In fact, all trajectories approach the origin as $t \to \infty$.

Liénard's Equation and Existence of a Limit Cycle

In the following theorem, we present the result on the existence of a limit cycle of Liénard's equation [28].

Theorem 1.9 (Liénard's Theorem)

Suppose that the functions $f(x)$ and $g(x)$ satisfy the following conditions:

i. $f(x)$ and $g(x)$ are continuously differentiable for all x.
ii. $f(x)$ and $g(x)$ are even and odd functions, respectively, for all x.
iii. $g(x) > 0$ for $x > 0$.
iv. The odd function $F(x) = \int_0^x f(u)\,du$ has the following properties:
    a. It has exactly one positive zero at $x = \alpha$.
    b. It is negative for $0 < x < \alpha$.
    c. It is positive and nondecreasing for $x > \alpha$.
    d. $F(x) \to \infty$ as $x \to \infty$.

Then, Liénard's equation

$$\ddot{x} + f(x)\dot{x} + g(x) = 0, \tag{1.23}$$

or the equivalent system

$$\dot{x} = y, \dot{y} = -g(x) - f(x)y, \tag{1.24}$$

has a unique, stable limit cycle surrounding the origin in the plane.

Example 1.24

Show that the van der Pol equation $\ddot{x} + \mu(x^2 - 1)\dot{x} + x = 0$ has a unique stable limit cycle for $\mu > 0$.

SOLUTION

Comparing the van der Pol equation with Liénard's equation, we get $f(x) = \mu(x^2 - 1)$ and $g(x) = x$. Conditions (i) through (iii) are automatically satisfied. Now, $F(x) = \int_0^x \mu(u^2 - 1)du = 1/3\,[\mu\,x(x^2 - 3)]$.

Condition (iv) is satisfied for $\alpha = \sqrt{3}$. By Theorem 1.9, the van der Pol equation has a unique stable limit cycle.

Energy Balance Method for Limit Cycles

The energy balance method was discussed by Jordan and Smith [13]. Consider the second-order nonlinear equation of the form

$$\ddot{x} + \varepsilon\,h(x,\dot{x}) + x = 0, \quad \text{where } |\varepsilon| \ll 1. \tag{1.25}$$

The contribution of the nonlinear term is small. Writing it as a system, we get

$$\dot{x} = y, \quad \dot{y} = -[\varepsilon\,h(x,y) + x], \tag{1.26}$$

which describes the phase plane. Assume that $h(0, 0) = 0$, so that the origin is an equilibrium point.

When $\varepsilon = 0$, Equation 1.25 reduces to the linear equation $\ddot{x} + x = 0$. In this case, the general solution is given by $x(t) = A \cos t + B \sin t$, or as $x(t) = a \cos (t + \theta)$, where a and θ are arbitrary constants. The family of phase paths is given by $x = a \cos t$, $y = -a \sin t$, where a is an arbitrary constant. That is, the family of phase paths is the system of circles $x^2 + y^2 = a^2$. The period of the closed paths is 2π. Therefore, for small values of ε, we may expect that the limit cycles or periodic paths will be close to the circular motions $x^2 + y^2 = a^2$ and approach it as $\varepsilon \to 0$. Hence, on the limit cycle, we may approximate

$$x(t) \approx a \cos t, y(t) \approx -a \sin t, T \approx 2\pi, \tag{1.27}$$

where a is the amplitude of the limit cycle.

Write Equation 1.25 as $\ddot{x} + x = -\varepsilon\,h(x,\dot{x})$.

A potential energy function is defined by $P.E. = \int x\,dx = (x^2/2)$.
The kinetic energy function is defined by $K.E. = (\dot{x}^2/2) = (y^2/2)$.
Hence, the total energy is given by $E = P.E. + K.E. = (x^2/2) + (y^2/2)$.
In the phase plane, the rate of change in energy, over one period $0 \le t \le T$ is given by

$$\frac{dE}{dt} = x\dot{x} + y\dot{y} = xy - y[\varepsilon\, h(x,y) + x] = -\varepsilon\, y h(x,y).$$

The change in energy over one period $0 \le t \le T$ is given by

$$E(T) - E(0) = -\varepsilon \int_0^T y(t)h(x(t), y(t))\,dt. \tag{1.28}$$

Since the periodic path is closed, $E(T) - E(0) = 0$ (the starting point is the same as the endpoint). Therefore, on the limit cycle

$$-\varepsilon \int_0^T y(t)h(x(t), y(t))\,dt = 0.$$

However, an approximation on the limit cycle is given by Equation 1.27. Hence, on the limit cycle, we have

$$\varepsilon\, a \int_0^T h(a\cos t, -a\sin t)\sin t\,dt = 0. \tag{1.29}$$

This equation can be solved for the unknown amplitude a of a limit cycle. Denote the approximation (1.29) as

$$g(a) = (\varepsilon\, a)\int_0^{2\pi} h(a\cos t, -a\sin t)\sin t\,dt. \tag{1.30}$$

Let $a \approx a_0 > 0$ on the limit cycle. That is, $g(a_0) = 0$. If the limit cycle is stable, then along the nearby interior spiral paths $(a < a_0)$, energy is gained and along the exterior spiral paths $(a > a_0)$, energy is lost. Therefore, for some $\alpha > 0$, $g(a) > 0$, for $a_0 - \alpha < a < a_0$, and $g(a) < 0$, for $a_0 < a < a_0 + \alpha$.
This implies that $g'(a_0) < 0$. Therefore, the limit cycle is

Stable when $g(a_0) = 0$, and $g'(a_0) < 0$,

$$\tag{1.31}$$

unstable when $g'(a_0) > 0$.

Example 1.25

Using the energy balance method, find the amplitude and discuss the stability of limit cycles for

$$\ddot{x} + \varepsilon(x^4 - 2)\dot{x} + x = 0. \quad 0 < \varepsilon < 1.$$

SOLUTION

Writing it as a system, we get $\dot{x} = y, \quad \dot{y} = -[\varepsilon\,(x^4 - 2)y + x]$.

Here, $h(x, y) = (x^4 - 2)y$. On the limit cycle, we have the approximation $x(t) \approx a\cos t, y(t) \approx -a\sin t, T \approx 2\pi$. The energy balance Equation 1.29 becomes

$$(-\varepsilon\,a^2)\int_0^{2\pi} [(a^4\cos^4 t - 2)\sin t]\sin t\,dt = 0.$$

$$\int_0^{2\pi} [a^4(\cos^4 t - \cos^6 t) + (2\cos^2 t - 2)]dt = 0.$$

Integrating, we obtain $(\pi/8)(a^4 - 16) = 0$. The positive solution of this equation is $a_0 = 2$.

Therefore, the amplitude is 2.

From Equation 1.30, we obtain

$$g(a) = (-\varepsilon\,a^2)\int_0^{2\pi} [(a^4\cos^4 t - 2)\sin^2 t]dt = -\frac{\pi\varepsilon\,a^2}{8}(a^4 - 16).$$

We get

$$g'(a_0) = -\frac{\pi\varepsilon}{8}(6a_0^5 - 32a_0) = -16\pi\varepsilon < 0.$$

By Equation 1.31, the limit cycle is stable when $\varepsilon > 0$ and unstable when $\varepsilon < 0$.

Focus

Consider the 2D real linear autonomous system

$$x' = Ax,$$

where

$$A = \begin{pmatrix} a_{11} & a_{12} \\ a_{21} & a_{22} \end{pmatrix}, \quad \det A = a_{11}a_{22} - a_{12}a_{21} \neq 0, \tag{1.32}$$

where $a_{11}, a_{12}, a_{21}, a_{22}$ are constants. We find that $(x_1, x_2) = (0, 0)$ is the only critical point of the system.

The solution of this system can be obtained [12, pp. 5.58–5.61] by assuming it to be in the form $x = e^{\lambda t} y$. Substituting into Equation 1.32 and canceling the factor $e^{\lambda t}$, we obtain the algebraic eigenvalue problem $Ay = \lambda y$. The characteristic equation is given by

$$\lambda^2 - (a_{11} + a_{22})\lambda + (a_{11}a_{22} - a_{12}a_{21}) = 0,$$

or

$$\lambda^2 - p\lambda + q = 0,$$

where $p = a_{11} + a_{22}$, and $q = a_{11}a_{22} - a_{12}a_{21}$.

Depending on the values of p and q, we have the following observations:

$p > 0, q > 0$: Two real positive roots or a complex pair of roots.

$p > 0, q < 0$ or $p < 0, q < 0$: One real positive root and one real negative root.

$p < 0, q > 0$: Two real negative roots or a complex pair of roots.

The discriminant of this equation is given by $\Delta = p^2 - 4q$. If $\Delta > 0$, we have real and distinct eigenvalues $2\lambda_{1,2} = p \pm \Delta^{1/2}$. If $\Delta = 0$, we have real and repeated eigenvalue $2\lambda_{1,2} = p$. If $\Delta < 0$, we have the complex pair of eigenvalues $2\lambda_{1,2} = p \pm i(-\Delta)^{1/2}$. Then, the system has the two linearly independent eigenvectors $y^{(1)}, y^{(2)}$ corresponding to the eigenvalues λ_1, λ_2. The general solution of the system is given by

$$x(t) = Ce^{\lambda_1 t}y^{(1)} + De^{\lambda_2 t}y^{(2)}. \tag{1.33}$$

Let $y^{(1)} = [y_{11}\ y_{21}]^T$, $y^{(2)} = [y_{12}\ y_{22}]^T$. Component-wise, we obtain

$$x_1(t) = Ce^{\lambda_1 t}y_{11} + De^{\lambda_2 t}y_{12}, \quad x_2(t) = Ce^{\lambda_1 t}y_{21} + De^{\lambda_2 t}y_{22}.$$

For our discussion, consider the case when the eigenvalues are a complex pair. In this case, $\lambda_2 = \bar{\lambda}_1 = \alpha - i\beta$, where $\alpha = p/2$, and $\beta = (-\Delta)^{1/2}/2$. $y_{11}, y_{21}, y_{12}, y_{22}$ are now complex numbers. Inserting into Equation 1.33, we obtain

$$x(t) = e^{\alpha t}[Ce^{i\beta t} y^{(1)} + De^{-i\beta t} y^{(2)}]. \tag{1.34}$$

The eigenvector corresponding to λ_2 is $y^{(2)} = \bar{y}^{(1)}$, that is, complex conjugate of $y^{(1)}$, $[A\bar{y} = \bar{\lambda}\bar{y}$ or $A\bar{y} = \bar{\lambda}\bar{y}$ since A is real]. Therefore,

$$x(t) = e^{\alpha t}[Ce^{i\beta t}y^{(1)} + De^{-i\beta t}\bar{y}^{(1)}],$$

where C and D are complex numbers. We choose $D = \bar{C}$, to obtain real solutions. Then, Equation 1.34 is of the form

$$x(t) = e^{\alpha t}[(Cy^{(1)})e^{i\beta t} + (\bar{C}\bar{y}^{(1)})e^{-i\beta t}]$$
$$= 2e^{\alpha t} \text{Re}[(Cy^{(1)})e^{i\beta t}], \tag{1.35}$$

where Re denotes the real part of the complex quantity. In component form, we get

$$x_1(t) = 2e^{\alpha t} \text{Re}[Cy_{11}e^{i\beta t}].$$

$$x_2(t) = 2e^{\alpha t} \text{Re}[Cy_{21}e^{i\beta t}].$$

First, consider the case $\alpha = 0$, that is, $p = 0$. If $q > 0$, the roots are purely imaginary.

In polar form, set $x_1 = r\cos\theta$, $x_2 = r\sin\theta$, $C = \rho e^{i\phi}$, $y_{11} = |y_{11}|e^{i\gamma}$, $y_{21} = |y_{21}|e^{i\psi}$.

Then,

$$x_1(t) = 2 \text{Re}[\rho|y_{11}|e^{i(\beta t + \phi + \gamma)}] = 2\rho|y_{11}|\cos(\beta t + \phi + \gamma),$$

$$x_2(t) = 2 \text{Re}[\rho|y_{21}|e^{i(\beta t + \phi + \psi)}] = 2\rho|y_{21}|\cos(\beta t + \phi + \psi).$$

Now, eliminate $\cos(\beta t + \phi)$ and $\sin(\beta t + \phi)$ after expanding the right-hand sides. In the phase plane, the trajectories (phase paths) of $(x(t), y(t))$ are inclined ellipses (inclined at a constant angle with the axes). The equilibrium point $(0, 0)$ is called a *center*. Therefore, conditions under which the origin $(0, 0)$ is a center are $p = 0$, $q > 0$.

Now, consider the case $\alpha \neq 0$. The solutions have the multiplying factor $e^{\alpha t}$. The ellipses are transformed as contracting spirals if $\alpha < 0$, that is, $p < 0$, and as expanding spirals if $\alpha > 0$, that is, $p > 0$. The equilibrium point $(0, 0)$ is called a *focus* or a *spiral*. The focus is stable when $\alpha < 0$, and unstable when $\alpha > 0$.

In summary, we have the following conditions:

The critical point is a *stable focus* if $\Delta = p^2 - 4q < 0$, $p < 0$, $q > 0$. The origin is asymptotically stable.

The critical point is an *unstable focus* if $\Delta = p^2 - 4q < 0$, $p > 0$, $q > 0$.

The critical point is a *center* if $p = 0$, $q > 0$. The origin is stable but not asymptotically stable.

Typical plots of a stable focus, an unstable focus, and a center are given in Figures 1.6a, b, c, respectively.

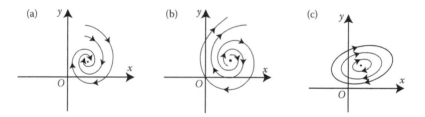

FIGURE 1.6
(a) A stable focus. (b) Unstable focus. (c) A center.

Example 1.26

Discuss the stability of critical point $(0, 0)$ for the following linear system [12, pp. 5.58–5.61]:

$$x' = -ax + ay, \quad y' = -ax - ay, \quad a > 0.$$

SOLUTION

We have the system $\mathbf{x'} = A\mathbf{x}$, where $A = \begin{pmatrix} -a & a \\ -a & -a \end{pmatrix}$.

The characteristic equation is $\lambda^2 + 2a\lambda + 2a^2 = 0$. Here, $p = -2a < 0$, $q = 2a^2 > 0$, $\Delta = p^2 - 4q = -4a^2 < 0$.

Therefore, the origin $(0, 0)$ is a stable focus and is asymptotically stable.

Let us analyze the solution in order to understand the formation of spirals.

The eigenvalues are $\lambda_1 = -a - ia$ and $\lambda_2 = -a + ia$. The corresponding eigenvectors are $y^{(1)} = [1 - i]^T$, and $y^{(2)} = [1\ i]^T$. Note that $\lambda_2 = \bar{\lambda}_1$, and $y^{(2)} = \bar{y}^{(1)}$.

The general solution is

$$x = Ce^{-a(1+i)t}y^{(1)} + De^{-a(1-i)t}y^{(2)} = e^{-at}[Ce^{-iat}y^{(1)} + De^{iat}y^{(2)}].$$

Choosing $D = \bar{C} = c_1 - ic_2$, we obtain

$$x = e^{-at}\left\{ (c_1 + ic_2)(\cos at - i\sin at)\begin{bmatrix} 1 \\ -i \end{bmatrix} + (c_1 - ic_2)(\cos at + i\sin at)\begin{bmatrix} 1 \\ i \end{bmatrix} \right\}$$

$$= 2e^{-at}\begin{bmatrix} c_1 \cos at + c_2 \sin at \\ c_2 \cos at - c_1 \sin at \end{bmatrix} = 2e^{-at}\ \text{Re}[(Cy^{(1)})e^{-iat}].$$

In polar form, set $x_1 = r\cos\theta$, $x_2 = r\sin\theta$, $C = \rho e^{i\phi}$, $y_{11} = 1$, $y_{21} = e^{-i\pi/2}$.

Then,

$$x_1(t) = 2e^{-at} \, \mathrm{Re}[\rho \, e^{-i(at-\phi)}] = 2\rho e^{-at} \cos(at - \phi).$$

$$x_2(t) = 2e^{-at} \, \mathrm{Re}[\rho \, e^{-i(at-\phi+(\pi/2))}]$$

$$= 2\rho e^{-at} \cos(at - \phi + (\pi/2)) = -2\rho e^{-at} \sin(at - \phi).$$

Therefore, $r = 2\rho e^{-at}$, $\theta = \phi - at$. Eliminating t, we get $r = 2\rho k e^\theta$, where $k = e^{-\phi}$. Thus, the phase portrait is a family of spirals. The critical point $(0, 0)$ is called a focus. It was shown earlier that the focus is stable.

Now, consider the autonomous system

$$x' = f(x). \tag{1.36}$$

Rewrite system (1.36) in polar coordinates. Set

$$x = r \cos\theta, \quad y = r \sin\theta, \quad x^2 + y^2 = r^2, \quad \theta = \tan^{-1}(y/x).$$

Then,

$$\dot{x} = \dot{r} \cos\theta - r(\sin\theta)\dot{\theta}, \; \dot{y} = \dot{r}\sin\theta + r(\cos\theta)\dot{\theta},$$

$$r\dot{r} = x\dot{x} + y\dot{y}, \quad \text{and} \quad \dot{\theta} = (x\dot{y} - y\dot{x})/r^2.$$

In terms of polar coordinates, we have the following definitions [23]:

Stable focus: The origin is called a stable focus for Equation 1.36 if there exists a $\delta > 0$ such that for $0 < r_0 < \delta$ and $\theta_0 \in R$, $r(t, r_0, \theta_0) \to 0$ and $|\theta(t, r_0, \theta_0)| \to \infty$ as $t \to \infty$.

Unstable focus: The origin is called an unstable focus if $r(t, r_0, \theta_0) \to 0$ and $|\theta(t, r_0, \theta_0)| \to \infty$ as $t \to -\infty$.

Stable node: The origin is called a stable node if there exists a $\delta > 0$ such that for $0 < r_0 < \delta$ and $\theta_0 \in R$, $r(t, r_0, \theta_0) \to 0$ as $t \to \infty$ and $\lim_{t\to\infty}\theta(t, r_0, \theta_0)$ exists; that is, each trajectory in a deleted neighborhood of the origin approaches the origin along a well-defined tangent line as $t \to \infty$.

Unstable node: The origin is called an unstable node if $r(t, r_0, \theta_0) \to 0$ as $t \to -\infty$ and $\lim_{t\to-\infty}\theta(t, r_0, \theta_0)$ exists for all $r_0 \in (0, \delta)$ and $\theta_0 \in R$.

The origin is called a proper node for the system (1.36) if it is a node and if every ray through the origin is tangent to some trajectory of the system (1.36).

Example 1.27

Determine the characteristic of the focus in the following problem:

$$\dot{x} = -3y - (1/2)x(x^2 + y^2), \quad r(0) = r_0,$$

$$\dot{y} = 3x - (1/2)y(x^2 + y^2), \quad \theta(0) = \theta_0.$$

SOLUTION

We have

$$\dot{r}\cos\theta - r(\sin\theta)\dot{\theta} = -3r\sin\theta - (1/2)r^3\cos\theta,$$

$$\dot{r}\sin\theta + r(\cos\theta)\dot{\theta} = 3r\cos\theta - (1/2)r^3\sin\theta.$$

Solving, we obtain $\dot{r} = -(1/2)r^3$, and $\dot{\theta} = 3$. The origin is an equilibrium point of the system.

Integrating and using the initial conditions, we obtain

$$r = \frac{r_0}{(1 + t\, r_0^2)^{1/2}}, \quad \text{for } t > -\frac{1}{r_0^2}. \, \theta = 3t + \theta_0.$$

We find that $r \to 0$, and $\theta \to \infty$ as $t \to \infty$. Therefore, the origin is a stable focus.

Dynamical System and Its Mathematical Model

A dynamical system gives a functional description of the mathematical model describing the physical problem. It is a concept where a fixed rule describes the time dependence of a point in a geometrical space. For example, the mathematical models that describe the swinging of a clock pendulum, the flow of water in a pipe, and the number of fish in a lake during each spring are dynamical systems. The *Lorenz attractor* is an example of a nonlinear dynamical system.

The dynamical system model is a means of describing how one state develops into another state over the course of time. Technically, a dynamical system is a smooth action of the reals or the integers of one object on another object (usually a manifold). When the reals are acting, the system is called a continuous dynamical system (it is also called a *flow*). When the nonnegative reals are acting, the dynamical system is called a *semi-flow*. When the integers are acting, the system is called a discrete dynamical system (it is also called a *cascade* or a *map*). When nonnegative integers are acting, it is called a *semi-cascade*. At any given time, a dynamical system has a state given by a set of real numbers (a vector), which can be represented by a point in an appropriate state space (a geometrical manifold). Small changes in the state of the system correspond to small changes in the numbers. The *evolution rule* of a dynamical system is a fixed rule that describes what future states follow from the current state.

A dynamical system is specified by the state vector $x \in U \subset \mathbf{R}^n$, $t \in \mathbf{R}^1$ and $\alpha \in V \subset \mathbf{R}^p$, where U and V are open sets in $\mathbf{R}^n$ and $\mathbf{R}^p$, respectively, and a function $f: \mathbf{R}^n \to \mathbf{R}^n$ which describes how the system evolves over time. The system $\dot{x} = f(t, x; \alpha)$ and $x \to f(x; \alpha)$ both represent a dynamical system.

A dynamical system has an associated mathematical model. The actual form of the mathematical model of the dynamical system depends on the method of description chosen to develop it. Depending on the degree of approximation and on the problem to be studied, the real system can be associated with different mathematical models, for example, a simple pendulum with or without friction. From a qualitative point of view, we often introduce a dynamical system, but it is not always possible to define its mathematical model. An example is the case of a cardiovascular system of a living organism.

In general, dynamical systems are classified as (i) linear or nonlinear, (ii) lumped or distributed, (iii) autonomous or nonautonomous, and (iv) conservative or dissipative systems. A special class includes the so-called *self-sustained systems*. Nonlinear dissipative systems in which nondecaying oscillations appear and are sustained without any external force are called self-sustained systems. The energy lost as dissipation in a self-sustained system is compensated from an external source. Among a wide class of dynamical systems, a special place is occupied by systems which can demonstrate oscillations. These systems are called oscillatory systems [2].

Systems with finite dimensional phase space are referred to as those having lumped parameters, because the parameters involved are not functions of spatial coordinates. Such systems are described by ordinary differential equations (ODEs) or return map, and are called *lumped systems*. However, there is a wide class of systems with infinite dimensional phase space. If the dynamical variables x_i of a system are functions of spatial coordinates r_k, $k = 1, 2, \ldots, M$, the system phase space is called infinite dimensional. Such systems are called distributed parameter systems or simply *distributed systems*. Such systems are described by partial differential equations or integral equations.

Several classes of dynamical systems can be distinguished depending on the properties of the evolution operator. If the evolution operator obeys the property of superposition, then the corresponding system is linear. Otherwise, it is nonlinear. If the system state and the evolution operator vary with time, then we are dealing with a time continuous system. If the system state is defined only at discrete time values, then we have a system with discrete time (map or cascade). If the evolution operator depends implicitly on time, then the corresponding system is *autonomous*, that is, it contains no additive or multiplicative external forces depending explicitly on time. If the evolution operator explicitly depends on time, then the corresponding system is called *nonautonomous*. For a *conservative system*, the volume or area in phase space is conserved as time evolves.

For a nonconservative system, the volume is usually contracted. The contraction of phase volume in mechanical systems corresponds to loss of energy as a result of dissipation. The growth of phase volume implies a supply of energy to the system that can be named as negative dissipation. Therefore, dynamical systems in which the phase area or phase volume or energy varies or is not preserved are called *dissipative systems*. For example, the linearized undamped pendulum conserves energy and its trajectories preserve phase area. On the other hand, the trajectories of the linearized damped pendulum decay to a single point. The phase area is not preserved and the system is said to be dissipative. Using these phase-space characteristics, Baker and Gollub [3] developed a method for determining from the equations of motion whether the system is conservative or dissipative. For a 3D system, the equations of motion of the system can be written in terms of phase-velocity components, as $X' = F(X)$ with $X = (x_1, x_2, x_3)$ and $F = (F_1, F_2, F_3)$. The logarithmic rate of volume change is expressed as $(1/V)(dV/dt) = divF$. The logarithmic derivative is therefore independent of the particular volume chosen and depends only on $divF$. If $divF = 0$, then the system is termed as conservative. If $divF < 0$, then the system is called dissipative. If $divF > 0$, then the system is called a volume expanding dynamical system. For dissipative dynamical systems, the state of the system evolves on an invariant geometrical object known as *attractor* whose dimension is less than the dimension of the original state space. For example, for a three-dimensional dissipative dynamical system involving three first-order ODEs, only three distinct possibilities of physical interest exist. They are represented by a stable focus, a stable limit cycle, and a strange chaotic attractor. In a nonlinear system, these possibilities very often coexist.

The kinematic properties of the flux in phase space for a conservative system are analogous to the flow of an incompressible fluid in hydrodynamics. The term Hamiltonian is sometimes used in connection with phase-volume-preserving systems [3]. Many dynamical systems obey Hamilton's equations of motion and such systems are called *Hamiltonian systems*. These systems preserve volume in phase space according to Liouville's theorem, and therefore Hamiltonian systems are a subset of the set of conservative systems [10].

The following are some examples of conservative, dissipative systems.

1. For an undamped pendulum, the equation of motion is $\ddot{x} + x = 0$. Write it as the system $\dot{x} = \omega, \dot{\omega} = -x$. Therefore, $F = (\omega, -x)$, and $div(F) = (\partial\omega/\partial x) + (\partial(-x)/\partial\omega) = 0$, ($\omega, x$ are independent variables). The phase area is preserved and the system is conservative.

2. For a damped pendulum, the equation of motion is $\ddot{x} + \dot{x} + x = 0$. Write it as the system $\dot{x} = \omega, \dot{\omega} = -\omega - x$. Therefore, $F = (\omega, -\omega - x)$ and $div(F) = (\partial\omega/\partial x) + (\partial(-\omega - x))/\partial\omega = -1$. The phase area diminishes with time and the system is dissipative.

3. For damped cubic anharmonic oscillator, the equation of motion is $\ddot{x} + d\dot{x} - ax + bx^3 = 0$, where a, b, and d are positive constants. The equation can be written as the system $\dot{x} = \omega$, $\dot{\omega} = -d\omega + ax - bx^3$. We obtain, $F = (\omega, \ -d\omega + ax - bx^3)$, $div(F) = \partial\omega/\partial x + (\partial(-d\omega + ax - bx^3))/\partial\omega = -d$. Therefore, the system is dissipative.

Like continuous time dynamical systems, a map can also be conservative or dissipative. If a 3D map $X_{n+1} = F(X_n)$ with $X = (x_1, x_2, x_3)$ and $F = (F_1, F_2, F_3)$ preserves the 3D phase space volume in each iteration, then it is called a conservative or volume-preserving map. A discrete time analog of the equation of a vertical pendulum called standard map (see [26]) is an example of conservative map.

The Jacobian matrix $J(X)$ of a map is defined by

$$J(X) = \begin{bmatrix} \partial F_1/\partial x_1 & \partial F_1/\partial x_2 & \partial F_1/\partial x_3 \\ \partial F_2/\partial x_1 & \partial F_2/\partial x_2 & \partial F_2/\partial x_3 \\ \cdots & \cdots & \cdots \\ \partial F_N/\partial x_1 & \partial F_N/\partial x_2 & \partial F_N/\partial x_3 \end{bmatrix}.$$

The map is *conservative* if $|\det(J(X)| = 1$. The map is *dissipative* if $|\det(J(X)| < 1$. The map is *volume (area) expanding* if $|\det(J(X)| > 1$. This result follows from the following fact of multivariable calculus: Let F be a two-dimensional map and J be its Jacobian matrix. F maps an infinitesimal rectangle at (x, y) with area $dx\,dy$ into an infinitesimal parallelogram with area $|\det J(x,y)|dxdy$. If $|\det J(x, y)| < 1$ everywhere, the map is area contracting or dissipative. If $|\det J(x, y)| = 1$ everywhere, the map is conservative. If $|\det J(x, y)| > 1$ everywhere, the map is area expanding.

Remark 1.14

All chaotic attractors of dissipative systems possess the *self-similar structure*. An attractor with a self-similar structure is called a *strange attractor*, a term coined by Ruelle [27]. For example, the chaotic attractor of the logistic map

$$x_{n+1} = ax_n(1 - x_n) = f(x_n), \quad n = 0, 1, 2, \ldots$$

possesses the self-similar structure.

Example 1.28

Find the characteristics of the Hénon 2D map described by the following set of two first-order difference equations.

$$x_{n+1} = 1 - ax_n^2 + y_n, \quad y_{n+1} = bx_n, \quad a > 0, \quad b > 0.$$

SOLUTION

The Jacobian matrix of the system is given by

$$J(X) = \begin{bmatrix} \dfrac{\partial x_{n+1}}{\partial x_n} & \dfrac{\partial x_{n+1}}{\partial y_n} \\[2mm] \dfrac{\partial y_{n+1}}{\partial x_n} & \dfrac{\partial y_{n+1}}{\partial y_n} \end{bmatrix} = \begin{bmatrix} -2ax_n & 1 \\ b & 0 \end{bmatrix}.$$

Now, $|\det(J(X))| = b$. Therefore, the cases $b < 1, b = 1$ and $b > 1$ correspond to dissipative, conservative, and area expanding maps, respectively.

Example 1.29

Find the characteristics of the standard map (the Taylor–Chirikov map [7])

$$x_{n+1} = x_n + y_{n+1}, \quad (\text{mod } 2\pi)$$

$$y_{n+1} = y_n + k \sin x_n, \quad (\text{mod } 2\pi),$$

where k is a positive parameter which determines the dynamics of the map and mod 2π corresponds to modulo 2π.

SOLUTION

The first equation is given by $x_{n+1} = x_n + y_n + k \sin x_n$, (mod 2π).
  The Jacobian matrix of the system is given by

$$J(X) = \begin{bmatrix} \dfrac{\partial x_{n+1}}{\partial x_n} & \dfrac{\partial x_{n+1}}{\partial y_n} \\[2mm] \dfrac{\partial y_{n+1}}{\partial x_n} & \dfrac{\partial y_{n+1}}{\partial y_n} \end{bmatrix} = \begin{bmatrix} 1 + k\cos x_n & 1 \\ k\cos x_n & 1 \end{bmatrix}.$$

Now, $|\det(J(X))| = 1$. Therefore, the map is conservative or area preserving.

Example 1.30

Find the characteristics of the map defined by [17]

$$x_{n+1} = x_n + y_{n+1}, \; y_{n+1} = Bx_n + (k/2\pi)\sin 2\pi x_n.$$

SOLUTION

The first equation is $x_{n+1} = x_n + Bx_n + (k/2\pi) \sin 2\pi x_n$. The Jacobian matrix is given by

$$J(X) = \begin{bmatrix} \dfrac{\partial x_{n+1}}{\partial x_n} & \dfrac{\partial x_{n+1}}{\partial y_n} \\ \dfrac{\partial y_{n+1}}{\partial x_n} & \dfrac{\partial y_{n+1}}{\partial y_n} \end{bmatrix} = \begin{bmatrix} 1 + B + k \cos 2\pi x_n & 0 \\ B + k \cos 2\pi x_n & 0 \end{bmatrix}.$$

Therefore, $|\det(J(X)| = 0 < 1$. The map is dissipative.

Hamiltonian Systems

Let $H(x, y)$ be a smooth real-valued function of two variables. A system of the form

$$\dot{x} = \frac{\partial H}{\partial y} = X(x, y), \quad \dot{y} = -\frac{\partial H}{\partial x} = Y(x, y) \tag{1.37}$$

is called a Hamiltonian system. The function H is called the Hamiltonian function for the system. Equations 1.37 are called Hamiltonian equations. The curves $X(x, y) = 0$, $Y(x, y) = 0$ are called isoclines. A necessary and sufficient condition for Equation 1.37 to be Hamiltonian is that

$$\frac{\partial X}{\partial x} + \frac{\partial Y}{\partial y} = 0. \tag{1.38}$$

If $x(t)$, $y(t)$ is any particular solution, then along the corresponding phase path, we get

$$\frac{dH}{dt} = \frac{\partial H}{\partial x}\frac{dx}{dt} + \frac{\partial H}{\partial y}\frac{dy}{dt} = -XY + XY = 0.$$

Therefore, along the phase path, H is independent of t, that is, $H = H(x, y)$. Geometrically, $H(x, y) = c$ represents level curves on the surface $z = H(x, y)$.

Suppose that the system has an equilibrium point at (x_0, y_0) so that $\partial H/\partial x = \partial H/\partial y = 0$, that is, $X(x, y) = 0$, $Y(x, y) = 0$ at this point.

Denote $\partial^2 H/\partial x^2 = r$, $\partial^2 H/\partial y^2 = t$, $\partial^2 H/\partial x \partial y = s$ evaluated at the equilibrium point (x_0, y_0). Then,

    i. If $rt - s^2 > 0$, $H(x, y)$ has a maximum or a minimum at (x_0, y_0) and this equilibrium point is a center.

    ii. If $rt - s^2 < 0$, $H(x, y)$ has a saddle point at (x_0, y_0).

A Hamiltonian system contains only centers and various types of saddle points but no nodes or focus.

Example 1.31

Show that the following system is Hamiltonian and obtain the Hamiltonian function. Also obtain the equilibrium points and classify them.

$$\dot{x} = \frac{4y^2}{1 + 3y^2} - x, \quad \dot{y} = -x + y.$$

SOLUTION

We have

$$X(x,y) = \frac{\partial H}{\partial y} = \frac{4y^2}{1 + 3y^2} - x, \quad \text{and} \quad Y(x,y) = -\frac{\partial H}{\partial x} = -x + y.$$

Since

$$\frac{\partial X}{\partial x} + \frac{\partial Y}{\partial y} = -1 + 1 = 0,$$

the given system is Hamiltonian.

Integrating the above equations for X and Y partially with respect to y and x, respectively, we get

$$H(x,y) = \int \left[\frac{4}{3}\left(1 - \frac{1}{1 + 3y^2} \right) - x \right] dy$$

$$= \left(\frac{4}{3} - x \right)y - \frac{4}{3\sqrt{3}} \tan^{-1}(\sqrt{3}y) + f(x).$$

$$H(x,y) = \int (x - y)dx = \frac{x^2}{2} - xy + g(y).$$

The required solution for $H(x, y)$ is given by

$$H(x,y) = \left(\frac{4}{3} - x \right)y + \frac{x^2}{2} - \frac{4}{3\sqrt{3}} \tan^{-1}(\sqrt{3}y).$$

To obtain the equilibrium points, set

$$X(x,y) = \frac{4y^2}{1 + 3y^2} - x = 0,$$

and

$$Y(x,y) = -\frac{\partial H}{\partial x} = -x + y = 0.$$

The second equation gives $x = y$. The first equation becomes $y(3y^2 - 4y + 1) = y(y - 1)(3y - 1) = 0$, which gives $y = 0$, 1, 1/3. The three equilibrium points are $(0, 0)$, $(1, 1)$, and $(1/3, 1/3)$.

Now,

$$r = \frac{\partial^2 H}{\partial x^2} = 1, \quad t = \frac{\partial^2 H}{\partial y^2} = \frac{8y}{(1 + 3y^2)^2}$$

and

$$s = \frac{\partial^2 H}{\partial x \, \partial y} = -1.$$

At $(0, 0)$, $rt - s^2 = -1 < 0$. $H(x, y)$ has a saddle point at $(0, 0)$.
At $(1, 1)$, we get $rt - s^2 = -(1/2) < 0$. $H(x, y)$ has a saddle point at $(1, 1)$.
At $(1/3, 1/3)$, we get $rt - s^2 = (1/2) > 0$. $H(x, y)$ has a maximum or a minimum at $(1/3, 1/3)$ and is a center.

Numerical Tools and Software Used

The numerical simulations in this book are carried out using the following software.

MATLAB®

MATLAB is a popular software package for high-performance numerical computation and visualization. It integrates the modules of computation, visualization, and programming in an easy-to-use environment where problems and solutions are expressed in a familiar mathematical notation. It provides an interactive environment with hundreds of built-in functions for technical computation, graphics, and animation. Best of all, it also provides easy extensibility with its own high-level programming language. In this book, we have generated space series, time series, two-dimensional snapshots, spatiotemporal patterns, chaotic attractors, and bifurcation diagrams using MATLAB 7.0.

We demonstrate the use of MATLAB through the following examples.

Example 1.32

Draw the phase plot and the corresponding time series using MATLAB for the following system of equations (see Example 1.22):

$$\dot{x} = -y + x(1 - x^2 - y^2), \ \dot{y} = x + y(1 - x^2 - y^2).$$

SOLUTION

```
%MATLAB code for Example 1.22
g = inline('[-u(2) + u(1).*(1-u(1)^2-u(2)^2); u(1) + u(2).*(1-u(1)^2-u(2)^2)];
    ','t','u')
[t ua] = ode45(g,[0 100],[0.1 0.1])
figure;
    plot(t,ua(:,1),'b');
    hold on;
    plot(t,ua(:,2),'r');
    xlabel('time');
    ylabel('state variable');
    legend('x','y');
    figure;
    plot(ua(:,1),ua(:,2));
    xlabel('x');
    ylabel('y');
```

The phase plot is given in Figure 1.7 and the time series is plotted in Figure 1.8.

Example 1.33

Draw the phase plot and the corresponding time series using MATLAB for the following system of equations $x' = 3x - 2y$, $y' = 9x - 3y$. (Exercise 1.1, Problem 30).

SOLUTION

```
% MATLAB code for Problem 30
g = inline('[3*u(1)-2*u(2); 9*u(1)-3*u(2)]; ','t','u')
[t ua] = ode45(g,[0 50],[0.01 0.02])
figure;
    plot(t,ua(:,1),'b');
    hold on;
    plot(t,ua(:,2),'r');
```

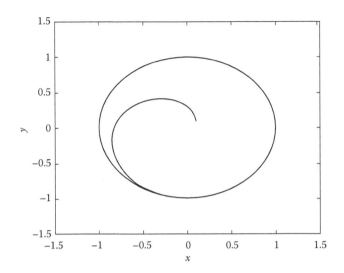

FIGURE 1.7
Phase plot in Example 1.32.

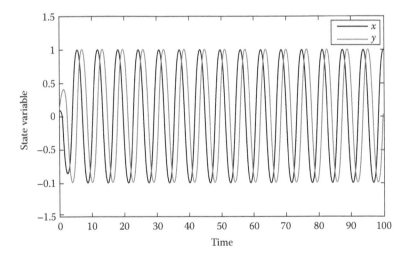

FIGURE 1.8
Time series in Example 1.32.

```
xlabel('time');
ylabel('state variable');
legend('x','y');
figure;
plot(ua(:,1),ua(:,2));
xlabel('x');
ylabel('y');
```

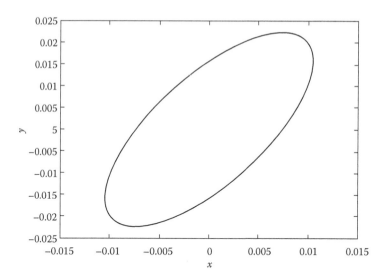

FIGURE 1.9
Phase plot in Example 1.33.

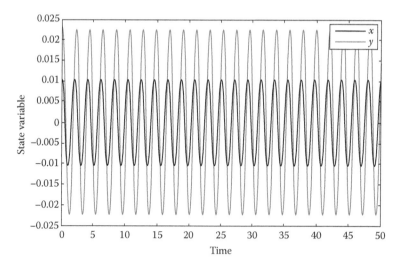

FIGURE 1.10
Time series in Example 1.33.

The phase plot is given in Figure 1.9 and the time series is plotted in Figure 1.10.

Dynamics: Numerical Explorations

This is a user-friendly software tool developed by the Maryland Chaos group, University of Maryland. The software is available in a disk provided along with a book titled *Dynamics: Numerical Explorations* by Nusse and Yorke [22]. The book provides all the details and applications of the software. It has advanced capabilities for maps and flows. These advanced capabilities include determination of the basin boundary structures, computation of bifurcation diagrams, dimension and Lyapunov exponents, periodic orbits, computations of the stable and unstable manifold, straddle trajectory, chaotic transients, following the unstable periodic orbits, and so on. In the following chapters of this book, we have generated the basin boundary structures and the bifurcation diagrams using this software, which exhibits the changes in the dynamical behavior when the system parameters are varied.

We now mention briefly how basin boundary calculations are performed. First, we define the *basin of infinity*. Let SD denote the diameter of the computer screen. It may be possible that the point at infinity is an attractor. Since we cannot examine rigorously whether the trajectory of a point goes to infinity, we conclude that a trajectory diverges or is diverging if it leaves the computer screen area; that is, it goes to the left or to the right of the screen by more than one SD width of the screen, or goes above or below the screen area by more than one SD screen height. The basin of infinity is the set of initial points whose trajectories are diverging.

Now, the basin of attraction of an attractor is the set of all initial conditions in the phase space whose trajectories go to that attractor. In the case of coexisting attractors, we have boundaries formed by the two basins. These boundaries may have very simple or very complicated geometries. The geometry of the basin boundary and the nature and strength of the external perturbations together decide how frequent would be the jumps in the system dynamics. The frequency of these jumps can sometimes be unpredictable, for example, when the basin boundary has a fractal structure.

We have used the *BAS* (basins and attractors structure) method for the computations. This method divides the basin into the following two groups: (i) the basin of attraction A whose points will be plotted and (ii) the basin B whose points will not be plotted. A generalized attractor is the union of finitely many attractors, and a generalized basin is the basin of a generalized attractor. The *BAS* routine does not plot the bowl lying outside. It considers a 100×100 grid of boxes covering the screen. The strategy is to test each grid point which is the center of the grid box. In the event that the center of a grid box is in basin A, then the same is plotted (colored). In the default case, basin A is the set of points whose trajectories are diverging, while basin B is empty. Therefore, the *BAS* routine will plot a grid box if the trajectory of its center is diverging. The important aspects of the basin boundary calculations are to specify the basins A, B and to find the radius RA, where RA stands for the radius of attraction for storage vectors which help to specify the basins A and B. The value of RA will be different for different dynamical systems. It must be set appropriately to avoid any misleading basin picture.

The basin boundary structures are presented as color diagrams. In the figures, the meanings of various colors are as follows:

Dark blue: color of points that diverge from the screen area

Green: color of the first attractor

Sky blue: basin of the first attractor

Red: color of the second attractor

Maroon: basin of the second attractor

Brown: color of the third attractor

White: basin of the third attractor

We demonstrate the use of the above-mentioned package through the following example.

Example 1.34

Draw the basin boundary for the limit cycle and chaos using the package *Dynamics: Numerical Exploration* for the following food-chain model [30].

$$\frac{dx}{dt} = a_1 x - b_1 x^2 - \frac{wxy}{(x + D)},$$

$$\frac{dy}{dt} = -a_2 y + \frac{w_1 xy}{(x + D_1)} - \frac{w_2 yz}{(y + D_2)},$$

$$\frac{dz}{dt} = cz^2 - \frac{w_3 yz}{y + D_3},$$

when $a_1 = 1.93$, $b_1 = 0.06$, $a_2 = 1.0$, $c = 0.029$, $w = 1.0$, $w_1 = 2.0$, $w_2 = 0.045$, $w_3 = 1.0$, $D = 10.0$, $D_1 = 10.0$, $D_2 = 10.0$, $D_3 = 20$.

SOLUTION

The $y - z$ plane ($-50 \le y \le 125; -470 \le z \le 70$) and $x - y$ plane ($-50 \le x \le 50$; $-50 \le y \le 50$) views of the basin boundary structure for coexisting attractors for the given set of parameter values are shown in Figures 1.11 and 1.12 [29].

EXERCISE 1.1

1. A birthday cake is to be packed in a rectangular box with an open top and has a volume of 20 m³. The length of its base is twice its width. The material for the base costs $15 per square meter and the material for the sides costs $5 per square meter. Determine the dimensions of the box if the total cost of the material is to be a minimum. Find the minimum cost.

FIGURE 1.11
y–z plane view of the basin boundary structure in Example 1.34. (Reprinted from *Chaos, Solitons and Fractals*, 16, Upadhyay, R. K. Multiple attractors and crisis route to chaos in a model food chain. 737–747. Copyright 2003, with permission from Elsevier.)

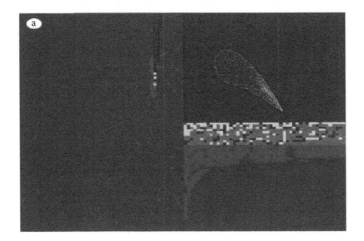

FIGURE 1.12
x–y plane view of the basin boundary structure in Example 1.34. (Reprinted from *Chaos, Solitons and Fractals*, 16, Upadhyay, R. K. Multiple attractors and crisis route to chaos in a model food chain. 737–747. Copyright 2003, with permission from Elsevier.)

2. A mall in a city normally sells 2000 electronic music systems a week at $3500 per unit. A market survey indicates that the sale of the systems can be increased by 200 units in a week if a rebate of $100 is offered to customers. Find the demand function and the revenue function. How large a rebate should the store offer in order to maximize its revenue?

3. (Cooling coffee problem) If a cup of coffee cools from 90°C to 60°C in 10 min in a room whose temperature was 20°C, find out how long it will take for the cup of coffee to cool to 35°C?

Test the stability of the zero solution in the following problems

4. $x_1' = -2x_2 + r\,x_1, \quad x_2' = 3x_1 + r\,x_2, \quad r = (x_1^2 + x_2^2)^{1/2}$.

5. $x_1' = -4x_1 + 8x_1x_2^2, \quad x_2' = -6x_2 - 12x_1^2x_2$.

6. $x_1' = -x_1 - x_2^2, \quad x_2' = 2x_1x_2 - x_2^3$.

7. $x_1' = -x_1^3 + x_1^4, \quad x_2' = x_1^4 - x_2^3$.

8. $x_1' = -x_1x_2^{2p}, \quad x_2' = x_1^{2q}x_2$, (*p, q* are positive integers).

Test the stability of the zero solution in the following problems using Krasovskii's method

9. $x_1' = -4x_1 - x_1x_2^2, \quad x_2' = -6x_2 - x_1^2x_2$.

10. $x_1' = -2x_1 - 3x_2 - 5x_1^5, \quad x_2' = 3x_1 - 2x_2 - 2x_2^5$.

Reduce the following differential equations to three-dimensional systems of the form x′ = Ax. Hence, construct a Lyapunov function, if possible, for the system. Is the origin asymptotically stable?

11. $z''' + 9z'' + 26z' + 24z = 0$.

12. $z''' - 4z'' - 4z' + 16z = 0$.

Test or find the conditions under which the roots of the following characteristic equations are negative or have negative real parts

13. $P(\lambda) = \lambda^3 + 9\lambda^2 + 6\lambda + 1$.

14. $P(\lambda) = \lambda^3 + 3\lambda^2 + \lambda + 8$.

15. $P(\lambda) = (2 - 3p)\lambda^3 + (4 + p)\lambda^2 + 2(1 + p)\lambda + p$.

Test the linear stability of the zero solution $x_1(t) \equiv 0$, $x_2(t) \equiv 0$, in the following problems using the linearization method

16. $x_1' = -x_1 + 3e^{x_2} - 3\cos x_1, \quad x_2' = -1 + e^{x_2} - 6x_2 - \sin x_1$.

17. $x_1' = -1 - x_2 + e^{x_1}, \quad x_2' = 4x_1 - 2\sin x_2$.

18. $x_1' = -x_1 + 3e^{x_2} - 3\cos x_1, \quad x_2' = 1 + 6x_2 - e^{x_2} - \sin x_1$.

19. $x_1' = -x_2 - 4x_1^3, \quad x_2' = 9x_1 - 2x_2^3$.

20. Show that the following logarithmic model is globally asymptotically stable:

$$x_1' = x_1(11 - 3\ln x_1 - 2\ln x_2); \; x_2' = x_2(13 - 4\ln x_1 - 3\ln x_2).$$

Find out whether the limit cycles in the following problems are stable?

21. $x_1' = -x_2 + 6x_1(x_1^2 + x_2^2 - 3), \quad x_2' = x_1 + 6x_2(x_1^2 + x_2^2 - 3)$.

22. $x_1' = -3x_2 + 2x_1(1 - x_1^2 - x_2^2), \quad x_2' = 3x_1 + 2x_2(1 - x_1^2 - x_2^2).$

23. $x_1' = -x_2 + x_1(1 - x_1^2 - x_2^2)^2, \quad x_2' = x_1 - x_2(1 - x_1^2 - x_2^2).$

Show that the following systems have no closed orbits/periodic solutions

24. $x_1' = -2x_1 + 3x_2, \quad x_2' = -x_1 - 6x_2^3.$

25. $x_1' = -2x_1 + 4x_2^3 - 4x_2^4, \quad x_2' = -x_1 - x_2 + x_1 x_2.$

Show that the following systems have a unique stable limit cycle using Liénard's theorem

26. $\ddot{x} + \mu(x^4 - 1)\dot{x} + x = 0, \quad \mu > 0.$

27. $\ddot{x} + \mu(x^2 - 1)\dot{x} + x^3 = 0, \quad \mu > 0.$

28. Apply the energy balance method to find the amplitude and discuss the stability of limit cycles for the system $\ddot{x} + \varepsilon(x^6 - 1)\dot{x} + x = 0, 0 < \varepsilon \le 1.$

Determine the characteristic of the focus in the following problems

29. $x' = (x/2) - y, \quad y' = x + (y/2).$

30. $x' = 3x - 2y, \quad y' = 9x - 3y.$

Determine the characteristic of the focus using polar coordinates in the following problems

31. $\dot{x} = -5y - x\sqrt{(x^2 + y^2)}, \quad r(0) = r_0,$

$\dot{y} = 5x - y\sqrt{(x^2 + y^2)}, \quad \theta(0) = \theta_0. \quad x^2 + y^2 \ne 0.$ Define $f(0) = 0.$

32. $\dot{x} = -3x - \dfrac{2y}{5\ln(x^2 + y^2)}, \quad r(0) = r_0,$

$\dot{y} = -3y + \dfrac{2x}{5\ln(x^2 + y^2)}, \quad \theta(0) = \theta_0. \quad x^2 + y^2 \ne 0.$ Define $f(0) = 0.$

Find out when the following systems are dissipative/area expanding

33. van der Pol oscillator $\ddot{x} - \varepsilon(1 - x^2)\dot{x} + x = 0$, where $\varepsilon > 0$, is a damping constant.

34. Lorenz system (a simplified model of atmospheric convection).

$\dfrac{dx}{dt} = \sigma(y - x); \dfrac{dy}{dt} = -xz + rx - y; \dfrac{dz}{dt} = xy - bz.$ b and σ are positive constants.

35. Rössler equation:

$\dot{x} = -(y + z), \dot{y} = x + ay, \dot{z} = b + xz - cz$, with $a = b = 0.2$ and $c = 5.7.$

36. Find out when the Burgers map $x_{n+1} = (1 - a)x_n - y_n^2$, $y_{n+1} = (1 + b + x_n)y_n$ is area expanding.

37. Find out when the coupled logistic map

$$x_{n+1} = 1 - (a + c)x_n^2, \quad y_{n+1} = 1 - (a - c)y_n^2 + 2b(x_n - y_n)$$

is dissipative/conservative/area expanding.

Find out whether the following systems are Hamiltonian. Obtain the Hamiltonian function. Also, find the equilibrium points and classify them

38. $\dot{x} = y - x$, $\dot{y} = 6 + y + x(4 - 2x - x^2)$.
39. $\dot{x} = y(4 - x^2)$, $\dot{y} = x(x^2 + y^2 + 5)$.
40. $\dot{x} = 4 - x^2 - y^2$, $\dot{y} = x(x^2 + y^2 + 3)$.
41. $\dot{x} = y(4 + 5x + x^2)$ $\dot{y} = [x(1 - 2y^2) - 5y^2]/2$.
42. $\dot{x} = y(4 - 5x - x^2)$, $\dot{y} = [(5 + 2x)y^2 - x]/2$.

Using MATLAB, solve the following problems

43. The Lotka–Volterra population model governing the growth/decay of species

$$\dot{x} = ax - bxy, \quad \dot{y} = cxy - dy, \text{ for } a = 1.0, b = 0.2, c = 0.04, d = 0.5.$$

44. The Rössler model (a model system of chemical reactions in a stirred tank)

$$\dot{x} = -(y + x), \quad \dot{y} = x + ay, \quad \dot{z} = b + z(x - c), \text{ for } a = 0.2, b = 0.2, c = 5.7.$$

45. Draw the phase plot and the corresponding time series for the following mathematical model described by the system of equations:

$$\frac{dX}{dt} = a_1 X - b_1 X^2 - \frac{wXY}{(\alpha + \beta Y + \gamma X)},$$

$$\frac{dY}{dt} = -a_2 Y + \frac{w_1 XY}{(\alpha + \beta Y + \gamma X)}.$$

when $a_1 = 2.5$, $b_1 = 0.05$, $w = 0.85$, $\alpha = 0.45$, $\beta = 0.2$, $\gamma = 0.6$, $a_2 = 0.95$, and $w_1 = 1.65$.

Using the Dynamics: Numerical Exploration package, solve the following problem

46. Draw the basin boundary for the limit cycle and chaos for the following food-chain model [24].

$$\frac{dx}{dt} = a_1 x - b_1 x^2 - \frac{wxy}{(x + D)},$$

$$\frac{dy}{dt} = -a_2 y + \frac{w_1 xy}{(x + D_1)} - \frac{w_2 yz}{(y + D_2)},$$

$$\frac{dz}{dt} = -cz + \frac{w_3 yz}{y + D_3}$$

when $a_1 = 1.75$, $b_1 = 0.05$, $a_2 = 1.0$, $c = 0.7$, $w = 1.0$, $w_1 = 2.0$, $w_2 = 1.5$, $w_3 = 3.75$, $D = 10.0$, $D_1 = 10.0$, $D_2 = 10.0$, $D_3 = 20$.

References

1. Ahmad, S., Rao, M. R. M. 1999. *Theory of Ordinary Differential Equations*. Affiliated East-West Press Private Limited, New Delhi.
2. Anishchenko, V. S., Astakhov, V., Neiman, A., Vadivasova, T., Schimansky-Geier, L. 2007. *Nonlinear Dynamics of Chaotic and Stochastic Systems*. Springer-Verlag, Berlin.
3. Baker, G. L., Gollub, J. P. 1990. *Chaotic Dynamics: An Introduction*. Cambridge University Press, Cambridge.
4. Barnes, B., Fulford, G. R. 2002. *Mathematical Modeling with Case Studies*. Taylor & Francis, Boca Raton, FL.
5. Berg, M. T., Shuman, L. J. 1995. A three-dimensional stochastic model of the behavior of radionuclides in forests III. Cs-137 uptake and release by vegetation. *Ecol. Model.* 83(3), 387–404.
6. Berry, J., Houston, K. 1995. *Mathematical Modeling*. Edward Arnold, a division of Hodder Headline PLC, London.
7. Chirikov, B. V. 1979. A universal instability of many dimensional oscillator systems. *Phys. Rep.* 52, 263–379.
8. Gertsev, V. I., Gertseva, V. V. 2004. Classification of mathematical models in ecology. *Ecol. Model.* 178, 329–334.
9. Hale, J. K. 1969. *Ordinary Differential Equations*. Wiley, Interscience, New York.
10. Helleman, R. H. G. 1983. Self-generated chaotic behavior in nonlinear systems. In: *Universality in Chaos*, P. Cvitanovic, ed. Adam Hilger, Bristol.
11. Hirsch, M. W., Smale, S. 1974. *Differential Equations, Dynamical Systems and Linear Algebra*. Academic Press, Elsevier Science, USA.
12. Jain, R. K., Iyengar, S. R. K. 2007. *Advanced Engineering Mathematics*, 3rd ed. Narosa Publishing House, New Delhi.
13. Jordan, D. W., Smith P. 2009. *Nonlinear Ordinary Differential Equations*. Oxford University Press, New York.
14. Jørgensen, S. E., Bendoricchio, G. 2001. *Fundamentals of Ecological Modeling*, 3rd ed. Elsevier, Amsterdam.

15. Kapur, J. N. 1994. *Mathematical Modeling*. Wiley Eastern Limited. (Presently) New Age International (P) Ltd., New Delhi.
16. Krasovskii, N. N. 1963. *Some Problems in the Theory of Stability of Motion* (translated from Russian). Standard University Press, Stanford.
17. Lakshmanan, M., Rajasekar, S. 2003. *Nonlinear Dynamics: Integrability, Chaos and Pattern*. Springer-Verlag, Germany.
18. Mauersberger, P. 1983. General principles in deterministic water quality modeling. In: *Mathematical Modeling of Water Quality: Streams, Lakes and Reservoirs*, Orlob, G.T., ed. Wiley, New York.
19. May, R. M. 1973. *Stability and Complexity in Model Ecosystems*. Princeton University Press, Princeton.
20. May, R. M. 1974. Ecosystem pattern in randomly fluctuating environments. *Prog. Theor. Biol.* 3, 1–50.
21. Neumaier, A. 2004. Mathematical model building, Chapter 3. In: *Modeling Language in Mathematical Optimization*, J. Kallrath, ed. Kluwer, Boston.
22. Nusse, H. E., Yorke, J. A. 1994. *Dynamics: Numerical Explorations*. Springer-Verlag, New York.
23. Perko, L. 2001. *Differential Equations and Dynamical Systems*. Springer-Verlag, New York.
24. Rai, V., Upadhyay, R. K. 2004. Chaotic population dynamics and biology of the top-predator. *Chaos, Solitons and Fractals* 21, 1195–1204.
25. Rao, V. S. H., Rao, P. R. S. 2009. *Dynamic Models and Control of Biological Systems*. Springer, Dordrecht.
26. Rasband, S. N. 1990. The standard map. In: *Chaotic Dynamics of Nonlinear Systems*. Wiley, New York.
27. Ruelle, D. 1980. Strange attractors. *Math. Intell.* 2, 126.
28. Strogatz, S. H. 2007. *Nonlinear Dynamics and Chaos*. Westview Press, USA.
29. Upadhyay, R. K. 2003. Multiple attractors and crisis route to chaos in a model food chain. *Chaos, Solitons and Fractals* 16, 737–747.
29a. Upadhyay, R. K. 2009. Observability of chaos and cycles in ecological systems: Lessons from predator–prey models. *Int. J. Bifu. Chaos* 19(10), 3169–3234.
30. Upadhyay, R. K., Rai, V. 1997. Why chaos is rarely observed in natural populations? *Chaos, Solitons and Fractals* 12, 1933–1939.
31. Wiggins, S. 1990. *Introduction to Applied Nonlinear Dynamical Systems and Chaos*. Springer-Verlag, New York.

2

Modeling of Systems from Natural Science

Introduction

Science attempts to understand how natural systems function. However, all natural sciences are not at the same level of development. For example, physical sciences are more developed than biological/ecological sciences. One branch of science benefits from the developments in the other branches. A classic example is the contribution of Crick [40] to solve the structure of the DNA molecule. The building blocks of molecular biology were created by a practitioner of solid state physics. The Lotka–Volterra models originated from chemical kinetics.

Gause [58] conducted some experiments in microbiology. He had grown bacteria (*Paramecia aurelia* and *Paramecia caudatum*) in a test tube and studied the variations in the population over a few days. Initially, the population increased very fast and then more slowly; by the fourth day, the population attained a maximum value. In later days, the population did not increase. The results obtained by Gause for the separated and combined cultivation of these bacteria are normally used to demonstrate a good correspondence between theory (Verhulst's model of isolated population dynamics and Lotka–Volterra's model of competing species) and experiment. It is also used for the demonstration of the validity of use of such mathematical models for the approximation of real datasets [114].

The world is experiencing serious global environmental problems such as climate change, global warming, and rise in sea levels. Rapid urbanization, industrialization, and globalization have enhanced the pace of these changes and have exerted severe ecological stress on the Earth and its life-supporting systems. Sustainability can only be assured with proper understanding of the complex ecological interactions among environmental factors and with careful planning and management based on sound ecological principles. Ecology is one of the tools we may use in our efforts to make our world a better place. The study of ecology is concerned with how one life form (more precisely a species) influences the lives of other species in the same ecosystem. Any given life form interacts with only a few other life forms. The network between various life forms (species) through which interactions take place (physical environment) and the life forms constitute an ecological unit known as an *ecosystem*. In another sense, ecosystem is a modeling concept. The concept of an ecosystem is the modeling of interactions between the constituent species and the study of the influence of a species over those which belong either to the same family or to other families. The study of the fragile nature of these interactions may help us to find ways of conserving the ecosystems. An ecosystem can be defined as the smallest self-contained ecological functional unit. The four basic ecological interactions, predation, competition, mutualism, and interference among species, weave the structure of ecosystems. It is too complex to be handled by most of the existing scientific methods. The application of mathematical methods/techniques to problems in ecology has resulted in a branch of ecology known as *mathematical ecology*. The complete process that attempts to describe, model, and solve an ecological phenomenon and interprets the results in ecological terms is called *ecological modeling*.

Even the simplest of ecosystems can be complex. Most ecosystems have evolved over thousands of years. Schaffer and Kot [137] pointed out that deterministic chaos may be an important component of a system's behavior in ecology. These observations were made from the application of dynamical system tools to field data. All ecosystems or geophysical events are nonlinear. For example, landslides, earthquakes, tsunamis, and floods are all complex nonlinear geophysical events. The type of behavior that nonlinear systems generate can sometimes be chaotic, that is, chaos is also a possible result of nonlinear system dynamics. Chaos may be defined as an aperiodic long-term behavior in a deterministic system that exhibits sensitive dependence on initial conditions [145]. However, there is an "order" within chaos, which can be estimated by studying the attractors of the system. Ecologists have the opinion that deterministic chaos is a mathematical possibility not realized by the real world. Deterministic chaos (sensitive dependence of system's dynamics on initial conditions) which serves as a working model for turbulent motion and dynamics of interacting species creates hierarchical structure of organization. In turn, these hierarchies maintain deterministic chaos. Deterministic chaos, in the temporal evolution of species' population densities, is created by forces of diffusion when individuals and populations move from one place to another to fulfill various requirements [113,117]. It should be noted here that two degrees of freedom are not enough to generate deterministic chaos. There must be an additional degree of freedom, that is, an additional dimension to generate deterministic chaos. The objects of ecological research are populations, communities, and ecosystems. Conducting experiments on such objects is not possible as it can lead to changes or even destruction of the ecological objects. Therefore, mathematical modeling plays a key role in ecological research. Ecological objects are very complex and it is impossible to reflect all the features of such objects in the model. The population model, for instance, cannot take into consideration linear-weight characteristics, productivity, and ecological–physiological reactions of every organism in the population. Therefore, it is necessary to resort to some assumptions and group characteristics.

Natural scientists have made great strides in understanding dynamical processes in the physical and biological worlds using a synthetic approach that combines mathematical modeling with statistical and dynamical analysis [152].

Models with Single Population

The single population models can be categorized as continuous time models and discrete time models.

Continuous Time Models

Malthusian Model

Let $P(t)$ represent the total number of individuals in a population or density of a population in an environment. The Malthus law states that *the rate of growth of population is proportional to the population present at that instant.* The rate of change of the entire population is given by dP/dt, while $(1/P)\,(dP/dt)$ represents the per capita rate of change in the population. Assume that the change in the population is caused due to two processes only and they are births and deaths. If the per capita birth and death rates are given by b and d, respectively, then we can represent per capita rate of change by the difference between the birth and death rates

$$\frac{1}{P}\frac{dP}{dt} = b - d,$$

or

$$\frac{dP}{dt} = rP, \tag{2.1}$$

where $r = b - d$ is called the intrinsic growth rate or reproduction rate for the population. If the initial population at time $t = t_0$ is P_0, then the mathematical model representing the dynamics of the population is given by the initial value problem

$$\frac{dP}{dt} = rP, \quad P(t_0) = P_0. \tag{2.2}$$

The solution of this model is given by $P = P_0 e^{r(t-t_0)}$.

This model is called the *Malthusian model* or an *exponential model*. It is restrictive due to the following assumptions.

 i. The per capita growth rate $(1/P)\,(dP/dt)$ is always a constant. That is, contribution due to an average individual is always a constant given by r and this contribution does not depend on the density of the population.

 ii. The growth rate of the population dP/dt is always either increasing (if $r > 0$) or decreasing (if $r < 0$). The process is exponential growth (or exponential decay).

 iii. The population grows (decays) exponentially from the initial value P_0. The population will remain constant when the births and deaths balance each other, that is, $b = d$ or $r = 0$.

The exponential model is often used in the following applications: (a) Plant or insect quarantine (population growth of introduced species). (b) Fishery (prediction of fish dynamics). (c) Conservation biology (restoration of disturbed populations). (d) Microbiology (growth of bacteria). (e) Insect rearing (prediction of yield).

Verhulst–Pearl Logistic Model

The logistic model was originally proposed by Pierre–François Verhulst in 1838 as a model for world population growth. The logistic law has useful applications to human and animal populations [30]. The model was discussed by Verhulst [160] and Pearl [121], and it is referred to as the *Verhulst–Pearl model* in population dynamics. Gotelli [62] gave detailed explanation of the logistic model. It is a useful starting point for modeling density-dependent population dynamics [150].

We have observed in the previous section that the Malthusian model is not a realistic model. It is more realistic to assume that the per capita growth rate is a function of the total population, and it decreases with the total population (as the population increases they have to share the limited food resources available which limit their growth). Verhulst observed that populations normally grow exponentially due to abundance of food, then the population gets crowded and there shall be competition for food. This is called the *crowding effect*. Due to this effect, the population reduces, then again grows, and so on. He argued that the statistical average of the number of encounters of two members of population per unit time is proportional to the square of the population. The encounters between the members reduce the population.

Now, we assume that the per capita growth rate is a linearly decreasing function of the total population. If the initial population at time $t = t_0$ is P_0, then the mathematical model representing the dynamics of the population is given by the initial value problem

$$\frac{1}{P}\frac{dP}{dt} = r\left(1 - \frac{P}{K}\right), \quad P(t_0) = P_0, \tag{2.3}$$

where K is called the carrying capacity of the environment, which is determined by the available sustaining resources, and $r = [(\text{per capita birth rate}) - (\text{per capita death rate})]$ is called the intrinsic growth rate or reproduction rate for the population. K represents the total number of population the environment can support in the long run. Observe that the per capita growth rate continuously reduces from r as the population P increases from zero and it becomes zero when the population reaches K. This seems reasonable as the resources are always limited and the population is controlled by these resources. This model representing growth (decay) in a single species is called the Verhulst–Pearl logistic model. We write the initial value problem as

$$\frac{dP}{dt} = rP\left(1 - \frac{P}{K}\right), \quad P(t_0) = P_0. \tag{2.4}$$

The solution of the problem can be obtained by separation of variables. Rewriting the equation and integrating, we get the solution as

$$P(t) = \frac{K}{1 - Ae^{-rt}}. \tag{2.5}$$

Taking the initial condition as $P(0) = P_0$, we get

$$A = (P_0 - K)/P_0. \tag{2.6}$$

The solution is given by Equation 2.5 where A is given by Equation 2.6. As $t \to \infty, P \to K$.

Let us analyze the model to understand the qualitative behavior of the solutions. Setting $f(P) = rP[1 - (P/K)] = 0$, we obtain two equilibrium points, $P_1 = 0$ and $P_2 = K$.

Now,

$$f'(P) = r[1 - (2P/K)], \quad f'(P_1) = r > 0, \quad f'(P_2) = -r < 0.$$

Hence, the equilibrium point $P = 0$ is unstable and the equilibrium point $P = K$ is asymptotically stable. Thus, solutions initiating in a neighborhood of K approach K as $t \to \infty$, while no solution starting in a neighborhood of $P = 0$

remains close to zero in the future. When the population P approaches the carrying capacity K, we find that $dP/dt \to 0$. This means that the population levels off. As long as $r > 0$ and $K > 0$, this model is always stable.

Now,

$$\frac{d^2P}{dt^2} = \frac{r}{K}(K - 2P)\frac{dP}{dt} = \frac{r^2}{K^2}P(K - 2P)(K - P).$$

$P''(t) = 0$, when $P = 0$, $(K/2)$, and K. We find that $P''(t)$ changes sign from negative to positive at $P = 0$, from positive to negative at $P = K/2$, and from negative to positive at $P = K$. Therefore, the solution curve is concave upward (convex) in $0 < P < (K/2)$ and concave downward in $(K/2) < P < K$. The point of inflection is at $P = K/2$.

Remark 2.1

When the population is small, that is, $P(t)$ is small, P^2 can be neglected in Equation 2.4. The population grows exponentially. When $P(t)$ becomes large, the second term on the right-hand side in Equation 2.4 containing $-P^2$ dominates and the population decreases, thus providing a natural balancing of population. The solution of this equation is the S-shaped curve called the sigmoid curve (see Figure 2.1).

Turchin [151] gave the following reasons for the possible failure of the above logistic model.

i. The realized per capita rate of change is linearly related to P.

ii. The rate of population change responds to variations in density instantaneously without a time lag.

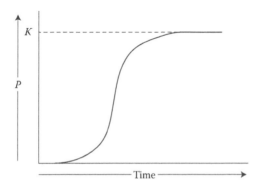

FIGURE 2.1
Sigmoid curve.

iii. The model does not incorporate the effects of exogenous influences.

iv. The model ignores the effects of the population structure.

Gompertz Growth Model

The Gompertz equation was originally formulated as a law of decreasing survivorship while carrying out the mortality analysis of senior citizens [46,60]. It was also used for modeling the growth of plants [27], tumor growth [167,168], and in fishery ecology as a surplus yield model [53]. The Gompertz model is one of the most popular nonlinear models for self-limiting cell population growth. Using experimental data, Laird [87] fitted Gompertz curves for a variety of primary and transplanted tumors of mouse, rat, and rabbit.

The main feature of the Gompertz model is the incorporation of exponentially decreasing growth coefficient r in the Malthusian model. Here, the growth coefficient r varies with time according to the relationship $dr/dt = -\alpha r$. If $r(0) = r_0$, we obtain the expression for the growth coefficient as $r(t) = r_0 e^{-\alpha t}$. Substituting this expression in the Malthusian model, we obtain

$$\frac{dP}{dt} = r_0 e^{-\alpha t} P. \quad P(0) = P_0. \tag{2.7}$$

The solution of Equation 2.7 under the given initial condition is $P(t) = P_0 \exp[r_0(1 - e^{-\alpha t})/\alpha]$.

Note that by expanding $e^{-\alpha t}$ and neglecting the higher-order terms, we obtain the Malthusian model $P(t) = P_0 e^{r_0 t}$. Also, as $t \to \infty$, $P \to P_0 e^{r_0/\alpha} = P^*$. This relation offers an alternative form for the solution of the Gompertz equation as

$$P(t) = P_0 \exp\left[\frac{r_0}{\alpha}(1 - e^{-\alpha t})\right] = P^* \exp\left(-\frac{r_0}{\alpha}e^{-\alpha t}\right). \tag{2.8}$$

Thus, P^*, which is the asymptotic value of the model, can be viewed as the carrying capacity of the system. However, it should be noted that this carrying capacity is not due to the crowding effect, but due to the effort of exponentially decaying coefficient $r(t)$.

From Equation 2.8, we get

$$\frac{P^*}{P} = \exp\left[\frac{r_0}{\alpha}e^{-\alpha t}\right], \quad r_0 e^{-\alpha t} = \alpha \ln\left(\frac{P^*}{P}\right).$$

The model equation can be written in terms of the asymptotic value P^*. Inserting in Equation 2.7, we get

$$\frac{1}{P}\frac{dP}{dt} = r_0 e^{-\alpha t} = \alpha \ln\left(\frac{P^*}{P}\right). \tag{2.9}$$

Now,

$$\frac{d^2P}{dt^2} = \alpha\left[\ln\left(\frac{P*}{P}\right) - 1\right]\frac{dP}{dt} = \alpha^2 P\ln\left(\frac{P*}{P}\right)\left[\ln\left(\frac{P*}{P}\right) - 1\right].$$

$P''(t) = 0$ when $P = (P*/e)$ and $P*$. We find that $P''(t)$ changes sign from positive to negative at $P = (P*/e)$. Therefore, the solution curve has a point of inflection at $P = (P*/e)$.

Remark 2.2

It was postulated that the Gompertz law should be used only for sufficiently large populations, while for the small populations one should just consider the exponential growth.

Remark 2.3

A generalization of the Verhulst–Pearl logistic equation and the Gompertz equation is given by

$$\frac{dP}{dt} = \frac{rP}{\alpha}\left[1 - \left(\frac{P}{K}\right)^\alpha\right].$$

For $\alpha = 1$, the model reduces to the logistic Equation 2.4. In the limit, as $\alpha \to 0$, we get

$$\frac{dP}{dt} = rP\ln\left(\frac{K}{P}\right),$$

or

$$\frac{dx}{dt} = r(\ln K - x),$$

where $x = \ln P$.
  Solving the equation, we get

$$x = \ln K + ae^{-rt},$$

or

$$\ln P = \ln K + ae^{-rt},$$

or

$$P = K\exp\left[ae^{-rt}\right],$$

where a is an arbitrary constant. Applying the initial condition $P(0) = P_0$, we get the Gompertz growth law

$$P = P_0 \exp\left[\ln\left(\frac{K}{P_0}\right)(1 - e^{-rt})\right].$$

Theta-Logistic Model

The logistic model (Equation 2.4) can be modified so that it allows various types of nonlinear relationships between the realized rate of population change and density. The model equation is written as

$$\frac{dP}{dt} = rP\left[1 - \left(\frac{P}{K}\right)^\theta\right]. \tag{2.10}$$

Introduction of the factor θ controls the shape of the relationship between the realized per capita rate of increase $r(t) = (1/P)(dP/dt)$ and the population density. It also decides the significance of the carrying capacity term. For example, if θ is large ($\theta \gg 1$), the model behaves in the following way:

i. *If $P < K$:* The system is going to look like having an exponential growth for a while.

ii. *If $P > K$:* The rate of the population change is negative and hence the population decreases rapidly.

iii. *If $P = K$:* The rate of the population change is zero, so that the population never changes.

Biologically, the third case is realistic because as P increases from zero, the per capita rate of change should stay nearly constant until P gets near K, where density dependence finally sets in. The relationship between $r(t)$ and P is convex initially, and then near K we observe a rapid decrease [151]. When $\theta \neq 1$, the exact solution to the theta-logistic equation cannot be obtained.

Model with Allee Effect

An important model, which does not follow the logistic-like population growth, is a model with *Allee effect*. In 1931, Warder Clyde Allee studied biological species from an ecological perspective and played a major role in the establishment of ecology as an independent biological science. The origin of the above study was based on an observation by Allee in which he identified that survival and reproduction are often limited by the lack of conspecifics, and this can lead to a population decline at low population densities [6,7]. Thus, the growth rate of a population can reach zero, or even negative values, because of a decrease in reproduction or survival when conspecific individuals are not numerous enough.

The presence of the Allee effect indicates that there is a minimal population size necessary for a population to maintain itself in nature. The Allee effect may arise from a number of sources such as difficulties in finding mates, reproduction of cooperative interaction in social species, inbreeding depression, food exploitation, predator avoidance, or defense [38,144]. The Allee effect can be observed in different organisms including vertebrates, invertebrates, insects, and plants [144]. There are several mechanisms that generate Allee effects and a classification of these effects was presented in a book by Courchamp et al. [39]. A review article by David and Berec [41] presents a classification of single species models, which are subjected to various Allee effects with emphasis on a nonspatial, deterministic approach. There are a number of real-world examples exhibiting the presence of such Allee effects [21,54]. As a consequence, the analysis of systems involving the Allee effect has gained a lot of importance in real-world problems associated with various fields such as conservation biology [36,143], sustainable harvesting [90], biological control [76], population management [19], meta population dynamics [172,174], interacting species [42,173], and so on.

When the population growth rate becomes negative of the population density, the phenomenon is termed as the Allee effect. The population is said to have an Allee effect if the per capita growth rate is initially an increasing function, and then decreases to zero at a higher density. The Allee effect has been known to strongly affect not only local population dynamics [44], but also the population interactions in space [89,125,126]. The effect has attracted much attention from researchers due to its strong impact potential on population dynamics [8,65,119]. The Allee effect significantly increases the system's spatiotemporal complexity and may enhance chaos in population dynamics. For example, in the predator–prey system with the Allee effect, chaotic temporal population oscillations can appear in the case when the spatial distribution of species remains regular [139]. In other words, a spatially regular species distribution in the system without the Allee effect can only cause periodic or quasi-periodic population oscillations [112,113,124,139].

There are two concepts that are useful in the classification of Allee effects [18,19,144]; they are (i) component Allee effects and (ii) demographic Allee effects, depending on whether the decline in individual fitness at low population densities concerns a fitness component or the overall fitness, respectively. Component Allee effects need not always give rise to demographic Allee effects. Demographic Allee effects are further categorized into two types—strong and weak. Let the growth equation be written as $(dP/dt) = Pf(P)$, where $f(P)$ is the per capita rate of growth of the population. If the growth rate of the population is initially positive, $f(0) > 0$, it is called a *weak Allee effect*. When the growth rate of the population is initially negative,

$f(0) < 0$, it is called a *strong Allee effect*. A reduction in the per capita population growth rate takes place in small or sparse populations if a demographic Allee effect is present. Strong Allee effects give rise to negative per capita population growth rates once the population size or density falls below the Allee threshold. A strong Allee effect results in a critical population threshold (or Allee threshold) below which the population crashes to extinction. On the other hand, the (overall) fitness decline due to a weak Allee effect still ensures a positive, yet reduced, per capita growth rate of small or sparse populations [18]. A model with Allee effect can be written as [84]

$$\frac{dP}{dt} = rP\left(\frac{P}{K_0} - 1\right)\left(1 - \frac{P}{K}\right) = f(P), \tag{2.11}$$

where r, K_0, and K represent the intrinsic growth rate, threshold population, and carrying capacity, respectively. This equation models the well-known fact that for small populations that live in a large environment in which individuals cannot easily find each other, reproduction becomes difficult.

Equating the right-hand side $f(P)$ to zero, we obtain three equilibrium points, $P_1 = 0$, $P_2 = K_0$, and $P_3 = K$. We find

$$f'(P_1) = f'(0) = -r < 0,$$

$$f'(P_2) = f'(K_0) = r[1 - (K_0/K)] > 0,$$

$$f'(P_3) = f'(K) = -r[(K/K_0) - 1] < 0.$$

Hence, the equilibrium points $P = 0$ and $P = K$ are asymptotically stable nodes and the equilibrium point $P = K_0$ is an unstable node. The solutions with initial conditions in the neighborhood of K approach the carrying capacity K, while those with initial conditions in the neighborhood of the origin approach zero. The model exhibits the phenomena called *critical dispensation* of net growth rate being negative at low population levels. Also, the per capita growth rate is no longer a monotonically decreasing function of density, but shows an Allee effect or an increase in the per capita growth rate over certain ranges of density. When the population P approaches the carrying capacity K, we find that $dP/dt \to 0$. This means that the population levels off.

Example 2.1

Find the equilibrium solutions and the solution of the following model with Allee effect

$$\frac{dP}{dt} = rP\left(1 - \frac{P}{K}\right)\left(1 - \frac{m}{P}\right),$$

where m is the minimum population level below which the species tends to become extinct.

If P_0 is the initial value, discuss the cases $P_0 > m$ and $P_0 < m$.

SOLUTION

Equating the right-hand side $f(P)$ to zero, we obtain the two equilibrium points, $P_1 = m$ and $P_2 = K$. We find

$$f'(P_1) = f'(m) = r[1 - (m/K)] > 0, \quad f'(P_2) = f'(K) = -r[1 - (m/K)] < 0.$$

Hence, the equilibrium point $P_1 = m$ is unstable and the equilibrium point $P_2 = K$ is asymptotically stable. The solutions with initial conditions in the neighborhood of K approach the carrying capacity K. When the population P approaches the carrying capacity K, we find that $dP/dt \rightarrow 0$. This means that the population levels off.

Simplifying the given equation, we obtain

$$\frac{dP}{dt} = \frac{r}{K}[-P^2 + P(m + K) - mK].$$

This equation is a *Riccati equation*, which is of the form $y' = p(x)y^2 + q(x) y + r(x)$.

One of the solutions of the differential equation is $P = K$. Substituting, $P = K + (1/z)$ in the differential equation and simplifying, we get

$$z' - bz = (r/K), \quad \text{where } b = r(K - m)/K.$$

The integrating factor is e^{-bt}. The solution is given by

$$z = \frac{1}{m - K} + ce^{bt}.$$

Therefore,

$$P = K + \frac{1}{z} = K + \frac{m - K}{1 + c(m - K)e^{bt}} = \frac{cK(m - K)e^{bt} + m}{1 + c(m - K)e^{bt}}.$$

Applying the initial condition $P(0) = P_0$, and simplifying, we obtain $c(m - K) = (P_0 - m)/(K - P_0)$. The solution is given by

$$P(t) = \frac{m(K - P_0) + K(P_0 - m)e^{[r(K-m)/K]t}}{(K - P_0) + (P_0 - m)e^{[r(K-m)/K]t}}$$

$$= \frac{m(K - P_0)e^{-[r(K-m)/K]t} + K(P_0 - m)}{(K - P_0)e^{-[r(K-m)/K]t} + (P_0 - m)}$$

$$= \frac{me^{-\alpha t} + KC}{e^{-\alpha t} + C},$$

where

$$\alpha = \frac{r(K-m)}{K} > 0, \quad C = \frac{P_0 - m}{K - P_0}.$$

If $P(0) = P_0 = m$, then $C = 0$ and $P(t) = m$, which means that the population is stationary.

If $P(0) = P_0 > m$, then $0 < C < 1$. As $t \to \infty$, $P(t) \to K$, that is, the carrying capacity. All the solution trajectories starting with initial values $P_0 > m$ converge to the carrying capacity.

If $P(0) = P_0 < m$, then $C < 0$ and $P(t)$ becomes negative for small values of P_0. This implies that for $0 < P_0 < m$, $P(t) \to 0$, that is, the population becomes extinct.

Limited Growth Model

Populations cannot continue growing exponentially over time due to limited resources and/or competition for food with other species. If the populations are observed over long periods, they often appear to reach a limit or to stabilize [15]. We modify the exponential growth model to include the death rate due to the resource limitation or from competition and allow the population to stabilize. As in the Malthusian model, let the per capita birth and death rates be given by b and d, respectively. The model for the density-dependent growth is given by

$$\frac{dP}{dt} = bP - dP - \gamma P^2 = P(r - \gamma P) = f(P), \tag{2.12}$$

where $b - d = r$ is the reproduction rate and γ is the per capita dependence of deaths on the population size (called death due to crowding).

Equilibrium points of the model are $P_1 = 0$, and $P_2 = r/\gamma$. Now,

$$f'(P) = r - 2\gamma P, \quad f'(P_1) = f'(0) = r > 0, \quad f'(P_2) = f'(r/\gamma) = -r < 0.$$

Hence, the equilibrium point $P_1 = 0$ is unstable and the equilibrium point $P_2 = r/\gamma$ is asymptotically stable.

Example 2.2

Find the solution of the limited growth model

$$\frac{dP}{dt} = P(r - \gamma P). \quad P(0) = P_0.$$

When does the population increase or decrease with time?

SOLUTION

The differential equation is a Bernoulli equation. Setting $(1/P) = x$, we obtain the linear equation $(dx/dt) + rx = \gamma$. Solution of the linear equation is

$$x = (\gamma/r) + Ae^{-rt},$$

or

$$P(t) = \frac{1}{(\gamma/r) + Ae^{-rt}}.$$

Using the initial condition, $P(0) = P_0$, we obtain

$$A = \frac{r - \gamma P_0}{rP_0}.$$

The solution of the model is given by

$$P(t) = \frac{rP_0}{\gamma P_0 + (r - \gamma P_0)e^{-rt}}.$$

As $t \to \infty$, $P(t) \to (r/\gamma)$. The solutions with initial conditions in the neighborhood of the equilibrium point (r/γ) approach the carrying capacity (r/γ).

From the given differential equation, we observe the following:

i. When $P < r/\gamma$, $(dP/dt) > 0$. This implies that the population is increasing with time and approaches $P = r/\gamma$.
ii. When $P > r/\gamma$, $(dP/dt) < 0$. This implies that the population is decreasing with time toward $P = r/\gamma$.

Remark 2.4

The logistic Equation 2.4 is also referred to as the limited growth model or density-dependent model. It is obtained by putting $K = r/\gamma$ in Equation 2.12.

Harvest Model

Consider the dynamics of a single species resource with harvesting. The study of population dynamics with harvesting is a vital topic of mathematical bioeconomics and is related to the optimal management of renewable resources [31]. The management of renewable resources is based on the concept of maximum sustainable yield, which suggests exploiting the surplus production on the basis of biological growth model. The effect of harvesting on fisheries management using ecological and economic models was examined and reviewed by Colin W. Clark in his famous book [31]. Optimal management of a resource over time is very important in the context of net economic benefits from the fishery (resource).

For example, assume that we have a resource, say a fish stock, which is growing as per the logistic equation, and the fish stock is subjected to harvesting. Let E represent the effort used for harvesting (E may represent the number of boats, fishing nets, etc.). Assume that catch per unit effort is proportional to the density of the stock or stock level, N. If C and E represent the total catch and effort, respectively, then $(C/E) \propto N$, or $C = qEN$ where q is the proportionality constant. q is called the *catchability coefficient* and qE is called the *fishing mortality*. Thus, C represents the forced reduction in the fish stock per unit time due to harvesting activity. It is also called the rate of harvesting or harvest rate. The net growth of the resource is reduced by C due to harvesting. Hence, in the presence of harvesting, the dynamics of the resource (the logistic Equation 2.4) gets modified to (see Kot [84])

$$\frac{dN}{dt} = rN\left(1 - \frac{N}{K}\right) - qEN = N\left[r\left(1 - \frac{N}{K}\right) - qE\right] \equiv f(N), \qquad (2.13)$$

where r is the intrinsic growth rate of the stock and K is the natural carrying capacity.

Equation 2.13 is a Bernoulli equation

$$\frac{dN}{dt} = aN - \frac{r}{K}N^2,$$

where $a = r - qE$.

By setting $(1/N) = x$, we obtain the linear equation $(dx/dt) + ax = (r/K)$. The solution of the equation is given by

$$x = \frac{r}{Ka} + Be^{-at},$$

or

$$N(t) = \frac{1}{(r/Ka) + Be^{-at}}.$$

Using the initial condition, $N(0) = N_0$, we obtain

$$B = \frac{Ka - rN_0}{KaN_0}.$$

The solution of the model is given by

$$N(t) = \frac{KaN_0}{rN_0 + (Ka - rN_0)e^{-at}}, \quad \text{where } a = r - qE. \tag{2.14}$$

As $t \to \infty$, $N(t) \to [(r - qE)K/r]$.

By setting $f(N) = 0$, we obtain the equilibrium points as $N_1 = 0$ and $N_2 = [(r - qE)K/r]$. The interior equilibrium point N_2 exists when $r > qE$. Also, N_2 is a linear decreasing function of the fishing mortality qE. Note from solution (2.14), N_2 is the asymptotic value of N.

We have, $f'(N) = r - qE - (2Nr/K)$, $f'(N_1) = f'(0) = r - qE$.

$$f'(N_2) = [r - qE - (N_2r/K)] - (N_2r/K) = -(N_2r/K) = qE - r.$$

If $r > qE$, that is, the intrinsic growth rate r is greater than the fishing mortality qE, then we have $f'(N_1) > 0$ and $f'(N_2) < 0$, (N_2 exists in this case). Hence, the trivial equilibrium point N_1 is unstable and the interior equilibrium point N_2 is asymptotically stable. Thus, the stock is sustained and approaches $N_2 = [(r - qE)K/r]$ asymptotically.

If $r < qE$, that is, the intrinsic growth rate r is smaller than the fishing mortality qE (indicating over harvesting), then we have $f'(N_1) < 0$, (N_2 does not exist in this case). Hence, the trivial equilibrium point is asymptotically stable. The stock gets extinct due to over harvesting. Thus, $qE = r$ is the critical value where the two equilibrium points collide and exchange their stability nature.

The fishing effort E, satisfying the condition $r > qE$, plays a major role in the sustenance of the stock. The yield Y at the equilibrium, which is also called as sustainable yield (catch), is defined as

$$Y(qE) = qEN_2 = qEK[1 - (qE/r)]. \tag{2.15}$$

This yield equation gives us important information regarding the amount of the fishing effort to be made available to obtain maximum sustainable yield. The yield curve (2.15) is a parabola. Setting $Y(qE) = y$ and $qE = x$, we can write the equation as

$$y = xK\left(1 - \frac{x}{r}\right) = \frac{K}{r}(rx - x^2) = -\frac{K}{r}\left(x - \frac{r}{2}\right)^2 + \frac{Kr}{4},$$

or

$$y - \left(\frac{Kr}{4}\right) = -\frac{K}{r}\left(x - \frac{r}{2}\right)^2. \tag{2.16}$$

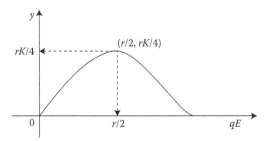

FIGURE 2.2
Yield curve. (From Kot, M. *Elements of Mathematical Ecology.* Cambridge University Press, Cambridge, UK, 2001. With permission.)

The curve is an inverted parabola with vertex at the point $(r/2, rK/4)$ and passes through the origin (see Figure 2.2). Obviously, the maximum occurs at the vertex, that is, when $x = qE = r/2$. The maximum value is given by $Y(qE) = y = (rK/4)$. Therefore, increasing the fishing effort from zero increases the sustainable yield up to the value $qE = r/2$. Further increase in the fishing effort lowers the sustainable yield as a result of over exploitation and the stock gets depleted. Thus, the optimal level of effort is given by $E_{opt} = r/(2q)$ and the corresponding maximum sustainable yield is given by $Y_{opt} = (rK/4)$.

Case of Constant Rate Harvesting

Consider a population that is growing logistically at a constant rate of harvesting h. From Equation 2.13, the dynamics of the model for constant rate harvesting is governed by the equation

$$\frac{dN}{dt} = rN\left(1 - \frac{N}{K}\right) - h \equiv f(N). \tag{2.17}$$

Solving $Kf(N) = (rNK - rN^2 - hK) = 0$, we obtain the equilibrium points as

$$N = \frac{1}{2r}[rK \pm \sqrt{r^2K^2 - 4rhK}] = \frac{K}{2}[1 \pm \sqrt{1 - p}],$$

where

$$p = \frac{4h}{rK}.$$

Let,

$$N_1 = K[1 - \sqrt{1 - p}]/2,$$

and

$$N_2 = K[1 + \sqrt{1-p}]/2.$$

Real solutions are obtained only when $p < 1$, or $h < (rK/4)$.

To determine the nature of the equilibrium points N_1 and N_2, we compute $f'(N_1)$ and $f'(N_2)$. Now,

$$f(N) = rN - (r/K)N^2 - h, \quad f'(N) = r[1 - (2/K)N],$$

$$f'(N_1) = r\sqrt{1-p} > 0, \quad f'(N_2) = -r\sqrt{1-p} < 0.$$

The equilibrium point N_1 is unstable, whereas the equilibrium point N_2 is asymptotically stable. Now, $f''(N) = [-2r/K] < 0$. Hence, the solution graph is concave downward.

The two equilibrium points coincide when $p = 1$, that is, when $4h = rK$, or $h = rK/4$, and $N = K/2$. The point $N = K/2$ is called a critical point. When the harvest $h > rK/4$, there are no equilibrium points and the entire system collapses. (In this case, N_1 and N_2 are a complex pair.) Over exploitation of the fishing stock reduces the sustainable yield and the stock gets completely depleted.

If the harvest $h = 0$, then Equation 2.17 reduces to the logistic Equation 2.4 and admits two equilibrium points $N_1 = 0$ and $N_2 = K$.

The amount of harvesting h plays an important role in sustaining the stock. If the harvest is made at, say, $h = h^*$, we obtain two equilibrium points N_1 and N_2. If the stock is to the left of N_1, then slowly the density of population approaches zero as time progresses. If the stock lies between N_1 and N_2, the density of population grows with time and approaches N_2. If the stock is N_1, it remains the same. If the stock is to the right of N_2, the density of population decreases and reaches N_2 as the time progresses. If the harvest is made at $h = rK/4$, then the density of the population remains at $K/2$. Note that in fisheries models, the level of the harvesting effort at the stable equilibrium gives the maximum sustainable yield of exploitation. In Refs. [24,97], it was shown that for an autonomous population model with constant rate harvesting, there is either an asymptotically stable equilibrium or the extinction of population depending on the harvest rate. For periodic population models, they have examined some special cases in which there are asymptotically stable periodic solutions. Wang and Wang [165] presented an analysis of a harvested logistic equation oriented to obtain maximum harvesting yield in a rigid time interval. Pradhan and Chaudhuri [129] discussed the optimal harvesting policy of a single species fishery with the Gompertz law of population growth. A single species dynamics governed by the periodic Gompertz equation was investigated by Zeng [170]. Dong et al. [45] studied the periodic Gompertz equation with continuous and impulsive harvesting to obtain the maximum annual sustainable yield.

A general single harvesting population model is written as

$$\frac{dN}{dt} = r(N) - h(N) \equiv f(N),$$ (2.18)

where $r(N)$ is the growth term and may follow the Malthusian, logistic, Gompertz, or Allee growth laws, and $h(N)$ is the harvesting term or exploitation term. The harvesting term is generally chosen in one of the following ways:

i. $h(N) = h$: Constant rate of harvesting.
ii. $h(N) = hN$: Holling type I harvesting/exploitation. It leads to the original Malthus model with a different growth function.
iii. $h(N) = hN/(N + D)$: Holling type II harvesting, where h and D are constants.
iv. $h(N) = hN^2/(N^2 + D^2)$: Holling type III harvesting.
v. $h(N) = hN/[(N^2/i) + N + D]$: Holling type IV harvesting, where h, i, and D are constants.

Now, consider the case when the growth term $r(N)$ follows the Allee growth law subjected to a strong Allee effect (also known as critical dispensation-type growth equation) and the resource is subjected to proportionate harvesting $h(N) = qEN$, that is, the catch is proportional to the stock and effort [30]. The growth equation is written as

$$\frac{dN}{dt} = aN(N - b)(c - N) - qEN, \quad a > 0, \quad 0 < b < c,$$ (2.19)

where the constants a, c, and b represent the intrinsic growth rate, carrying capacity of the resource, and the threshold value below which the growth rate of the resource is negative, respectively. E represents the effort and q stands for the catchability coefficient. For $h(N) = 0$, this equation is also referred to as an equation exhibiting a multiplicative Allee effect [2,3] and admits two positive solutions, $N_1 = b$ and $N_2 = c$, and one trivial solution as its equilibrium solutions. We observe that the solutions with initial values greater than b approach the carrying capacity c, while those with initial value less than b decline to zero. The behavior of the solutions was established by Kot [84].

Consider the case when $q = 1$. To reduce the number of parameters, we use the following transformations $N = cx$, $ac^2t = T$, and $E/(ac^2) = e$. Equation 2.19 gets transformed as

$$\frac{dx}{dT} = x(x - K)(1 - x) - ex = G(x),$$ (2.20)

where

$$K = \frac{b}{c} < 1.$$

Here, K represents the threshold population level below which the population has negative growth rate and e stands for the normalized effort. Setting $G(x) = 0$, we obtain the equilibrium points as

$$x = 0, \quad x_{1,2} = [(K + 1) \pm \sqrt{s}]/2, \quad \text{where } s = (K - 1)^2 - 4e.$$

The positive equilibrium points exist when $(K - 1)^2 > 4e$, or $0 < e < [(K - 1)^2/4]$. Now, $G'(x) = -3x^2 + 2(K + 1)x - (K + e)$.

At $x = 0$: $G'(0) = -(K + e) < 0$. At $x = x_1$: $G'(x_1) = -(1/2)\sqrt{s}[\sqrt{s} + (K + 1)] < 0$.

At $x = x_2$: $G'(x_2) = (1/2)\sqrt{s}[(K + 1) - \sqrt{s}] = (1/2)\sqrt{s}[(K + 1) - \sqrt{(K-1)^2 - 4e}] > 0$.

Hence, $x = 0$ and $x_1(e)$ are locally stable and $x_2(e)$ is unstable.

If $e > [(K - 1)^2/4]$, then Equation 2.20 admits only the trivial equilibrium point, which is globally asymptotically stable. Under this condition, we have $dx/dT < 0$. All solutions with a positive initial value decrease monotonically and approach the trivial solution asymptotically. Therefore, the trivial solution is globally asymptotically stable. It is observed that the positive equilibrium points are always greater than the threshold value K.

If $e = [(K - 1)^2/4]$, then $x_1 = x_2 = (K + 1)/2$, and the equilibrium solution is

$$x(e = (K - 1)^2/4) = (K + 1)/2.$$

Integrating the differential equation in this case and using the initial condition as $x(0) = \alpha$, we obtain

$$\left[\ln\left(\frac{x}{x - x_1} \right) - \frac{x_1}{x - x_1} \right] - \left[\ln\left(\frac{\alpha}{\alpha - x_1} \right) - \frac{x_1}{\alpha - x_1} \right] = -Tx_1^2.$$

For $x(0) = \alpha < x_1$, the solutions do not exist. All the solutions with initial conditions $x(0) < [(K + 1)/2]$ are repelled and $x(t) \to 0$ in the limit.

For $x(0) = \alpha > x_1$, we write the solution as

$$\frac{x(\alpha - x_1)}{\alpha(x - x_1)} = e^{-Tx_1^2 + q},$$

where

$$q = \frac{x_1(\alpha - x)}{(x - x_1)(\alpha - x_1)}.$$

In the limit, all the solutions with the initial condition $x(0) > (K + 1)/2$ are attracted to the equilibrium solution.

Example 2.3

In a fish farm, a population of fish is introduced into a pond and harvested regularly. Find the nonzero equilibrium population for the harvesting model

$$\frac{dN}{dt} = rN\left(1 - \frac{N}{K}\right) - hN.$$

i. What value of dN/dt corresponds to a stable population?
ii. If a pond can sustain 10,000 fish, the birth rate is 5%, and harvesting rate is 4%, find the stable population level. What happens if harvesting rate is raised to 5%?
iii. At what critical harvesting rate can extinction occur?

SOLUTION

We follow the solution procedure as given in Equations 2.13 through 2.15. The solution of the model is

$$N(t) = \frac{KaN_0}{rN_0 + (Ka - rN_0)e^{-at}}, \quad \text{where } a = r - h.$$

$$\text{As } t \to \infty, \quad N(t) \to [(r - h)K/r].$$

The equilibrium points are $N_1 = 0$ and $N_2 = [(r - h)K/r]$. The equilibrium point N_2 exists when $r > h$. The case $r < h$ is biologically incorrect. N_2 is the asymptotic value of N. We have shown in the text that the equilibrium point N_1 is unstable and the equilibrium point N_2 is stable.

i. The stable population level (*SPL*) is $SPL = hN_2 = (hK/r)(r - h)$.
   The value of dN/dt corresponding to the stable population is

$$\left[\frac{dN}{dt}\right]_{hN_2} = rhN_2\left(1 - \frac{hN_2}{K}\right) - h^2N_2 = hN_2\left[r\left(1 - \frac{hN_2}{K}\right) - h\right]$$

$$= \frac{hK}{r}(r - h)^2(1 - h).$$

ii. We have $r = 0.05$, $h = 0.04$, $K = 10{,}000$, and

$$N_2 = \frac{(r - h)K}{r} = \frac{(0.05 - 0.04)10{,}000}{0.05} = 2000.$$

The model becomes $(dN/dt) = 0.05N(1 - (N/10,000)) - 0.04N$.
The stable population level (SPL) is $SPL = hN_2 = 0.04(2000) = 80$.
If the harvesting rate is increased to 5%, then $r - h = 0.0$ and
the population level becomes zero.

iii. The critical value of the harvesting rate is $r = h$, (growth
rate = harvesting rate). If the harvesting rate $h \geq r$, extinction of
the population of fish occurs, that is, when the harvesting rate
is $\geq 5\%$.

Models with Delay

From the discussions in the above subsections, we observe that the simplest
models may not capture the rich variety of dynamics observed in natural systems. To have a better understanding of the complexities, we can construct
larger systems of ODE or PDE containing more parameters, where some of
the parameters cannot be determined experimentally. A second approach that
is gaining prominence is the inclusion of time delay terms in the differential equations. The delays or lags often represent gestation times, incubation
periods, transport delays, or can simply lump complicated biological processes
together, accounting only for the time required for these processes to occur.
Such models have the advantage of combining a simple, intuitive derivation
with a wide variety of possible behavior regimes for a single system. However,
these models may hide much of the detailed workings of the complex biological systems, which is sometimes precisely what we are looking for. Delay
models are becoming popular in various application areas. For example, delay
models are being used for describing several aspects of infectious disease
dynamics like primary infection [29], drug therapy [115], immune response
[34], and so on. Delay models have also appeared in the study of chemostat
models [171], circadian rhythms [141], epidemiology [35], respiratory system
[161], tumor growth [162], neural networks [25], and so on.

In this subsection, we present very briefly a class of delay differential equation models for a single species describing population dynamics. The goal is
to determine whether the introduction of time delays enriches the dynamics
of these models.

Consider a single species model described by the following differential
equation with time delay [52]

$$\frac{dP}{dt} = b[P(t - \tau)] \, P(t - \tau) - d[P(t)] \, P(t). \tag{2.21}$$

We assume the following:

i. $b[P(t - \tau)]$ is a positive, continuous, decreasing function of the argument representing the per capita growth rate of the population which
may decrease with increased population levels. This is an instance
of density-limited growth of which the logistic model is another

example. The delay in this case may represent a gestation or maturation period so that the number of individuals entering the population depends on the levels of the population at a previous instance of time.

ii. $d[P(t)]$ is a positive nondecreasing function representing the per capita death rate, which may increase due to intraspecific competition.

We state the following three theorems due to Forde [52], which describe the possible behavior regimes for model Equation 2.21.

Theorem 2.1

If b is a positive function in Equation 2.21 and sup $b(P) < $ inf $d(P)$, then the zero steady state is globally asymptotically stable.

Theorem 2.2

Let b and d be positive functions. Suppose that there exists a $\bar{P}$ such that (i) $\mathrm{sign}(b(P) - d(P)) = -\mathrm{sign}(P - \bar{P})$ and (ii) $b'(\bar{P}) < d'(\bar{P})$, (where dash denotes derivative with respect to P). Then, $\bar{P}$ is a positive steady state and the trivial steady state is unstable. (iii) If $b'(\bar{P})\bar{P} > -2d(\bar{P}) - d'(\bar{P})\bar{P}$, then $\bar{P}$ is linearly stable for all τ. Otherwise, there exists a $\tau_c > 0$, such that $\bar{P}$ is stable for $\tau < \tau_c$, and unstable for $\tau > \tau_c$.

Theorem 2.3

If $\lim_{P \to \infty} b(P) \geq \lim_{P \to \infty} d(P)$, then all the solutions of Equation 2.21 with positive initial value are unbounded and a solution $P(t)$ is bounded if and only if $\lim_{t \to \infty} P(t) = 0$.

We consider a particular form of the general model (2.21), with constant per capita death rate and exponentially decaying per capita birth rate, which was studied by many authors [23,63], especially those dealing with Nicholson's blowflies data [63]. These studies sparked much debate about the possibility of chaotic dynamics in natural populations [23,64]. The form of the equation studied is the following:

$$\frac{dP}{dt} = pe^{-aP(t-\tau)}P(t - \tau) - qP(t), \tag{2.22}$$

where p and q are constants. Comparing with Equation 2.21, we have $b[P(t - \tau)] = pe^{-aP(t-\tau)}$, and $d[P(t)] = q$.

The nontrivial steady state occurs when

$$pe^{-a\bar{P}} = q,$$

or

$$\bar{P} = \frac{1}{a}\left(\ln\frac{p}{q}\right) = \text{a constant.}$$

Since per capita death rate is constant, $d'(P) = q' = 0$ and also since $b = pe^{-aP(t-\tau)}$ is a positive decreasing function, $b'(P) < 0$. The requirement (ii) of Theorem 2.2, $b'(P) < d'(P)$, is satisfied. Now, we check the condition $b'(\bar{P})\bar{P} > -2d(\bar{P}) - d'(\bar{P})\bar{P}$. We obtain

$$\bar{P}\left[\frac{d}{dP}\left(pe^{-aP}\right)\right]_{P=\bar{P}} > -2q,$$

or

$$\bar{P}[p(-a)e^{-a\bar{P}}] > -2q,$$

or

$$\bar{P}(-a)q > -2q,$$

or

$$\ln(p/q) < 2, \quad \text{or } p < qe^2.$$

By Theorem 2.2, $\bar{P}$ is linearly stable for all τ when $p < qe^2$.

Now, consider the case when $p > qe^2$. Let $\alpha = \ln(p/q) - 1$. Then, $\alpha > 1$.

Substituting $P(t) = e^{\lambda t}$ into Equation 2.22 and simplifying, we get the characteristic equation as $\lambda = -q(\alpha e^{-\lambda\tau} + 1)$. When $\tau = 0$, $\lambda = -q(\alpha + 1)$. For $\tau > 0$, let $\lambda = i\sigma$, $\sigma > 0$ be a purely imaginary root. Then, the real and imaginary parts of the characteristic equation give the relations $\alpha\cos(\sigma\tau) + 1 = 0$, and $\alpha q\sin(\sigma\tau) = \sigma$. Eliminating $\cos(\sigma\tau)$ and $\sin(\sigma\tau)$, we obtain

$$\left(-\frac{1}{\alpha}\right)^2 + \left(\frac{\sigma}{\alpha q}\right)^2 = 1,$$

or

$$\sigma^2 = q^2\,(\alpha^2 - 1).$$

We find that $\cos\sigma\tau = -(1/\alpha) < 0$ and $\sin\sigma\tau = [(\sqrt{\alpha^2 - 1})/\alpha] > 0$. The critical delay τ_c, at which the eigenvalue crosses into the left half-plane, $\sigma\tau_c \in (\pi/2, \pi)$, is $\tau_c = \cos^{-1}(-1/\alpha)/[q\sqrt{\alpha^2 - 1}]$. For $\tau > \tau_c$, the equilibrium point is unstable. We now discuss some one-dimensional (1D) discrete time models.

Discrete Time Models

Mathematical modeling of an ecological problem started with Volterra [163], a mathematician by profession, who was approached by his friend to advise him on the management of the fish population in the pond he owned. By constructing a model and devising an ingenious method to solve it, Volterra [164] arrived at a solution to the problem. The recent history of mathematical ecology can be traced to May [102] who made pioneering contributions to this field. Performing computations using a pocket calculator, May [100,101] showed that a 1D difference equation can exhibit as complicated a dynamics as chaos, that is, it is able to support a temporal dynamics that consists of many periods at a definite value of the critical parameter of the system. The finding has been instrumental in dismantling the belief that several degrees of freedom are required to interact with each other to give rise to chaotic dynamics. Many authors [17,118] carried out similar investigations on different difference equation models of one and two dimensions. Most of these authors found that simple mathematical models can support chaotic dynamics in reasonably large parameter regimes.

In discrete growth models, we assume that the population does not change except at discrete intervals of time.

Linear Map

First, consider a linear map $x_{n+1} = kx_n$, where k is an integer. This equation can be viewed as a linear difference equation. The evolution of the system is in discrete time. Starting with an initial value x_0, we find the successive approximations $x_1 = kx_0$, $x_2 = kx_1$, …. Obviously, $x_n = k^n x_0$. For $k > 1$, we have the case of stretching of the map. As $n \to \infty$, $x_n \to \infty$. That is, the trajectories are attracted to infinity. For $k < 1$, we have the case of contraction of the map. As $n \to \infty$, $x_n \to 0$. That is, the trajectories are attracted to the origin. When $k = 1$, $x_n = x_0$. The initial trajectory is not transformed. Hence, $k = 1$ is a bifurcation point. The behavior of the map is completely predictable and there is no loss of the qualitative behavior in successive iterations. However, when a model is nonlinear, the dynamical behavior cannot easily be predicted.

In ecological models, the discrete intervals may correspond to breeding seasons. Instead of the rate of change of the population at an arbitrary time in continuous time models, we consider the time interval as the time from one breeding season to the next breeding season. We take the measure of time in units of this period. Let P_n denote the population size in the nth breeding interval and P_{n+1} be the population in the next breeding interval. Here, n is an integer representing time, but taking only discrete values.

A nonlinear model representing nonoverlapping generations is of the form

$$P_{n+1} = F(P_n), \quad \text{or as} \quad P_{n+1} = P_n f(P_n), \tag{2.23}$$

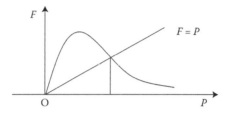

FIGURE 2.3
Population in nonoverlapping genereations.

where $P_n \geq 0$, $F(P_n) \geq 0$, $F(0) = 0$. Note the following:

i. F is a one-to-one self-mapping on $[0,\infty)$. $F(P)$ is an increasing function for small P.

ii. Owing to the limitations of resources, $F(P) \rightarrow 0$ as $P \rightarrow \infty$. $F(P)$ is not monotonic on the entire axis.

iii. The equilibrium point is the intersection of the graphs of $y = F(x)$ and $y = x$ (Figure 2.3).

Consider the following examples in ecology.

Model 1

We assume that the deaths are due to natural causes and also due to crowding effects or due to competition for limited resources. The per capita death rate is represented by $\alpha + \gamma P_n$. Here, the parameters γ and α describe the crowding effect on the population and per capita death rate due to natural attrition, respectively. If the per capita birth rate is β, then the total number of births is βP_n. Therefore, over a single breeding interval, we write

{Change in population size} = {number of births} − {number of deaths}

$$P_{n+1} - P_n = \beta P_n - (\alpha + \gamma P_n)P_n$$

or

$$P_{n+1} = P_n + (\beta - \alpha)P_n - \gamma\, P_n^2.$$

Setting $\beta - \alpha = r$ (per capita reproduction rate) and $\gamma = r/K$, we get

$$P_{n+1} = P_n + rP_n\left(1 - \frac{P_n}{K}\right) = P_n\left[(1 + r) - \frac{rP_n}{K}\right], \quad n = 0, 1, 2, \ldots, \quad (2.24)$$

where K is the carrying capacity of the population.

This equation is sometimes called the *discrete logistic equation*.
For $P_n > [K(1 + r)/r]$, $P_{n+1} < 0$ which is not biologically correct.
Substitute $P_n = [K(1 + r)/r]x_n$ in Equation 2.24 and cancel $[K(1 + r)/r]$. We get,

$$x_{n+1} = (1 + r)x_n(1 - x_n) = a\, x_n(1 - x_n), \quad a = r + 1, \tag{2.25}$$

which is the standard and conventional form of a *logistic map*. In a logistic map, x_n denotes the population in the nth year expressed as a fraction of its maximum possible value. When the population is small, it multiplies (boom) unmindful of the resource limitation. But, when it grows too large, the resource crunch is felt and this may lead to a decline in the population (bust). However, it may also lead to oscillations or even chaotic fluctuations depending on the logistics. The control parameter a in this case is called the *boom and bust parameter*. x_n is interpreted as the normalized population density and the control parameter a is interpreted as the reproduction rate.

We have noted that for $P_n > [K(1 + r)/r]$, $P_{n+1} < 0$, which is not biologically correct. The situation can be avoided by writing the following approximations to Equation 2.24.

i. Write an exponential approximation to Equation 2.24 as

$$P_{n+1} = P_n \exp\left[r\left(1 - \frac{P_n}{K}\right)\right]. \tag{2.26}$$

Expanding the exponential and neglecting the second and higher order terms, we get the approximation in Equation 2.24. P_n is never negative and it is a biologically correct equation.

ii. Write a rational approximation to Equation 2.24 as

$$P_{n+1} = P_n(1 + r)\left[1 - \frac{r}{K(1 + r)}P_n\right] \approx P_n(1 + r)\left[1 + \frac{r}{K(1 + r)}P_n\right]^{-1}. \tag{2.27}$$

P_n is never negative and it is also a biologically correct equation. Take the term in the denominator to the numerator and expand symbolically. Neglecting the second and higher-order terms, we obtain the approximation in Equation 2.24.

iii. More generally, we may write Equation 2.27 as

$$P_{n+1} = \frac{aP_n}{1 + bP_n}, \quad a > 0, \quad b > 0, \tag{2.28}$$

where the constants a, b are suitably defined.

Let us now analyze the model given in Equation 2.23 for stability using the linearization method.

Stability of model (2.23): The equilibrium solution, $P_n = P^* = $ constant, satisfies the equation $P^* = F(P^*)$. If the solution is stable, then P^* is a stable point. Conversely, equilibrium is possible if $P^* = F(P^*)$ has at least one positive root.

Write $P_n = P^* + x_n$ and linearize $P_{n+1} = F(P_n)$. We obtain

$$P^* + x_{n+1} = F(P^* + x_n) \approx F(P^*) + x_n \left(\frac{dF}{dP} \right)_{P^*}$$

or

$$x_{n+1} = Ax_n + \text{(higher-order terms)}, \quad \text{where } A = (dF/dP)_{P^*}.$$

Neglecting the higher-order terms, we obtain the solution of the difference equation as

$$x_{n+1} = A^n x_0, \tag{2.29}$$

where x_0 is the initial population. Therefore,

  i. *P^* is asymptotically stable when $|A| < 1$:* The solution monotonically approaches P^* when $0 < A < 1$ and approaches P^* with damped stable oscillations when $-1 < A < 0$.

 ii. *P^* is unstable when $|A| > 1$:* The solution monotonically increases when $A > 1$ and it oscillates with increasing amplitude when $A < -1$.

Model 2

If we assume that changes in the population size is balanced by purely birth and death rates and in the absence of crowding effect ($\gamma = 0$), we obtain

$$P_{n+1} - P_n = \beta P_n - \alpha P_n,$$

or

$$P_{n+1} = [1 + (\beta - \alpha)]P_n = (1 + r)P_n = \lambda P_n, \tag{2.30}$$

where $\lambda = 1 + r$. The equation is called a *discrete Malthus model or exponential model*. The constant λ is referred to as the finite growth rate of the population. The solution of the difference equation is $P_n = \lambda^n P_0$, where the initial population is P_0. The model exhibits the same drawback as its continuous counterpart; namely, for $\lambda < 1$ we find $P_n \to 0^+$ (the population becomes extinct) and for $\lambda > 1$ we have $P_n \to \infty$ (the population grows unboundedly). For $-1 < \lambda < 0$, the solution oscillates between negative and positive values (damped oscillations), which are not possible. In this case, a natural correction would require defining $P_{n+1} = \max \{\lambda P_n, 0\}$, [97].

Model 3

A more robust model similar to Equation 2.26 was discussed by May [103], which does not have a negative population for any value of r (see the exponential approximation (2.26))

$$P_{n+1} = e^{\lambda[1-(P_n/K)]} P_n, \quad \text{where } \lambda = \ln(r+1). \tag{2.31}$$

This model is called the *Ricker model* [131], which exhibits a behavior similar to the logistic model, with damped oscillations, 2-cycles, 4-cycles, and chaotic growth behavior. It is superior to the logistic model for discrete growth because it incorporates a constant probability of death for each individual. It is well known that the Ricker model is capable of stability, limit cycles, and chaos.

Setting $(P_n/K) = x_n$, we obtain the nondimensional form of model (2.31) as

$$x_{n+1} = x_n \exp[\lambda(1 - x_n)].$$

We note that x_n is nonnegative for all n, and it demonstrates logistic growth. Sharov [138] has also discussed the above model in his online lectures.

Ricker's model is stable for $0 < \lambda < 2$ (see Problem 16, Exercise 2.1). When the stability criterion is violated, we may have limit cycles; for larger values of λ, we may have chaotic behavior. For different values of λ and $K = 250$, plots of population dynamics are given below. Figure 2.4a shows monotonous increase of numbers for $\lambda = 0.45$, showing a stable equilibrium. Figure 2.4b displays a limit cycle with period 2 for $\lambda = 2.25$, and Figure 2.4c displays chaotic behavior for $\lambda = 3.05$.

Example 2.4

Determine the range of r for which model (2.31) is asymptotically stable.

SOLUTION

We find the solution of the equation $P^* = F(P^*)$, that is, of

$$P^* = P^* \exp\left[\lambda\left(1 - \frac{P^*}{K}\right)\right],$$

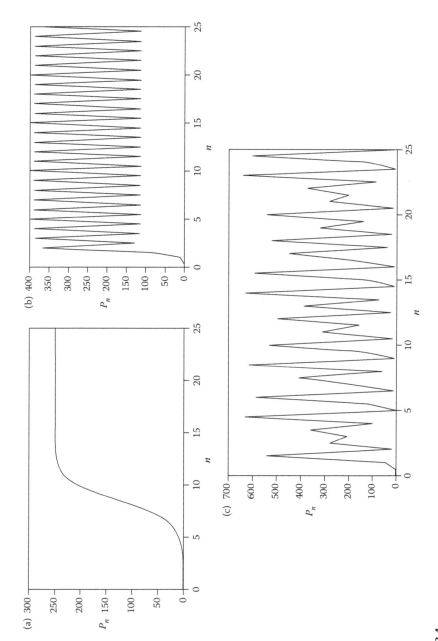

FIGURE 2.4
(a)–(c) Plots of population dynamics in Ricker model.

or

$$\exp\left[\lambda\left(1 - \frac{P*}{K}\right)\right] = 1.$$

The solution is given by $P* = K$, for all λ.
Now,

$$F(P) = P\exp\left[\lambda\left(1 - \frac{P}{K}\right)\right], \quad \frac{dF}{dP} = \exp\left[\lambda\left(1 - \frac{P}{K}\right)\right]\left[1 - \frac{\lambda P}{K}\right].$$

At $P = P* = K$, we get $(dF/dP) = [1 - \lambda]$.
The model is asymptotically stable for

$$|1 - \lambda| < 1, \text{ or } 0 < \lambda < 2, \text{ or } 0 < \ln(r + 1) < 2, \text{ or } 1 < r + 1 < e^2, \text{ or } 0 < r < e^2 - 1.$$

Neighboring trajectories approach the equilibrium point asymptotically for $0 < r < e^2 - 1$.
It is unstable for $\lambda > 2$, or $r > e^2 - 1$.

Model 4

The *Beverton–Holt model* (*BV*) is characterized by the fact that the solutions approach the stable equilibrium point exponentially for all parameter values. The model is given by (see the rational approximations (2.27), (2.28))

$$P_{n+1} = \frac{\lambda P_n}{1 + [(\lambda - 1)/K]P_n}, \quad \lambda \neq 1. \tag{2.32}$$

The curve is called the Beverton–Holt stock recruitment curve [22] and is a monotonically increasing function. The nonlinear difference equation has an exact, closed-form solution that increases monotonically, but with ever-decreasing slope. Write Equation 2.32 as

$$P_{n+1} = \frac{\lambda}{(1/P_n) + a},$$

where

$$a = \frac{\lambda - 1}{K}.$$

Substitute $x_n = 1/P_n$. We obtain

$$\frac{1}{x_{n+1}} = \frac{\lambda}{x_n + a},$$

or

$$x_{n+1} = \frac{1}{\lambda}(a + x_n).$$

Hence,

$$x_1 = \frac{1}{\lambda}(a + x_0), \quad x_2 = \frac{a}{\lambda}\left(1 + \frac{1}{\lambda}\right) + \frac{1}{\lambda^2}x_0, \cdots$$

$$x_n = \frac{a}{\lambda}\left(1 + \frac{1}{\lambda} + \cdots + \frac{1}{\lambda^{n-1}}\right) + \frac{x_0}{\lambda^n} = \frac{a}{\lambda}\left[\frac{1 - (1/\lambda)^n}{1 - (1/\lambda)}\right] + \frac{x_0}{\lambda^n}$$

$$= \frac{a}{\lambda^n}\left[\frac{\lambda^n - 1}{\lambda - 1}\right] + \frac{x_0}{\lambda^n} = \frac{1}{\lambda^n}\left[\frac{\lambda^n - 1}{K} + x_0\right],$$

or

$$\frac{1}{P_n} = \frac{1}{\lambda^n}\left[\frac{\lambda^n - 1}{K} + \frac{1}{P_0}\right] = \frac{1}{\lambda^n}\left[\frac{(\lambda^n - 1)P_0 + K}{KP_0}\right].$$

The solution of the model is

$$P_n = \frac{\lambda^n KP_0}{(\lambda^n - 1)P_0 + K} = \frac{\lambda^n P_0}{1 + [(\lambda^n - 1)/K]P_0}. \quad \lambda \neq 1.$$

The trajectories predicted by the Beverton–Holt model are numerically very near to the trajectories generated by the logistic map [64]. The model has equilibrium points at $P = 0$ and $P = K$.

We have

$$F(P) = \frac{\lambda P}{1 + [(\lambda - 1)/K]P}, \quad \frac{dF}{dP} = \frac{\lambda}{(1 + [(\lambda - 1)/K]P)^2}.$$

At $P = 0$, $F'(0) = \lambda$ and $F'(K) = 1/\lambda$. The trivial equilibrium point $P^* = 0$ is unstable if $\lambda > 1$ and the equilibrium point $P^* = K$ is asymptotically stable for $|\lambda| > 1$. For $\lambda > 1$, small perturbations from the trivial equilibrium grow at first geometrically. Small perturbations from the carrying capacity K decay geometrically [84]. The BV model has the same steady monotonic approach to carrying capacity K that we associate with the Verhulst–Pearl logistic model.

Model 5

Consider the simple nonlinear map

$$x_{n+1} = kx_n \text{ (mod 1)}. \quad k > 1 \text{ is an integer.} \quad 0 < x_n < 1. \qquad (2.33)$$

For illustration, let $k = 2$. That is, $x_{n+1} = 2x_n$ (mod 1). The map is called the Bernoulli shift (also shift map, doubling map, or dyadic transformation). We can imagine x_n as the population of a community at the end of the nth year. The first operation, multiplication by 2, stretches the map by the factor 2, that is, doubles the initial separation. The second operation (mod 1) gives a folding operation, that is, folds it back into the unit interval. The folding operation can send them to various destinations chaotically. Depending on the value of k, the folding operation can produce catastrophic results. The stretching out and folding back operation is also known as Baker's transformation, which can be written as

$$x_{n+1} = 2x_n (\text{mod } 1) = \begin{cases} 2x_n, & 0 \le x_n < 1/2 \\ 2x_n - 1, & 1/2 \le x_n \le 1. \end{cases}$$

The Bernoulli shift map typically generates *deterministic chaos*. This map is nonconservative and noninvertible with the Lyapunov exponent ln2. It has sensitive dependence on initial conditions, infinitely many periodic and aperiodic orbits, and a dense orbit.

For example, consider the following sequences of values obtained for two different but close initial values.

i. $x_0 = 0.2499, 0.4998, 0.9996, 0.9992, 0.9984, 0.9968, 0.9936, 0.9872, 0.9744, 0.9488, 0.8976, 0.7952, 0.5904, 0.1808, 0.3616, \ldots$

ii. $x_0 = 0.2501, 0.5002, 0.0004, 0.0008, 0.0016, 0.0032, 0.0064, 0.0128, 0.0256, 0.0512, 0.1024, 0.2048, 0.4096, 0.8192, 0.6384, \ldots$

An error of 0.0002 in initial approximation has produced a completely different phenomenon. The two orbits do not resemble at all (see Figure 2.5). The map is a simple example of a system exhibiting chaotic behavior.

Model 6

Consider the logistic model

$$x_{n+1} = Ax_n(1 - x_n), \quad 0 < x_n < 1. \qquad (2.34)$$

May [100,101,103] made pioneering studies of the logistic models. Other authors have also studied the logistic models very extensively [49,50,86] as it

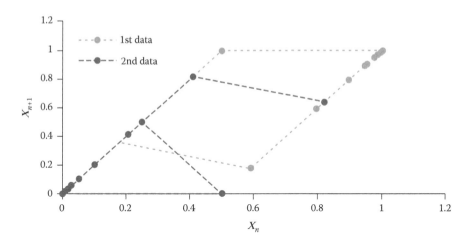

FIGURE 2.5
Map of Equation 2.33 for $k = 2$. First data are given in (i) and second data are given in (ii).

provides a simple example exhibiting all the situations of nonlinear behavior like period doubling, bifurcation, chaos, and so on, depending on the value of A.

When the population is small, it multiplies unmindful of resource limitations. When the population is large, the resource crunch is felt and this leads to population decline. Depending on the value of A, the map exhibits different types of behavior.

First, let us discuss the linear stability of Equation 2.34.

$x^* = F(x^*)$ gives the equilibrium points as $x^* = 0$, and $[1 - (1/A)]$. Now, $F' = A(1 - 2x)$.

At $x = x^* = 0$, $F' = A$, and $|F'| < 1$, gives $0 < A < 1$. The equilibrium point $x^* = 0$ is asymptotically stable for $0 < A < 1$.

At $x = x^* = [1 - (1/A)]$, $F' = 2 - A$, and $|F'| < 1$, gives $1 < A < 3$. The equilibrium point $x^* = [1 - (1/A)]$ is asymptotically stable for $1 < A < 3$.

Graphically, the equation $y = Ax(1 - x)$ is a parabola with vertex at $V(1/2, A/4)$

$$y - \frac{A}{4} = A\left[x - \frac{1}{2}\right]^2, \quad 0 < x < 1, \quad 0 < y < 1. \tag{2.35}$$

The point of intersection of the parabola and $y = x$ inside the unit square is the equilibrium point.

For $0 < A < 1$, the only solution is $x^* = 0$: All initial approximations x_0 in the interval $0 < x_0 < 1$ are attracted to the solution $x^* = 0$ and the equilibrium point is stable. For example, let $A = 0.8$. Then, the vertex of the parabola is at $(0.5, 0.2)$. Let an initial approximation be $x_0 = 0.4$. We obtain the solution values as $x_1 = 0.192, 0.1241, 0.0870, \ldots$ (see Figure 2.6). As $n \to \infty$, $x_n \to 0$,

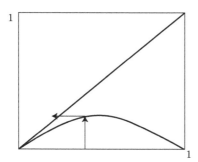

FIGURE 2.6
Stable equilibrium point. $A = 0.8$ (Not to scale.)

irrespective of any initial approximation. Nonlinearity has no effect in this case.

For $1 < A < 3$*, the solution at* $x^* = [1 - (1/A)]$ *appears:* All initial approximations x_0 in the interval $0 < x_0 < 1$ are attracted to the solution x^* and the equilibrium point is stable. The interval $1 < A < 3$ is called the *basin of attraction*. The solution at $x^* = 0$ becomes a repeller. For example, at the right end of the interval, $A = 3.0$, the equilibrium point is $x^* = 2/3$. The vertex of the parabola is at $(0.5, 0.75)$. Let an initial approximation be $x_0 = 0.3$. We obtain the sequence of values as 0.63, 0.6993, 0.6308, 0.6987, 0.6315, 0.6980, 0.6324, 0.6974, ... As $n \to \infty$, the oscillation decays and converges to the equilibrium point $x^* = 2/3$, for all initial approximations in $(0, 1)$. $A = 3$ is called the first *threshold* point (see Figure 2.7). That is, when $A = 3$, the first bifurcation takes place.

For $3 < A < 3.449 \ldots$*,* F' *is negative and* $|F'| > 1$*:* The attractor at $x^* = [1 - (1/A)]$ is also unstable and becomes a repellor. Bifurcation takes place for $3 < A < 3.449$. For example, let $A = 3.2$. The vertex of the parabola is at $V(0.5, 0.8)$. With $x_0 = 0.2$, we obtain the sequence of values as 0.512, 0.7995, 0.5129, 0.7995, 0.5129, 0.7995, ... The solution values oscillate between the two values $y_1 = 0.5129$ and $y_2 = 0.7995$. That is, $y_1 \to y_2 \to y_1 \to y_2 \ldots$. This

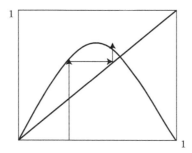

FIGURE 2.7
Stable equilibrium point. $A = 3.0$ (Not to scale.)

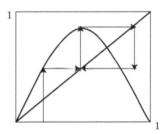

FIGURE 2.8
Stable equilibrium point. $A = 3.2$ (Not to scale.)

constitutes a stable attractor (see Figure 2.8). It is called a period-2 orbit, 2^1 cycle, period-doubling bifurcation.

Let us analyze the formation of this attractor. We have

$$x_{n+2} = Ax_{n+1}(1 - x_{n+1}) = A^2 x_n(1 - x_n)[1 - Ax_n(1 - x_n)].$$

Setting $x^* = F(x^*)$, we get $x^* = A^2 x^*(1 - x^*)[1 - Ax^*(1 - x^*)]$.

Factorizing, we get $x^*[Ax^* - (A - 1)][A^2 x^{*2} - (A^2 + A)x^* + (A + 1)] = 0$.

The solutions are $x^* = 0$, $x^* = [1 - (1/A)]$, and $x^* = [(A + 1) \pm \sqrt{(A + 1)(A - 3)}]/(2A)$.

For $A = 3.2$, we get the new solutions as $x^* = 0.7995$ and $x^* = 0.5130$: These are the values y_1 and y_2 between which the solution oscillates. The value $A = 3.449\ldots$ is called the next threshold point.

For $3.449\ldots < A < 3.5699\ldots$, F is negative and $|F'| > 1$: All the earlier attractors are unstable and become repellers. For example, let $A = 3.48$. The vertex of the parabola is at $V(0.5, 0.87)$. With $x_0 = 0.4$, we obtain the sequence of values as 0.8352, 0.4790, 0.8685, 0.3974, 0.8334, 0.4832, 0.8690, 0.3962, 0.8325, 0.4853, 0.8692, 0.3960, ... The solution values oscillate between the four values $z_1 = 0.396\ldots$, $z_2 = 0.832\ldots$, $z_3 = 0.485\ldots$, and $z_4 = 0.869\ldots$. That is, $z_1 \rightarrow z_2 \rightarrow z_3 \rightarrow z_4 \rightarrow z_1 \rightarrow z_2 \rightarrow z_3 \rightarrow z_4\ldots$. This constitutes a stable attractor. It is called a period-4 orbit, 2^2 cycle, period-doubling bifurcation.

Consider again the difference equation $x_{n+1} = Ax_n(1 - x_n)$. The solutions are in a unit square. When $A(1 - x_n) > 1$, $Ax_n(1 - x_n) > x_n$, and we have stretching. As x_n increases, at some stage, $A(1 - x_n) < 1$, and $Ax_n(1 - x_n)$ diminishes. Now, we have folding back of the solution trajectory. This stretching and folding is an important characteristic of the chaotic behavior.

For $3.5699\ldots < A < 4$, this period-doubling bifurcation continues, giving a 2^n cycle. As $n \rightarrow \infty$, all the threshold points accumulate at $A_c = 3.5699\ldots$ where the period becomes infinite. At this stage, the motion is aperiodic and the onset of chaos takes place.

The study of this periodic-doubling route to chaos gave rise to a new universal constant called *Feigenbaum constant*. Feigenbaum [49,50] showed that for all nonlinear 1D systems, which take the period-doubling route to chaos, the ratio of the successive parameter values is always a constant, independent of the actual details of the equations. Thus, the Feigenbaum constant is a new universal constant. The value of this constant is $F_\infty = 4.6692....$ Further, the ratios of the spacings of the points of successive attractors (y_1, y_2), (z_1, z_2, z_3, z_4), ... get smaller and smaller and in the limit the ratios converge to the number $\alpha = 2.5029....$ The number is called as *numerical universality* and also *metric universality*. We shall revisit this map again later while discussing deterministic chaos.

Applications of Logistic Maps and Their Variants

1. *Fashion dress created by Eri Matsui, Keiko Kimoto, and Kazuyuki Aihara:* Brilliant practical applications of the logistic model (2.34) and other models are the fashion dresses conceptualized by Professor Kazuyuki Aihara (University of Tokyo, Japan) for a fashion show held in Japan in March 2010. The design of the dress in Figure 2.9 was based on the bifurcation diagram of the logistic map $x_{n+1} = ax_n(1 - x_n)$. The dresses were designed and made by Eri Matsui, Keiko Kimoto, and Kazuyuki Aihara (Eri Matsui is a famous fashion designer in Japan) (see Kazuyuki A., private communication) [82,98].

2. *Model for the spread of avian influenza A/H5N1:* An interesting application of a variant of the logistic equation is a model of a one-parameter discrete dynamical model for the spread of avian influenza derived by Eifert et al. [48]. This model utilizes the Lindblad dissipation dynamics of quantum physics [5,61,91] for biological rate equation. It is known in epidemiology that the hygienic stress of the virus ensemble plays an important role in the prevention of this virus. The viruses are damped in their replication rate by hygienic stress and prevention methods, provided these methods are intense. On the other hand, if the hygienic stress is minimal, it will stimulate the virus replication. The plain infection model without hygienic stress is given by Equation 2.34, $x_{n+1} = Ax_n(1 - x_n)$, where x_n denotes the relative number infected at time step n and A is the infection rate. Utilizing the methods suggested in Refs. [5,61,91], Eifert et al. [48] have modified model (2.34) incorporating the hygienic stress and assuming a power-law relation [106]. The new model is a particular case of the following model:

$$x_{n+1} = Ax_n(1 - x_n) - y^\alpha A^{\beta y} x_n \ln(Ax_n), \tag{2.36}$$

where y is a fixed positive number, $\alpha > 0$, $\beta > 0$ are real positive parameters. In Ref. [48], the authors have considered model (2.36) with $\alpha = 1.5$ and $\beta = 50$. These numbers were chosen so that the

FIGURE 2.9
The design of the dress was based on the bifurcation diagram of the logistic map $x_{n+1} = ax_n(1 - x_n)$. (Courtesy Eri Matsui, Keiko Kimoto, and Kazuyuki Aihara [Kazuyuki Aihara, private communication].)

probabilities x_n stay positive and less than or equal to 1. The second term on the right-hand side indicates that the hygienic stress of the virus increases with the infection rate and this increase depends on $A^{\beta y}$. When the stress coefficient vanishes, model (2.36) reduces to the plain infection model (2.34). Since the true equation of state for the H5N1 virus is unknown, it was assumed that the hygienic stress coefficient is a monotone function of the infection rate A. The following conclusions were made from the computer experiments [48]:

i.   The chaotic attractor settles at $A = 1$ and finally vanishes as $y \to \infty$.

ii. The infection probability is reduced by the hygienic stress and the onset of chaos is earlier as without any hygienic stress. A maximum of hygienic stress extinguishes the virus.

iii. Vaccination (immunization) of the domestic animals and poultry is the best way to stop the spread of virus.

There are many other discrete time models for a single population such as the Gompertz model, Theta–Ricker model, Moran–Ricker map, sine map, tent map, and so on.

Model 7

Consider the discrete Hassell model [70]

$$P_{n+1} = \lambda P_n [1 + \alpha P_n]^{-\beta}. \tag{2.37}$$

where λ is the net growth rate, and α and β are constants defining the density-dependent feedback term. The model can be viewed as a generalization of Beverton–Holt model (2.32). The equilibrium points are the solutions of $P^* = F(P^*)$. We obtain $P^* = 0$ and $P^* = [(\lambda^{1/\beta} - 1)/\alpha]$, where $\lambda^{1/\beta} > 1$. Now,

$$\frac{dF}{dP} = \frac{\lambda[1 + \alpha P(1 - \beta)]}{(1 + \alpha P)^{\beta + 1}}.$$

At $P = P^* = 0$, we get $(dF/dP) = \lambda$. $P^* = 0$ is asymptotically stable if $|\lambda| < 1$. All the trajectories starting in the neighborhood of origin converge to the origin. $P^* = 0$ is monotonically stable for $0 < \lambda < 1$, and unstable for $\lambda > 1$.
At $P = P^* = [(\lambda^{1/\beta} - 1)/\alpha]$, or $1 + \alpha P^* = \lambda^{1/\beta}$, we obtain

$$\frac{dF}{dP} = \frac{\lambda}{(\lambda^{1/\beta})^{\beta+1}} [\lambda^{1/\beta} - \beta(\lambda^{1/\beta} - 1)] = 1 - \beta(1 - \lambda^{-1/\beta}).$$

The equilibrium point P^* is asymptotically stable when $|1 - \beta(1 - \lambda^{-1/\beta})| < 1$. The condition for equilibrium is $0 < \beta(1 - \lambda^{-1/\beta}) < 2$ [104]. Therefore, $\beta(1 - \lambda^{-1/\beta}) = 2$ defines the isocline that divides the stable and unstable regions. Simplifying, we get

$$\lambda^{-1/\beta} = 1 - \frac{2}{\beta},$$

or

$$-\frac{1}{\beta} \ln \lambda = \ln\left(1 - \frac{2}{\beta}\right),$$

or

$$\lambda = \exp\left[-\beta \ln\left(1 - \frac{2}{\beta}\right)\right].$$

Note that the stability condition is independent of the parameter α. The expression for λ allows us to represent the stability boundary in the parameter space.

The Hassell model was used by Hassell, Lawton, and May [71] to fit the census data of surviving larvae in insect populations. For this purpose, the model was rewritten as follows: Define $P_s = P_{n+1}/\lambda$. Here, P_s denotes the number of surviving larvae in generation n; the number of individuals surviving the density-dependent phase of the life cycle when the number entering this stage is P_n. From Equation 2.37, we write

$$\frac{\lambda P_n}{P_{n+1}} = \frac{P_n}{P_s} = (1 + \alpha P_n)^\beta,$$

or

$$\ln\left[\frac{P_n}{P_s}\right] = \beta \ln(1 + \alpha P_n).$$

The equation gives the relationship between mortality in the density-dependent stage and the initial density at that stage. The parameter β denotes the slope of the relationship between mortality $(\ln(P_n/P_s))$ and the population size $(\ln P_n)$ at high population levels and α is related to the point of inflection to the curve. To estimate the parameters α and β, least-squares technique to the data was used.

EXERCISE 2.1

1. Using the Malthus model, find how long it will take for the population to triple its size.

2. Solve the logistic equation $(dP/dt) = 0.09P[1 - (P/900)]$, $P(0) = 90$. Use it to find the population size at time $t = 90$ units.

3. A population is modeled by the logistic equation $(dP/dt) = 0.5P[1 - (P/500)]$.

    i. Find the equilibrium solutions and discuss their nature.

    ii. If $P(0) = 50$, find the solution of the model.

4. The population of a particular aquatic species in the Indian Ocean was about 8 billion in 1995. The birth and death rates of the species are approximately 40 million and 20 million, respectively. Assume that the carrying capacity of the Ocean is 100 billion. Using the Verhulst–Pearl logistic model, estimate the population in the years 2005 and 2100.

5. A model for the spread of a rumor is that the rate of spread is proportional to the product of the fraction R of the population who had heard the rumor and the fraction who have not heard the rumor. The

model is written as $(dR/dt) = aR(1 - R)$. A small town has a population of 10,000 persons. At 8.00 a.m., 800 people have heard a rumor. By noon half of the town has heard it. At what time 90% of the population would have heard the rumor?

6. Show that the model $dP/dt = rP/\alpha[1 - (P/K)^\alpha]$ has a point of inflection at $P = K(1 + \alpha)^{-1/\alpha}$. Also, show that as α varies from 1 to zero, the point of inflection moves from $K/2$ to K/e.

7. Assume that the population $P(t)$ at any time t in a suburban area of a large town is modeled by the initial value problem $(dP/dt) = P(10^{-3} - 10^{-9} P)$, $P(0) = 10^5$, t is measured in months. Find the limiting value of the population.

8. Suppose that the population $P(t)$ at any time t is given by the initial value problem $(dP/dt) = P(0.09 - 0.0009P)$, $P(0) = 2500$, where t is measured in months. Find the solution, limiting value of the population, and the equilibrium populations of the model.

9. A tumor may be regarded as a population of multiplying cells. It is found that the birth rate of the cells in a tumor decreases exponentially with time, so that $r(t) = r_0 e^{-\alpha t}$ and $(dP/dt) = r_0 e^{-\alpha t}P$, $P(0) = P_0$. Show that $P(t)$ approaches the finite limiting population $P_0 \exp (r_0/\alpha)$ as $t \to \infty$.

10. Suppose a student with severe viral fever returns to his residential campus of 2500 students. The rate at which the viral fever spreads is proportional not only to the number p of infected students but also to the number of noninfected students. Determine the number of students infected after 15 days if, after 5 days, it is observed that $p(5) = 50$.

11. Find the equilibrium points and discuss the local stability for the harvested logistic fish model $(dP/dt) = P(6 - P) - h$, $P(0) = P_0$. What should be the condition on P_0?

12. Find the nonzero equilibrium populations for the harvesting model [15]

$$\frac{dN}{dt} = rN\left(1 - \frac{N}{K}\right) - h.$$

   i. Find the critical value h_c, for which the harvesting rate $h > h_c$, the nonzero equilibrium values do not exist and the population tends to extinction.

   ii. Find the critical initial value for which the population may still become extinct if the initial value n_0 is below some critical value if $h < h_c$.

13. A population of fish is introduced into a pond and harvested regularly. Determine the nature of equilibrium points, the yield curve, and the maximum sustainable yield for the harvesting model

$$\frac{dN}{dt} = \alpha N \ln\left(\frac{K}{N}\right) - qEN.$$

14. For the time delay model

$$\frac{dP}{dt} = be^{-\mu\tau}P(t-\tau)e^{-aP(t-\tau)} - dP(t),$$

where b and d are constants, show that the trivial steady state is globally stable.

15. Derive the linear stability properties of the rational logistic Equation 2.27

$$P_{n+1} = P_n(1+r)\left[1 + \frac{r}{K(1+r)}P_n\right]^{-1}.$$

16. Derive the linear stability properties of the logistic Equation 2.26

$$P_{n+1} = P_n \exp\left[r\left(1 - \frac{P_n}{K}\right)\right].$$

Find the point of maximum value. Draw plots of the trajectories for $K = 500$, $r = 1$, $P_0 = 50$; $K = 500$, $r = 1$, $P_0 = 660$.

17. Constructing a suitable Lyapunov function, show that the following logistic equation [146] is globally asymptotically stable.

$$P_{n+1} = P_n \exp\left[r\left(1 - \frac{P_n}{K}\right)\right].$$

18. Discuss the behavior of the discrete exponential logistic growth model

$$x_{n+1} = x_n \exp[r(1 - x_n)].$$

19. Determine the equilibrium points and their eigenvalues for the model $P_{n+1} = \lambda P_n / (1 + b P_n^2)$. Show that the population is bounded by $\lambda / 2\sqrt{b}$.

Two-Dimensional (2D) Continuous Models (Modeling of Population Dynamics of Two Interacting Species)

In a two species ecological model system, one species is a predator and the other species is a prey. Understanding the relationship between the predator and prey is an important goal in ecology. One of the significant components of the predator–prey relationship is the predator's rate of feeding upon the prey. Among the four basic interactions (predator–prey, competition, interference, and mutualism), predator–prey interaction is the most common and is well known for generating oscillatory dynamics. One of the important observations is that these oscillations have the following property: the peak of the prey population always occurs some time before the natural enemy's population peak. There are four basic features of predator–prey interaction:

i. There should be a relationship between the prey's loss and the predator's gain.

ii. In the absence of the prey, the predator population should die out exponentially. We usually call such a predator as a specialist predator.

iii. An increase or decrease in the prey population is followed (with a delay) by an increase or decrease in the predator population, respectively.

iv. The number of prey and predator populations has an oscillatory character.

The dynamic relationship between preys and predators is an important area of research in ecology [47]. The framework of Lotka–Volterra prey–predator modeling of interactions is still popular today. It consists of two differential equations with a simple correspondence between prey consumption and predator production. The link between the dynamics of the two species in prey–predator models is based on the "trophic function," the function that describes the number of prey consumed per predator per unit time for given numbers of preys and predators.

Predator–prey interaction and competition are often viewed as the two main building blocks in mathematical population models. The species compete for the limited resources. In the case of two species competing for the same resource, in the long run, one species may survive and the other may become extinct. This is known as Gause's principle of competitive exclusion derived from a series of experiments conducted by microbiologists on the competing species of yeast cells. Further, in the predator–prey models, the populations do not appear to oscillate with time. When these resources are explicitly modeled, a simple competition interaction may be expanded as a multiple species predator–prey interactions. According to Berryman et al. [20], credible and simple predator–prey populations should in general possess some minimum biological and ecological properties. They are the following:

i. There is a negative effect of predators on prey.

ii. There is a positive effect of prey on predators.

iii. Predators must have finite appetites and finite per-capita reproductive rates.

iv. When resources are low relative to population density, the predator per-capita growth rate should decline with its density.

v. The solution of positive initial conditions should stay eventually uniformly bounded.

The predator-dependent models possess properties (i) through (v), while the prey-dependent ones lack property (iv).

Consider the two species models of the form

$$\frac{dx_1}{dt} = x_1 F_1(x_1, x_2), \quad \frac{dx_2}{dt} = x_2 F_2(x_1, x_2),$$

where F_1 and F_2 are linear functions of x_1 and x_2, respectively.

Suppose $F_1(x_1, x_2) = a_1 + b_{11}x_1 + b_{12}x_2$, $F_2(x_1, x_2) = a_2 + b_{21}x_1 + b_{22}x_2$, where a_i, b_{ij}, $i, j = 1,2$ are real constants.

Then, the above model where x_1 denotes the prey and x_2 denotes the predator is referred to as

i. *A prey–predator model:* if $a_1 > 0$, $a_2 < 0$, $b_{12} < 0$, and $b_{21} > 0$.

ii. *A competitive model:* if $a_1 > 0$, $a_2 > 0$, $b_{12} < 0$, and $b_{21} < 0$.

iii. *A cooperative model (symbiotic or mutualistic):* if $a_1 > 0$, $a_2 > 0$, $b_{12} > 0$, and $b_{21} > 0$.

A general predator–prey model in its classical form is represented by

$$\frac{dX}{dt} = Xf(X) - Yp(X,Y) = XF(X,Y), \tag{2.38}$$

$$\frac{dY}{dt} = -\delta Y + \theta Yp(X,Y) = YG(X,Y), \tag{2.39}$$

where

X: Density of preys (resource, victim) population. $X \geq 0$.

Y: Density of predators (consumer, exploiter) population. $Y \geq 0$.

$f(X)$: Per capita growth rate of prey in the absence of predation.

δ: Food-independent predator mortality, which is assumed to be a constant.

$p(X,Y)$: Trophic function. It is called the functional response in the prey equation and numerical response in the predator equation.

θ: Conversion efficiency or trophic efficiency.

F, G : Continuous and analytic functions in the domain $X \geq 0$, $Y \geq 0$.

The term $[\theta Y p(X,Y)]$ on the right-hand side of Equation 2.39 describes the per capita predator production. In population dynamics, a functional response of the predator to the prey density refers to the change in the density of prey attached per unit time per predator as the prey density changes. In chemical kinetics or microbial dynamics, it describes the uptake of the substrate by the microorganisms [134]. The simplest model of functional response is obtained by assuming that in the time available for searching, the total change in the prey density/substrate concentration is proportional to the prey density/substrate concentration.

Choice of a trophic function: The trophic function in some of the earlier works was only concerned with prey densities such as Holling type functions I, II, III, and IV, ignoring the effect of predator density, that is, $p(X,Y) = p(X)$, (prey dependent). The following are the Holling type trophic functions.

i. *Holling type I (linear):*

$$p(X) = X, \quad p(X) \rightarrow \infty \text{ as } X \rightarrow \infty. \tag{2.40}$$

The attack rate of the individual consumer increases linearly with prey density, but then suddenly reaches a constant value when the consumer is satiated. It is a simple form of the law of the mass action and is the component of the Lotka–Volterra predation model.

ii. *Holling type II (cyrtoid):*

$$p(X) = \frac{X}{X + D}, \quad D > 0, \text{ and } \lim_{X \to \infty} p(X) = 1. \tag{2.41}$$

The attack rate increases at a decreasing rate with prey density until it becomes satiated. Cyrtoid behavior responses are typical of predators that specialize on one or a few prey (see Figure 2.10).

iii. *Holling type III (sigmoid):*

$$p(X) = \frac{X^2}{X^2 + D^2}, \quad D > 0. \tag{2.42}$$

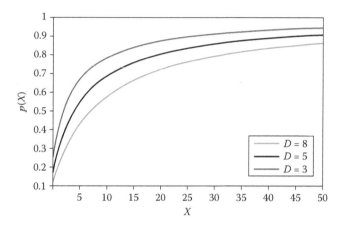

FIGURE 2.10
Holling type II functional response plotted for different values of $D = 3, 5,$ and 8.

$p(X)$ is a sigmoid and $\lim_{X \to \infty} p(X) = 1$ (see Figure 2.11). The attack rate accelerates at first and then decelerates toward satiation. Sigmoid-type functional responses are typical of generalist predators, which readily switch from one food species to another and/or which concentrate their feeding in areas where certain resources are most abundant.

Specialist predators should be characterized by hyperbolic or Holling type II response, while generalists are expected to exhibit a sigmoid-type response if they are characterized by switching behavior. For example, a house cat (a generalist predator) will hunt voles in the grassy patch when voles are abundant there. When voles become

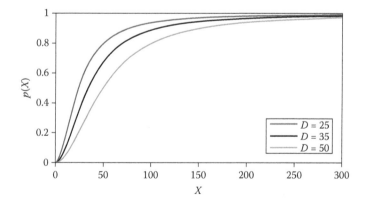

FIGURE 2.11
Holling type III functional response plotted for different values of D.

sparse, cats may shift to birds coming to the feeder. Meanwhile, cat numbers will not respond numerically to the changes in either vole or bird abundance, since they are regulated by the amount of cat food provided by their owners [151].

iv. *Holling type IV:*

$$p(X) = \frac{cX}{(X^2/i) + X + a}.$$ (2.43)

The Holling type IV functional response is relatively less studied in population ecology. It was first introduced by Haldane [66] in enzymology. Collings [33] used this response function in a mite predator–prey interaction model. Andrews [9] called it the Monod–Haldane function. The parameter i is a measure of the predator's immunity from the prey. The parameters c and a can be interpreted as the maximum per capita predation rate and the half-saturation constant in the absence of any inhibitory effect. As the value of the parameter i decreases, the predator's foraging efficiency decreases. In the limit of large i, it reduces to a type II functional response [84]. This response function describes a situation in which the predator's per capita rate of predation decreases at sufficiently high prey densities (see Figure 2.12).

Sokol and Howell [142] proposed a simplified Holling type IV function of the form

$$p(X) = \frac{cX}{X^2 + a}.$$

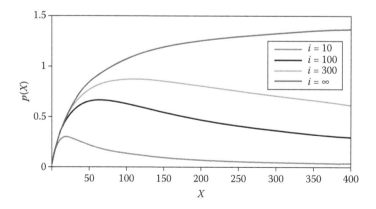

FIGURE 2.12
Holling type IV functional response plotted for fixed values of $c = 1.4$, $a = 40$ and different values of $i = 10, 100, 300$, and ∞.

Arditi and Ginzburg [11] have labeled $p(X)$ as prey-dependent functional response. They argued that the consumption rate decreases as the abundance of predator increases. They proposed that the resource-sharing mechanism could be modeled by a ratio-dependent trophic function (depends on both the prey and predator densities), in which the relevant variable is the per capita availability of resource of the consumer, that is, $p(X,Y) = p(X/Y)$.

Prey-dependent $p(X)$ and ratio-dependent $p(X/Y)$ responses offer simpler, less parameter-rich formulations. Such functions, therefore, are better starting points for modeling real predator–prey systems. If the ratio-dependent response fits the data better than the prey-dependent one with the same number of parameters, then we should consider constructing the model using the ratio-dependent form [150]. In the ratio-dependent model, we assume

$$p\left(\frac{X}{Y}\right) = \frac{(\alpha X/Y)}{1 + \alpha\beta(X/Y)} = \frac{\alpha X}{Y + \alpha\beta X},$$

where α is the total attack rate for the predator and β is the handling time. The predator density can have a direct effect on the trophic function. The first predator-dependent model was given by Hassell and Varley [72], who proposed that the attack rate should decrease with increasing predator density. They proposed the functional response in the form

$$P(X,Y) = P\left(\frac{X}{Y^m}\right) = \frac{aX}{bX + Y^m}, \quad m \in (0,1]. \tag{2.44}$$

Parameter m can be interpreted as an interference coefficient. This view is also plausible biologically [37]. When $m = 0$ or $Y = 1$, the Hassell–Varley functional response reduces to a Holling type II functional response. Arditi and Akcakaya [10] compared the two functional responses and found that $m > 0$ in each of the prey–predator systems they analyzed. They concluded that the Hassell–Varley functional response was a better descriptor of the data than the Holling type II functional response.

Beddington [16] and DeAngelis et al. [43] proposed a predator-dependent functional response (called the Beddington–DeAngelis functional response) in the form

$$p(X,Y) = \frac{aX}{bY + ahX + 1} = \frac{wX}{\alpha + \beta Y + \gamma X}, \tag{2.45}$$

where h is the prey handling time and the empirical constant b can be interpreted as the product of predator encounter rate and predator handling time. α and β represent the protection provided to the prey by its environment and intensity of interference between individuals of the specialist predator, respectively. γ determines the speed with which the per capita feeding rate approaches its saturation value w. The saturation value is defined as

$$\lim_{X \to \infty} p(X, Y) = \lim_{X \to \infty} \frac{w}{(\alpha/X) + \beta(Y/X) + \gamma} = \frac{w}{\gamma}.$$

For small values of α, $p(X,Y)$ is ratio dependent, whereas for very small values of β, $p(X,Y)$ becomes prey dependent. Ratio-dependence and prey-dependence functions are the limiting cases of the general Beddington–DeAngelis functional response or the Hassell–Varley-type functional response. Abrams and Ginzburg [1] have also proposed some predator-dependent models. Comparing with the empirical evidence from a different predator–prey model, Skalski and Gilliam [140] pointed out that the predator-dependent functional responses could provide a better description of the predator feeding over a range of predator–prey abundance, and in some cases the Beddington–DeAngelis-type functional response performed even better [92]. If predators do not waste time interacting with another or if their attacks are always successful and instantaneous, that is, $\beta = 0$, then it changes into Holling type II functional response. If $\beta < 0$, then the predators benefit from cofeeding.

We present below an important analytical tool to study the behavior of two species interacting population models.

Analytical Tool

Kolmogorov Theorem

An important tool for the analysis of model ecological systems is the Kolmogorov theorem [82a,99]. The Kolmogorov theorem assures the existence of either a stable equilibrium point or a stable limit cycle in the phase space of the 2D dynamical systems provided nine conditions (five constraints and four requirements) are satisfied. The amplitude and the period of stable limit-cycle oscillations depend on the values of the system parameters. The theorem is used to derive the conditions on system parameters which lead us to the period-doubling route to chaos and is based on the Poincaré–Benedixson theory [99]. However, it is not applicable to three-dimensional (3D) ODE models, which are commonly used to study the ecological problems. If a 3D ODE system can be subdivided into two 2D systems, then the complete system would either exhibit a stable limit cycle or chaotic dynamics when both the corresponding subsystems are in oscillatory mode. Kolmogorov theorem strongly suggests that those natural ecosystems which seem to exhibit a persistent pattern of reasonably regular oscillations [83] are in fact stable limit cycles.

Consider the following prey–predator model described by the system of equations

$$\frac{dH}{dt} = HF(H,P), \quad \frac{dP}{dt} = PG(H,P), \tag{2.46}$$

where H is the prey population at any instant of time and P is the predator population at the same instant of time. F and G are continuous functions of H and P with continuous first partial derivatives in the domain $H \geq 0, P \geq 0$.

Kolmogorov theorem: The theorem states that the 2D system (2.46) possesses either a stable equilibrium point or a stable limit cycle if the following five conditions and four requirements are satisfied.

Conditions:

$$\text{(i) } \frac{\partial F}{\partial P} < 0, \quad \text{(ii) } H\left(\frac{\partial F}{\partial H}\right) + P\left(\frac{\partial F}{\partial P}\right) < 0, \quad \text{(iii) } \frac{\partial G}{\partial P} < 0,$$

$$\text{(iv) } H\left(\frac{\partial G}{\partial H}\right) + P\left(\frac{\partial G}{\partial P}\right) > 0, \quad \text{(v) } F(0,0) > 0. \tag{2.47}$$

Requirements:

$$\text{(vi) } F(0, A) = 0, \ A > 0; \quad \text{(vii) } F(B, 0) = 0, \ B > 0;$$

$$\text{(viii) } G(C, 0) = 0, \ C > 0; \quad \text{(ix) } B > C. \tag{2.48}$$

Consider any given population sizes (as measured by numbers, biomass, etc.). Then, in biological terms, Kolmogorov conditions have the following meanings, respectively [169]:

  i. The per capita rate of change of the prey density is a decreasing function of the number of predators.
 ii. The rate of change of prey density is a decreasing function of both densities.
iii. The per capita rate of change of predator population is a decreasing function of the number of predators.
 iv. The rate of change of predator density is an increasing function of both the densities.
  v. When both the population densities are low, the prey has a positive rate of increase.

Kolmogorov requirements have the following meanings, respectively:

i. There is a predator population density sufficiently large to stop further prey growth, even when the prey is scarce.

ii. Even in the absence of predators, the prey will not grow without
bound.

iii. The predator population, when rare, cannot grow at arbitrarily small
prey densities.

iv. The minimum prey level that will permit an extremely sparse predator population to grow must be a level at which the prey is also
capable of growth.

It is important to note that the above original Kolmogorov constraints can
sometimes be relaxed in that an inequality can be replaced by the equality [99].

Local Stability Analysis

Linear stability of the state of a nonlinear dynamical system refers to the
response of the system in that state, to a small perturbation of the state. The
state may be a stable/unstable fixed point (equilibrium) or a limit cycle or
some reference trajectory, in general, for a flow or a map. If the perturbation
grows exponentially with time, the state is said to be unstable; otherwise, it
is stable. To perform the linear stability analysis, we linearize the nonlinear
dynamical equations about that reference state in the phase space.

Consider the dynamical evolution equations

$$\frac{dX_i}{dt} = F_i(X_1, X_2, ..., X_N), \quad i = 1, 2, ..., N.$$

Let $(X_1^0, X_2^0, ..., X_N^0)$ be a fixed point, that is, $F_i(X_1^0, X_2^0, ..., X_N^0) = 0$.

Let the equilibrium state be slightly perturbed. Set $X_i = X_i^0 + \delta x_i$, and write
down the time evolution for the perturbation δx_i by linearizing the nonlinear
equation about the fixed point. We have the tangent map $\delta x_i = \Sigma_1^N J_{ik} \delta x_k$ where
$J_{ik} = (\partial F_i/\partial x_k)_0$. The elements of the stability matrix or community matrix J are
given by the expression $J_{ik} = (\partial F_i/\partial x_k)^*$, $i, k = 1, 2, 3, ..., N$, where * signifies
that the elements of the community matrix are evaluated at a nondegenerate
steady state. The characteristic equation of the community matrix can be written and the Routh–Hurwitz criterion (see Chapter 1, section "Routh–Hurwitz
Criterion for Stability") is applied to obtain the conditions for the system to
possess local asymptotic stability. If the choices of the values of the parameters
are such that all the constraints are simultaneously satisfied, then the system
will be asymptotically stable in a local sense. On the other hand, violation of
any one of these conditions would result in the system jumping into the nonlinear regime where phenomena like bifurcation and chaos are observed [86].

It may be mentioned that asymptotic stability is distinct from the orbital stability. A limit cycle is asymptotically stable in the sense that neighboring trajectories converge to it and points on the orbits get arbitrarily close as $t \to \infty$. This can happen only for a dissipative system. The center, on the other hand, is only orbitally stable, that is, neighboring periodic orbits do not converge to any limiting orbit, and the closest points on the neighboring orbits need not stay closest. Here, a perturbation can lead to *jittering*. The center is possible only for nondissipative (conservative) systems.

In the next section, we present a few of the mathematical models which describe the population dynamics of the interacting species.

The solutions of the mathematical models, phase diagrams, and plotting of the time series in this chapter are done using MATLAB® 7.0.

Lotka–Volterra Model

In a system governed by the Lotka–Volterra (LV) model, both the prey (host) and the predator (parasite) undergo constant oscillations whose amplitudes bear no relation to the biology of the two species, but only to the initial sizes of their populations. This kind of behavior seems to be unlikely for real eco-systems because there has been no successful application of it to any field or laboratory population system.

One of the earliest prey–predator models was devised by Lotka [93] and independently by Volterra [163]. The model can be mathematically described by the following pair of coupled nonlinear differential equations:

$$\frac{dP}{dt} = (a_1 - b_1 Z)P = PF(Z), \quad \frac{dZ}{dt} = (-a_2 + b_2 P)Z = ZG(P), \quad (2.49)$$

where a_1, a_2, b_1, b_2 are positive constants and

   P: Density of the prey population at any time t.

   Z: Density of the predator population at any time t.

   a_1: Intrinsic rate of increase of the prey population.

   a_2: Intrinsic death rate of the predator population.

   b_1, b_2: Constants which express the effect of the density of one species on the rate of growth of the other.

The first equation states that the rate of change in the density of prey population with time is a function of the intrinsic rate of increase of the prey minus losses due to the density of the predator population. Similarly, the second equation states that the rate of change in the density of the predators is equal to a gain due to the density of prey minus the intrinsic rate of death. The assumptions implicit in the model are the following:

   i. Neither the prey nor the predator population inhibits its own rate of growth.

ii. The environment is completely closed and homogeneous.

iii. Every prey has equal probability of being attacked.

The equilibrium points of the system are $(0, 0)$, and $(P^*, Z^*) = (a_2/b_2, a_1/b_1)$.

Behavior of the system in the neighborhood of $(0, 0)$: In the neighborhood of $(0, 0)$, we neglect the nonlinear terms. We obtain the system as

$$\frac{dP}{dt} \approx a_1 P, \quad \frac{dZ}{dt} \approx -a_2 Z.$$

The solution of the system is $P(t) = P_0 e^{a_1 t}$, $Z(t) = Z_0 e^{-a_2 t}$. The prey population increases exponentially near the origin, while the predator population decreases exponentially. No solution with positive initial values approaches $(0, 0)$ in the phase plane. An equilibrium point which exhibits this type of nature is called a saddle point and is unstable.

Behavior of the system in the neighborhood of (P^*, Z^*): Using the transformation $u = P - P^*$, and $v = Z - Z^*$, in Equation 2.49, we obtain

$$\frac{du}{dt} = -b_1[vP^* + uv], \quad \frac{dv}{dt} = b_2[uZ^* + uv].$$

In a neighborhood of the interior equilibrium point (P^*, Z^*), we assume that u and v are sufficiently small and neglect the second-order terms. We obtain the linear system as

$$\frac{du}{dt} = -b_1 v P^*,$$

and

$$\frac{dv}{dt} = b_2 u Z^*.$$

Eliminating t, we obtain

$$\frac{du}{dv} = -\frac{b_1 v P^*}{b_2 u Z^*},$$

or

$$b_2 Z^* u \, du + b_1 P^* v \, dv = 0.$$

Integrating, we obtain $b_2 Z^* u^2 + b_1 P^* v^2 = c^2$.

In terms of the original variables, we have

$$b_2 Z^*(P - P^*)^2 + b_1 P^*(Z - Z^*)^2 = c^2.$$

These curves are a family of ellipses with centers at (P^*, Z^*).
To study the global stability analysis, we select the scalar function

$$V(u,v) = \int_0^u \frac{b_2 s\, ds}{s + P^*} + \int_0^v \frac{b_1 s\, ds}{s + Z^*}.$$

We find $V(u, v) > 0$ and $V(0, 0) = 0$. Now,

$$V^*(u,v) = \frac{b_2 u u'}{u + P^*} + \frac{b_1 v v'}{v + Z^*} = b_2 u[-b_1 v] + b_1 v[b_2 u] = 0.$$

Thus, the equilibrium point $E^*(a_2/b_2, a_1/b_1)$ is stable (known as marginally stable), but not asymptotically stable. Therefore, the equilibrium point $E^*(a_2/b_2, a_1/b_1)$ is a center. In fact, every trajectory of the LV model is a closed orbit except the equilibrium E^* and coordinate axes [75].

Another way of solving the above Equations 2.49 is to assume the parametric form of a solution curve in the P–Z plane as $P = P(t)$, $Z = Z(t)$. Dividing the two equations and separating the variables, we obtain

$$\frac{(a_1 - b_1 Z)dZ}{Z} = -\frac{(a_2 - b_2 P)dP}{P}.$$

Integrating this equation, we get,

$$a_1 \ln Z - b_1 Z = -a_2 \ln P + b_2 P + \ln K$$

or

$$Z^{a_1} e^{-b_1 Z} = K P^{-a_2} e^{b_2 P}, \tag{2.50}$$

where K is an arbitrary constant. Imposing the initial conditions $P(0) = P_0$, $Z(0) = Z_0$, we obtain $K = Z_0^{a_1} P_0^{a_2} e^{-(b_1 Z_0 + b_2 P_0)}$.

Although we cannot solve Equation 2.50 for either P or Z, we can determine the points on the curve by an ingenious method due to Volterra. Equating the left-hand and right-hand sides of Equation 2.50 to new variables H and W, we obtain

$$H = Z^{a_1} e^{-b_1 Z},$$

and

$$W = K P^{-a_2} e^{b_2 P}.$$

Thus, we find that the system is characterized by endlessly prolonged oscillations of constant amplitude which depend on the chosen initial population sizes, P_0 and Z_0. For the chosen initial population sizes P_0 and Z_0, the above equation represents an oval about the interior equilibrium point (P^*, Z^*) in the anticlockwise direction. In other words, the system possesses neutral stability, which means that the two populations undergo constant oscillations with amplitudes depending on the initial population sizes rather than on any intrinsic attributes of the two interacting species. Let T represent the period of the solutions. Now, consider the first equation in (2.49)

$$\frac{dP}{dt} = (a_1 - b_1 Z)P.$$

Integrating over one period T, we obtain

$$\int_{t_0}^{t_0+T} \frac{dP}{P} = \int_{t_0}^{t_0+T} (a_1 - b_1 Z)dt,$$

$$a_1 T - b_1 \int_{t_0}^{t_0+T} Z\,dt = \ln\left[\frac{P(t_0 + T)}{P(t_0)}\right] = 0,$$

or

$$\frac{1}{T} \int_{t_0}^{t_0+T} Z\,dt = \frac{a_1}{b_1}$$

since $P(t)$ is periodic. We conclude that the average of the predator density over a cycle is $Z_{average} = (a_1/b_1)$.

Similarly, from the second equation in Equation 2.49, we can show that $P_{average} = (a_2/b_2)$.

Another way of reaching the same conclusion, that is, the model possesses neutral stability, is by considering the behavior of the linearized model in the neighborhood of the positive equilibrium point $P^* = a_2/b_2$, $Z^* = a_1/b_1$.

Linearizing the equations about the equilibrium point (P^*, Z^*), we obtain the Jacobian matrix of the system as

$$\begin{bmatrix} F(Z) & PF'(Z) \\ ZG'(P) & G(P) \end{bmatrix} = \begin{bmatrix} a_1 - b_1 Z^* & -b_1 P^* \\ b_2 Z^* & (-a_2 + b_2 P^*) \end{bmatrix}$$

$$= \begin{bmatrix} 0 & -(a_2 b_1/b_2) \\ (a_1 b_2/b_1) & 0 \end{bmatrix}.$$

The characteristic equation is $\lambda^2 + a_1 a_2 = 0$. The eigenvalues are a pair of imaginary numbers $\lambda_{1,2} = \pm iw$, where $w = \sqrt{a_1 a_2}$. The perturbations to prey and predator populations are linear combinations of the factors $\exp(\lambda_1 t)$

and $\exp(\lambda_2 t)$, with the coefficients depending upon the initial disturbances. Therefore, we have linear combinations of the purely oscillatory factors $\exp(-iwt)$ and $\exp(iwt)$, or the combinations of $\cos(wt)$ and $\sin(wt)$. Hence, we obtain neutral stability with perturbations leading to undamped pure oscillations, of frequency w and period $(2\pi/w)$. The amplitudes depend on the initial conditions only. The behavior is not natural as the oscillations depend only on the initial conditions and not on the intrinsic biological attributes of the interacting system.

We find that model (2.49) has a limit-cycle solution for the following values of the parameters: $a_1 = 4.5$, $b_1 = 0.25$, $a_2 = 4.0$, and $b_2 = 0.5$. For this set of values, we obtain $P^* = 8$, $Z^* = 18$. Small variation in a control parameter, say a_1, shows shrinking in the size of limit-cycle attractor. For different values of the intrinsic growth rate parameter for prey population $a_1 = 4.5, 4.25, 3.5, 3.25$, and 3, we find endlessly prolonged oscillations of decreasing amplitude on the limit cycle (see Figure 2.13). A stable limit-cycle trajectory is such that any small perturbation from the trajectory decays to zero. Figure 2.14 shows the limit-cycle attractor for $a_1 = 4.5$ and its corresponding time series. The other values of the parameters are the same as for the limit-cycle solution. Every trajectory of the LV equation is a closed orbit.

Note that Kolmogorov theorem cannot be applied as the requirement (ii) $F(B, 0) = 0$ gives $a_1 = 0$, which is not true.

Generalized LV Model: The LV model was improved by Gause [58] who gave the following predator–prey model for consumer–resource interaction:

$$\frac{dx}{dt} = h(x) - yp(x), \quad \frac{dy}{dt} = cyp(x) - dy, \tag{2.51}$$

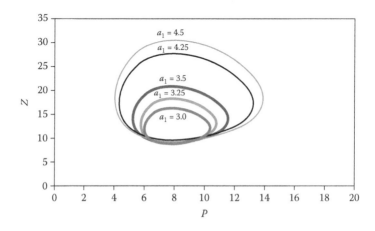

FIGURE 2.13
Closed (P, Z) phase plane trajectories for the LV model for different values of the parameter $a_1 = 4.5, 4.25, 3.5, 3.25$, and 3.0.

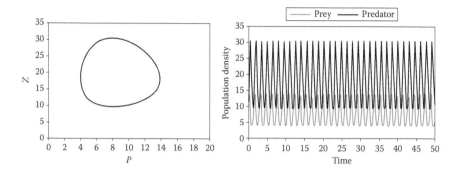

FIGURE 2.14
Phase plane trajectory and the corresponding time series for the LV model for the parameter values $a_1 = 4.5$, $b_1 = 0.25$, $a_2 = 4.0$, $b_2 = 0.5$.

where c and d are constants and

x, y: Prey and predator densities, respectively.

$h(x)$: Growth function of the prey, continuously differentiable, and is assumed to be positive in the interval $0 < x < K$ and negative for $x > K$, or $x < 0$.

K: Carrying capacity.

$p(x)$: Functional response. It is assumed to be an increasing function and is continuously differentiable. It has a unique zero at the origin.

c, d: Conversion factor and death rate of the predator population, respectively.

The physical conditions are such that the solution of system (2.51) remains positive and bounded. In the LV model, the functions $h(x)$ and $p(x)$ were assumed to be linear. For a generalized Gause-type model, the function $h(x)$ is taken as nonlinear. Model system (2.51) is a 2D system of ODEs and complicated oscillations are therefore excluded by the Poincaré–Bendixson theorem [75]. Depending on the parameter values and the initial conditions, the limit set is either a limit cycle or an equilibrium point. It is important to note that when the attractor of a Gause system is unique, the dynamical behavior of the system is practically independent of the initial condition. Since at most one of the fixed points of the system can be an attractor and the system has bounded solutions, the system has a unique attractor if it possesses at most one limit cycle.

A number of authors [55,59,73,74,108,109] have studied the following Gause-type predator–prey model:

$$\frac{dx}{dt} = rx(1 - x) - yp(x), \quad \frac{dy}{dt} = y[cp(x) - d], \tag{2.52}$$

where x, y are the prey and predator densities, respectively, and r, c, and d represent, respectively, the prey intrinsic growth rate, conversion efficiency rate of prey to predator, and predator death rate.

Most authors studied the dynamics of Equation 2.52 under the following assumptions: [73,110]: Let $p(x) \in C^2 [0, +\infty)$ and

$$(i)\ p(0) = 0, \quad (ii)\ p'(x) > 0, \quad \text{for } x \geq 0,$$

$$(iii)\ p''(x) \leq 0, \quad \text{for } x \geq 0, \quad (iv)\ \lim_{x \to +\infty} p(x) = C < \infty.$$

A functional response in which the consumption rate increases at a decreasing rate with prey density until it becomes constant at satiation satisfies the above assumptions. There are several functional responses that satisfy the above conditions [81]. Some simple examples are the responses (a) $p(x) = 1 - e^{-ax}$ [80], and (b) Holling type II response, which is also a most common type of functional response.

The positive interior equilibrium point (x^*, y^*) exists if $0 < x^* < 1$. Hesaaraki and Moghadas [74] have shown that the model has at least one limit cycle in the first quadrant if and only if [110]

$$2x^* + x^* (1 - x^*) \frac{p'(x^*)}{p(x^*)} < 1. \tag{2.53}$$

As a consequence of this result, it was shown in Ref. [74] that condition (iv) can be replaced by

$$(v)\ \lim_{x \to +\infty} p'(x) = 0. \tag{2.54}$$

They have also shown that if conditions (i), (ii), (iii), and (v) are satisfied and $p'''(x) > 0$, in $0 < x < 1$, then the model has at least one limit-cycle solution if Equation 2.53 holds.

Global analysis of the model was studied by several authors [12,28,73]. The analysis often led to some necessary and sufficient conditions for the existence of limit cycles, which are crucial to the survival of the ecosystem. The uniqueness of the limit cycle was proved by Cheng [28] and Hasik [69]. Ardito and Ricciardi [12] established the global stability of the positive critical point by constructing a Lyapunov function.

Consider the following examples.

Example 2.5

Consider the model (Rosenzweig [132], Hesaaraki and Moghadas [74], Moghadas and Corbett [110])

$$\frac{dx}{dt} = ax(1 - x) - yx^{\alpha}, \quad \frac{dy}{dt} = (cx^{\alpha} - d)y \quad 0 < \alpha < 1, \tag{2.55}$$

where a, c, and d are positive parameters with $0 < (d/c) < 1$.

Now, $p(x) = x^{\alpha}$ satisfies the assumptions (i) through (iii) and in place of (iv) it satisfies the weaker condition

$$\lim_{x \to \infty} p'(x) = \lim_{x \to \infty} \alpha\, x^{\alpha-1} = 0.$$

The model admits a unique positive critical point given by

$$x^* = (d/c)^{1/\alpha}, \quad y^* = [acx^*(1 - x^*)/d].$$

The model has at least one limit cycle if and only if Equation 2.53 is satisfied. That is,

$$2x^* + x^*(1 - x^*)\frac{\alpha(x^*)^{\alpha-1}}{(x^*)^{\alpha}} < 1,$$

or

$$x^*(2 - \alpha) < 1 - \alpha, \quad \text{or} \quad (d/c)^{1/\alpha} < (1 - \alpha)/(2 - \alpha).$$

Example 2.6

Consider the model described by

$$\frac{dx}{dt} = ax(1 - x) - y[5 - (x + 5)e^{-\alpha x}], \quad \frac{dy}{dt} = y[c\{5 - (x + 5)e^{-\alpha x}\} - d], \quad (2.56)$$

where all the parameters are positive, $0 < (d/c) < 5$ and $\alpha \geq 1$. In Moghadas and Corbett [110], the expression inside the brackets on the right-hand sides of the equations was taken as $[3 - (x + 3)e^{-\alpha x}]$ in the place of $[5 - (x + 5)e^{-\alpha x}]$.

We have $p(x) = 5 - (x + 5)e^{-\alpha x} > 0$, for $x > 0$, $p'(x) = e^{-\alpha x}[\alpha(x + 5) - 1] > 0$, for $x \geq 0$,

$$p''(x) = e^{-\alpha x}[2\alpha - \alpha^2(x + 5)] < 0, \quad \text{for } x \geq 0, \quad \text{and} \quad \lim_{x \to \infty} p(x) = 5 < \infty.$$

The model admits a unique critical point in the first quadrant such that

$$(x^* + 5)e^{-\alpha x^*} = 5 - (d/c), \quad y^* = cax^*(1 - x^*)/d.$$

The model has at least one limit cycle if and only if Equation 2.53 is satisfied. That is,

$$2x^* + x^* (1 - x^*)\frac{e^{-\alpha x^*}[\alpha(x^* + 5) - 1]}{5 - (x^* + 5)e^{-\alpha x^*}} < 1,$$

or

$$x^*(1 - x^*)[\alpha(x^* + 5) - 1]e^{-\alpha x^*} < (1 - 2x^*)[5 - (x^* + 5)e^{-\alpha x^*}].$$

Substituting $e^{-\alpha x^*} = [\{5 - (d/c)\}/(x^* + 5)]$ and simplifying, we get

$$\alpha < \frac{d(1 - 2x^*)(x^* + 5) + (5c - d)x^*(1 - x^*)}{(5c - d)x^*(1 - x^*)(x^* + 5)} \equiv Q.$$

But, $\alpha \geq 1$. This implies that $Q > \alpha \geq 1$. Hence, the system has no limit cycles if $Q \leq 1$. That is,

$$\frac{d(1 - 2x^*)(x^* + 5) + (5c - d)x^*(1 - x^*)}{(5c - d)x^*(1 - x^*)(x^* + 5)} \leq 1,$$

or

$$\frac{(1 - 2x^*)(x^* + 5)}{x^*(1 - x^*)(x^* + 4)} > \frac{5c - d}{d} = m.$$

Note that $m > 0$. Simplifying, we obtain

$$mx^*(x^* - 1)(x^* + 4) - 2(x^* + 5)[x^* - (1/2)] < 0. \qquad (2.57)$$

Let $f(x) = mx(x - 1)(x + 4) - 2(x + 5)[x - (1/2)]$.

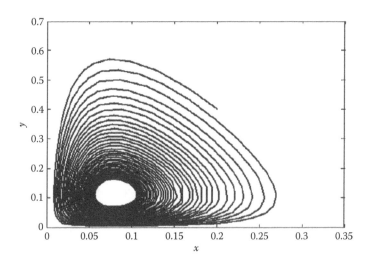

FIGURE 2.15
A limit cycle of system (2.56) for $a = 1, c = 3, d = 2, \alpha = 2$.

We find, $f(-5) = -30m < 0$, $f(-4) = 9 > 0$, $f(0) = 5 > 0$, $f(1/2) = -1.125m < 0$, and $f(x) > 0$ for large values of x. Therefore, $f(x)$ has a negative root $x_1 \in (-5, -4)$, a positive root $x_2 \in (0, 0.5)$, and a very large positive root x_3. Let $x_1 = -q$. Then Equation 2.57 can be written as $m(x + q)(x - x_2)(x - x_3) \leq 0$. For $0 < x < x_2$, the inequality is violated. Hence, a limit cycle may exist if $0 < x < x_2$, where $x_2 \in (0, 0.5)$. A limit cycle is drawn in Figure 2.15, when the parameters take the values $a = 1$, $c = 3$, $d = 2$, and $\alpha = 2$. For these values of the parameters, we get $x^* = 0.07943$, $y^* = 0.10968$.

Variation of the Classical LV Model

We shall study a variation of the classical LV model as the following example.

Example 2.7

Study the asymptotic stability of the Volterra model (density-dependent growth *Volterra model* [164]).

$$\frac{dP}{dt} = \left[a_1 \left(1 - \frac{P}{K} \right) - b_1 Z \right] P = PF(P, Z), \quad \frac{dZ}{dt} = (-a_2 + b_2 P)Z = ZG(P), \quad (2.58)$$

where a_1, a_2, b_1, b_2, and K (carrying capacity of the prey population) are positive constants and these parameters have the same meanings as defined in Equations 2.49.

SOLUTION

The LV model is based on the assumption that the prey grows exponentially in the absence of the predator which is unrealistic. Even in the absence of the predators, disease and/or food shortage will eventually curb the growth. The Volterra model assumes that in the absence of the predators, the prey grows as per the logistic equation. The second equation of this model is the same as in the LV model. The model has five constants. To reduce the number of constants, we nondimensionalize the model equations by using the transformations

$$x = \frac{P}{K}, \quad y = \frac{b_1 Z}{a_1}, \quad T = a_1 t, \quad \alpha = \frac{a_2}{b_2 K}, \quad \beta = \frac{b_2 K}{a_1}.$$

We obtain the nondimensional system as

$$\frac{dx}{dT} = x(1 - x - y), \quad \frac{dy}{dT} = \beta(x - \alpha)y.$$

The system admits three equilibrium points, $E_0(0, 0)$, $E_1(1, 0)$, and $E_2(\alpha, 1 - \alpha)$. The interior equilibrium point exists only if $\alpha < 1$.

Now, we study the local behavior of the system near the equilibrium points. The Jacobian matrix is given by

$$J = \begin{bmatrix} a_{11} & a_{12} \\ a_{21} & a_{22} \end{bmatrix}, \tag{2.59}$$

where $a_{11} = 1 - 2x - y$, $a_{12} = -x$, $a_{21} = \beta y$, and $a_{22} = \beta(x - \alpha)$.

At the equilibrium point (0, 0): We obtain $a_{11} = 1$, $a_{12} = 0$, $a_{21} = 0$, and $a_{22} = -\beta \alpha$. The eigenvalues of J are the diagonal elements $1, -\alpha\beta$. Since the eigenvalues are of the opposite sign, the equilibrium point (0, 0) is a saddle point.

At the equilibrium point (1, 0): We obtain $a_{11} = -1$, $a_{12} = -1$, $a_{21} = 0$, and $a_{22} = \beta(1 - \alpha)$.

The eigenvalues are the diagonal elements -1, $\beta(1 - \alpha)$. The nature of the equilibrium point depends on the value of the parameter α. For $\alpha < 1$, the equilibrium (1, 0) is unstable. For $\alpha > 1$, the equilibrium point is asymptotically stable. That is, the axial equilibrium point is asymptotically stable in the absence of the interior equilibrium and it turns unstable upon the emergence of the interior equilibrium. In terms of the original parameters, $\alpha > 1$ implies $a_2 > b_2K$, that is, the per capita death rate of the predators (a_2) exceeds the product of their per capita growth rate and the carrying capacity of the prey population (b_2K).

At the interior equilibrium point E_2 (α, $1 - \alpha$), $\alpha < 1$: We obtain $a_{11} = -\alpha$, $a_{12} = -\alpha$, $a_{22} = 0$, and $a_{21} = \beta(1 - \alpha)$. The characteristic equation is $\lambda^2 + \lambda \alpha + \alpha\beta(1 - \alpha) = 0$. The eigenvalues are

$$\lambda_1 = [-\alpha + \sqrt{\alpha^2 - 4\alpha\beta(1 - \alpha)}]/2,$$

and

$$\lambda_2 = [-\alpha - \sqrt{\alpha^2 - 4\alpha\beta(1 - \alpha)}]/2.$$

The real parts of the eigenvalues are negative. The interior equilibrium point is a stable node if the discriminant is positive, that is, $\alpha > 4\beta(1 - \alpha)$ or $\alpha > 4\beta/(1 + 4\beta)$ and a stable focus for $\alpha < 4\beta/(1 + 4\beta)$ [84].

The Volterra model is more stable than the LV model. In fact, the Volterra model is characterized by a globally stable equilibrium point—any trajectory starting at positive prey and predator density will be attracted to this equilibrium [150]. This is not surprising, since we expect that any density-dependent growth model should contribute to the stability of the system.

Leslie–Gower Model

The Leslie–Gower model [88] leads to asymptotic solutions tending to stable equilibrium, which is independent of the initial conditions and depends on the intrinsic factors governing the biology of the system. Although it marks a significant improvement over the LV model, it is limited in its explanatory capability. The equations governing the Leslie–Gower model are

$$\frac{dP}{dt} = (a_1 - c_1Z)P = PF(Z), \quad \frac{dZ}{dt} = [a_2 - c_2(Z/P)]Z = ZG(P,Z), \tag{2.60}$$

where the constants a_1, c_1, a_2, and c_2 are positive, and

a_1: Intrinsic growth rate of prey,

c_1: Effect of the density of the predator population on the population growth of the prey,

a_2: Intrinsic growth rate of the predator,

c_2: Number of prey necessary to support and replace each individual predator.

The factor $[-c_2(Z/P)]$ tells us that the rate of the growth of the predator population is limited and causes a decrease in the rate of increase of the predator population as Z increases.

In this model, the following assumptions are inherent:

i. The rate of increase of the predator (parasite) population has an upper limit.

ii. Intraspecific competition has negligible effect on prey's population growth.

The basic characteristic of the Leslie–Gower model is that it leads to a solution which is asymptotically independent of the initial conditions and depends only on the intrinsic attributes of the interacting system, that is, the parameters a_1, c_1, and so on.

Let us try to satisfy the conditions of the Kolmogorov theorem.

Conditions (i), (ii), (iii), (v), and (vi) are satisfied. Equality condition in (iv) is satisfied. The requirements (vii) and (viii) are violated. We find that $F(B, 0) = 0$ and $G(C, 0) = 0$ gives $a_1 = 0$, $a_2 = 0$, which violates the assumption that a_1, a_2 are positive. Hence, Kolmogorov theorem cannot be applied.

Linear stability of the system (2.60): The equilibrium points are the solutions of the equations

$$(a_1 - c_1 Z)P = 0, \quad [a_2 - c_2(Z/P)]Z = 0.$$

The positive equilibrium point is given by (P^*, Z^*), where $P^* = a_1 c_2/a_2 c_1$, and $Z^* = a_1/c_1$. Note that $(0, 0)$ is not an equilibrium point.

The Jacobian matrix of the system evaluated at an equilibrium point (P^*, Z^*) is given by Equation 2.59, where $a_{11} = 0$, $a_{12} = -c_1 P^* = -a_1 c_2/a_2$, $a_{21} = c_2 (Z^*/P^*)^2 = a_2^2/c_2$, and $a_{22} = a_2 - 2c_2(Z^*/P^*) = -a_2$.

The eigenvalues of J are the roots of $\lambda^2 + a_2\lambda + a_1 a_2 = 0$. The eigenvalues are

$$\lambda_1 = (-a_2 + \sqrt{a_2^2 - 4a_1 a_2})/2, \quad \text{and} \quad \lambda_2 = (-a_2 - \sqrt{a_2^2 - 4a_1 a_2})/2.$$

Therefore, the system is asymptotically stable for $a_2 > 4a_1$. The oscillations are damped and the solutions tend to a stable equilibrium level in both the populations.

Numerical simulation of model system (2.60) is done using the initial condition $(P_0, Z_0) = (10, 10)$, and $a_1 = 2.0$, $a_2 = 1.5$, $c_1 = 0.04$, $c_2 = 1$. The equilibrium point is obtained as $P^* \approx 33.33$, $Z^* = 50$. The parameter values violate the condition $a_2 > 4a_1$. We obtain a stable focus as presented in Figure 2.16. The time series is also given in Figure 2.16.

In the above model, we can also incorporate a self-interacting term like $(-b_1 P^2)$ in the equation for the time rate of change for the host. In this case, the equations governing the model become [155]

$$\frac{dP}{dt} = (a_1 - b_1 P - c_1 Z)P = PF(P,Z), \quad \frac{dZ}{dt} = [a_2 - c_2(Z/P)]Z = ZG(P,Z), \quad (2.61)$$

where the parameters are positive and b_1 measures the strength of intraspecific competition for prey.

We find that Kolmogorov theorem cannot be applied (see Problem 2, Exercise 2.2).

Linear stability analysis: Note that $(0, 0)$ is not an equilibrium point. The system has two equilibrium points $E_1(K,0)$, where $K = a_1/b_1$ and $E^*(P^*, Z^*)$ where

$$P^* = \frac{a_1 c_2}{a_2 c_1 + b_1 c_2}, \quad Z^* = \frac{a_2}{c_2} P^* = \frac{a_1 a_2}{a_2 c_1 + b_1 c_2}.$$

The Jacobian matrix of the system is given by Equation 2.59, where $a_{11} = a_1 - 2b_1 P^* - c_1 Z^*$, $a_{12} = -c_1 P^*$, $a_{21} = c_2(Z^*/P^*)^2$, and $a_{22} = a_2 - 2c_2(Z^*/P^*)$.

At the equilibrium point E_1 $(K, 0)$: We obtain

$$a_{11} = a_1 - 2b_1 K = -a_1, \quad a_{12} = -c_1 P^* = -a_1 c_1/b_1, \quad a_{21} = 0, \quad a_{22} = a_2.$$

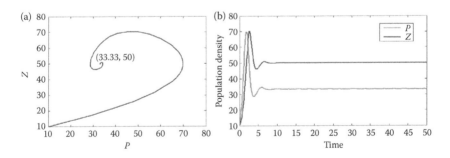

FIGURE 2.16
Stable focus: (a) Phase plot, (b) time series, for model (2.60).

The eigenvalues are $\lambda_1 = a_{11} = -a_1 < 0$, and $\lambda_2 = a_{22} = a_2 > 0$.

The equilibrium point is unstable. The fixed point $E_1(K, 0)$ is a hyperbolic saddle point, which attracts in the x-direction and repels in the y-direction. *At the equilibrium point $E^*(P^*, Z^*)$:* We obtain

$$a_{11} = a_1 - 2b_1P^* - c_1Z^* = \frac{-a_1b_1c_2}{(a_2c_1 + b_1c_2)}, \quad a_{12} = -c_1P^* = \frac{-a_1c_1c_2}{(a_2c_1 + b_1c_2)},$$

$$a_{21} = \frac{a_2^2}{c_2}, \quad a_{22} = a_2 - \frac{2c_2Z^*}{P^*} = -a_2.$$

The eigenvalues of J are the roots of $\lambda^2 - (a_{11} + a_{22})\lambda + (a_{11}a_{22} - a_{12}a_{21}) = 0$. The system is locally stable, if the eigenvalues are negative or have negative real parts. By the Routh–Hurwitz theorem, the necessary and sufficient conditions are $-(a_{11} + a_{22}) > 0$, $(a_{11}a_{22} - a_{12}a_{21}) > 0$. That is, $a_2 + [a_1b_1c_2/(a_2c_1 + b_1c_2)] > 0$ and $(a_1a_2b_1c_2 + a_1a_2^2c_1)/(a_2c_1 + b_1c_2) = a_1a_2 > 0$.

The conditions are satisfied as all the parameters are positive. The equilibrium point $E^*(P^*, Z^*)$ is locally asymptotically stable. The dynamical behavior of model system (2.61) is the same as the dynamical behavior of model system (2.60).

Numerical simulation of model system (2.61) is done using the initial condition $(P_0, Z_0) = (10,10)$, and $a_1 = 2.0$, $b_1 = 0.05$, $a_2 = 1.5$, $c_1 = 0.04$, $c_2 = 1$. The equilibrium point is obtained as $P^* \approx 18.18$, $Z^* \approx 27.27$. We obtain a stable focus as presented in Figure 2.17. The time series is also given in Figure 2.17.

This model also leads to solutions which are asymptotically independent of the initial conditions and depends only on the intrinsic attributes of the interacting system.

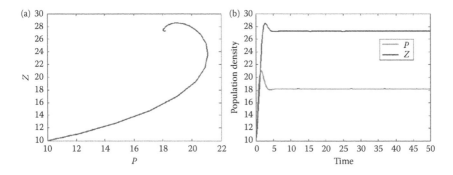

FIGURE 2.17
Stable focus: (a) Phase plot, (b) time series, for model (2.61).

The following two ecological principles form the basic structure of the model systems:

i.   A specialist predator (e.g., hawk owl, least weasel, stoat) dies out when its favorite food is absent or is in short supply. The predator population decays exponentially in the absence of its prey.

ii. The generalist predator (e.g., foxes, cats, common buzzards, hooded crow, etc.) switches to an alternative food option as and when it faces difficulty to find its favorite preys. The per capita growth of a generalist predator is limited by dependence on its favorite preys, and this limitation is inversely proportional to per capita availability of preys at any instant of time.

iii. Combining the two kinds of predation (specialist and generalist) in the same predator–prey model (discussed in a later section) demonstrates that the generalist predators have a stabilizing effect on the cycle driven by specialist predators. Sufficiently large numbers of generalist predator convert the limit cycle to a stable equilibrium point [67].

Rosenzweig–MacArthur Model

The Rosenzweig–MacArthur (RM) model [133] describes a prey-specialist predator dynamics. The assumptions underlying this formulation of predator–prey interaction are as follows: The life histories of each population involve continuous growth and overlapping generations. The predator Y dies out exponentially in the absence of its most favorite prey X. The predator's feeding rate saturates at high prey densities. The age structures of both the populations are ignored. It is assumed that the per capita growth rate of prey in the absence of predation defined by the function f grows logistically with the intrinsic growth rate a_1 (per capita rate of self-reproduction for the prey) and carrying capacity $K = a_1/b_1$ so that $f(X) = (a_1 - b_1 X)$. The functional response of the prey is taken as $p(x,y) = p_1(X) = wX/(X + D)$ (Holling type II functional response), and the functional response of the predator is taken as $p(x,y) = p_2(X) = w_1 X/(X + D_1)$.
The model system is given by

$$\frac{dX}{dt} = a_1 X - b_1 X^2 - \frac{wXY}{X + D} = XF(X,Y), \tag{2.62}$$

$$\frac{dY}{dt} = -a_2 Y + \frac{w_1 XY}{X + D_1} = YG(X,Y), \quad X(0) > 0, \quad Y(0) > 0, \tag{2.63}$$

where

$$F(X,Y) = a_1 - b_1 X - \frac{wY}{X + D}, \quad G(X,Y) = -a_2 + \frac{w_1 X}{X + D_1}.$$

The parameters have the following meanings:

b_1: Measures the intensity of competition among individuals of species X for space, food, and so on.

w: Maximum rate of per capita removal of prey species X due to predation by its predator Y.

D: Value of population density of X at which per capita removal rate is half of w.

a_2: Measures how fast the predator Y will die when there is no prey to capture, kill, and eat.

D_1: Population density of the prey at which per capita gain per unit time in Y is half of its maximum value, w_1.

$(w_1 X)/(X + D_1)$: Per capita rate of gain in predator Y, where w_1 is its maximum value.

The two species system (2.62), (2.63) qualifies as a Kolmogorov system when the conditions $w_1 > a_2$, $K(w_1 - a_2) > D_1 a_2$ are satisfied (see Problem 3, Exercise 2.2).

The system has three equilibrium points $E_0(0, 0)$, $E_1(K, 0)$, where $K = a_1/b_1$ and $E^*(X^*, Y^*)$ where

$$X^* = \frac{a_2 D_1}{w_1 - a_2}, \quad Y^* = \frac{b_1}{w}(K - X^*)(X^* + D).$$

Linear stability analysis shows that the equilibrium points $E_0(0, 0)$ and $E_1(K, 0)$ are hyperbolic saddle points. The equilibrium point $E^*(X^*, Y^*)$ is locally asymptotically stable if the following condition is satisfied (see Problem 4, Exercise 2.2).

$$2b_1\left(\frac{a_2 D_1}{w_1 - a_2}\right) + b_1 D - a_1 > 0. \tag{2.64}$$

Setting $x = a_2$ and $y = (a_1/b_1)$, we obtain the equation of the bounding curve (2.64) as $(2D_1 - D)x + Dw_1 - yw_1 + xy = 0$. The plot of the bounding curve is given in Figure 2.18 for $D = D_1 = 10$ and $w_1 = 2$. The bounding curve is given by $xy = 2(y - 5x - 10)$. Rai and Upadhyay [130] have erroneously taken the curve as a straight line.

We obtain a stable limit cycle in the phase space if inequality (2.64) is violated and the conditions from Kolmogorov analysis [99,169] are satisfied.

Aziz-Alaoui [13] carried out the stability analysis of $E^*(X^*, Y^*)$ and found the criteria for global asymptotic stability.

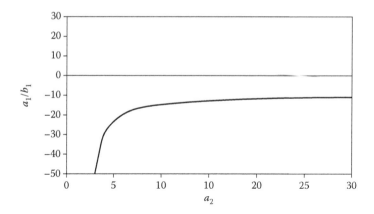

FIGURE 2.18
Plot of the bounding curve $xy = 2(y - 5x - 10)$.

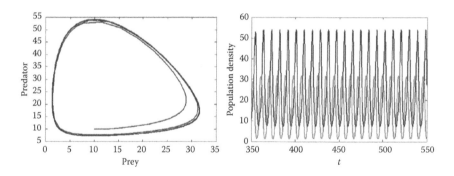

FIGURE 2.19
Phase plot and time-series displaying oscillatory dynamics in *RM* model (2.62) and (2.63).

Numerical simulation of model system (2.62) and (2.63) is done using the parameter values $a_1 = 2.0$, $b_1 = 0.05$, $w = 1.0$, $D = 10$, $a_2 = 1.0$, $w_1 = 2.0$, $D_1 = 10$, and initial condition $(X_0, Y_0) = (10, 10)$. The equilibrium point is obtained as $X^* = 20$, $Y^* = 30$. Condition (2.64) is violated. Phase plot and time series displaying oscillatory dynamics are given in Figure 2.19.

Variations of the RM Model

Model 1

Consider the following ecological problem: Modeling of phytoplankton–zooplankton interactions takes into account the zooplankton grazing with saturating functional response to phytoplankton abundance. These models are called Michaelis–Menten models of enzyme kinetics. The models can explain the phytoplankton and zooplankton oscillations and monotonous relaxation to one of the possible multiple equilibria [120,135,149]. The effect of fish predation stabilizes the predator dynamics relatively more than the prey dynamics. Fish density and nutrient concentration are environmental factors that can be manipulated. Phytoplankton and zooplankton are modeled dynamically. The phytoplankton equation is based on the logistic growth formulation with a Monod type of nutrient limitation. In case of biomanipulation, both nutrients and the third trophic level (fish) are manipulated variables. These conceptual models are keys to our understanding of non-Turing time-independent spatial patterns. The models are useful for studying the dynamical complexity of aquatic or oceanic systems for harvesting the natural resources in marine systems. The model equations are given by

$$\frac{dP}{dT} = \frac{\alpha NP}{H_N + N} - \beta P^2 - \frac{\upsilon PZ}{H_P + P} = PF(P, Z), \tag{2.65}$$

$$\frac{dZ}{dt} = \frac{e\upsilon PZ}{H_P + P} - \delta Z - \frac{FZ^2}{(H_Z^2 + Z^2)} = ZG(P, Z), \tag{2.66}$$

where

$$F(P, Z) = \frac{\alpha N}{H_N + N} - \beta P - \frac{\upsilon Z}{H_P + P}, \text{ and } G(P, Z) = \frac{e\upsilon P}{H_P + P} - \delta - \frac{FZ}{(H_Z^2 + Z^2)}.$$

The meanings of the parameters are as given below:

α: the maximum per capita growth rate of prey population,

β: the intensity of competition among individuals of prey,

υ: the rate at which the prey is grazed and it follows Holling type II functional response,

e: is the conversion coefficient from individuals of prey into individuals of predator,

δ: the mortality rate of the predator,

F: the constant predation rate by the predator population, which follows Holling type III functional response,

H_N: the prey density at which specific growth rate becomes half of its saturation value,

H_P, H_Z: half-saturation constants for prey and predator density, respectively,

N: the nutrient level of the system.

We assume that the system is defined on a bounded domain (habitat) and is augmented with appropriate initial and boundary conditions. The

individual species cannot leave the domain. Holling type II or type III pre-
dational forms are obvious choices for representing the behavior of predator
hunting [94]. Holling type II response function describes a situation in which
the number of prey consumed per predator initially rises quickly as the density of the prey increases and then levels off. Holling type III response function is sigmoid, rising slowly when preys are rare, accelerating when they
become more abundant, and finally reaching a saturated upper limit. Type
III functional response also levels off at some prey density. However, the
type II functional response curve behaves differently than the type III curve
when the prey density is low. Keeping the above-mentioned properties in
mind, the dependence of a predator's (e.g., zooplankton) grazing rate on prey
(e.g., phytoplankton) is taken as that of Holling type II response, whereas the
zooplankton predation by fish follows a sigmoidal functional response of
Holling type III as assumed in the Scheffer model [135,136]. The interactions
incorporated in the model system are given in Figure 2.20.

The interaction part of the model system investigated is depicted by solid
arrows. The interaction involving dotted arrows indicates that either of the
positive or negative effects involved are considered in the model. For example, the dotted arrow between fish and zooplankton indicates only the negative effect that fish have on the zooplankton.

The model was originally formulated as a system of ODEs by Scheffer [135]
and later studied by many authors [95–97,105,123,148,158].

We use the following variable transformations and rescaling of the
parameters:

$$u = \frac{P}{\omega}, \quad v = Z\left(\frac{v}{r\omega}\right), \quad t = rT, \quad r = \frac{\alpha N}{H_N + N}, \quad \frac{r}{\beta} = \omega.$$

$$h_P = \frac{H_P}{\omega}, \quad b = \frac{ev}{r}, \quad c = \frac{\delta}{r}, \quad f = \frac{Fv}{r^2\omega}, \quad h_Z = \frac{H_Z v}{r\omega}.$$

Substituting into Equations 2.65 and 2.66, we obtain the dimensionless
system

$$\frac{du}{dt} = u(1 - u) - \frac{uv}{(h_P + u)} = uF(u, v), \tag{2.67}$$

$$\frac{dv}{dt} = \frac{buv}{(h_P + u)} - cv - \frac{fv^2}{(h_Z^2 + v^2)} = vG(u, v), \tag{2.68}$$

FIGURE 2.20
Interactions incorporated in the model system.

where

$$F(u, v) = (1 - u) - v/(u + h_p),$$

and

$$G(u, v) = \frac{bu}{(h_p + u)} - c - \frac{fv}{(h_Z^2 + v^2)}.$$

Model system (2.67) and (2.68) has three nonnegative equilibrium points, $E_0(0, 0)$, $E_1(1,0)$, and $E^*(u^*, v^*)$. The other two equilibrium points are imaginary and are of no interest. $E^*(u^*, v^*)$ is the solution of the equations

$$(1 - u) - \frac{v}{(h_p + u)} = 0,$$

or

$$v = (1 - u)(h_p + u), \tag{2.69}$$

$$\frac{bu}{(h_p + u)} - c - \frac{fv}{(h_Z^2 + v^2)} = 0,$$

or

$$\frac{fv}{(h_Z^2 + v^2)} = \frac{bu}{(h_p + u)} - c. \tag{2.70}$$

The unique equilibrium point $E^*(u^*, v^*)$ exists in the positive quadrant of the u–v plane if

$$(h_Z^2 + v^2)(b - c) > fv, \quad (h_p + u)(f + 2cv) > 2buv,$$

and

$$0 < \frac{ch_p}{b - c} < 1, \quad 0 < h_p < 1.$$

The plot of the curves (2.69), (2.70), and the equilibrium point $E^*(u^*, v^*)$ is given in Figure 2.21 (Dubey et al., [45a]).

We find that the equilibrium point $E_0(0, 0)$ is a hyperbolic saddle point. The equilibrium point $E_1(K, 0)$ is also a saddle point. Sufficient conditions for the asymptotic stability of (u^*, v^*) are $v^* < (h_p + u^*)^2$, and $(v^*)^2 < h_Z^2$. If these conditions are satisfied, then both predator and prey species coexist, and they settle down at its equilibrium point (see Problem 5, Exercise 2.2).

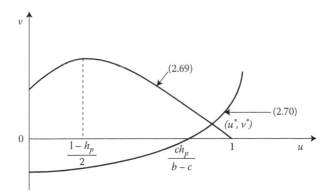

FIGURE 2.21
Equilibrium point $E^*(u^*, v^*)$ as the point of intersection of Equations 2.69 and 2.70. (From Dubey, B., Kumari, N., Upadhyay, R. K. 2009. *J. Appl. Math. Comput.* 31, 413–432. Copyright 2008, Springer. Reprinted with kind permission from Springer Science + Business Media B.V.)

The limit-cycle solution (phase plot and time series) of the model is plotted in Figure 2.22 for the following values of the parameters: $b = 1.9$, $h_p = 0.3$, $c = 0.8$, $f = 0.0001$, $h_Z = 2.5$, and the initial values $(u, v) = (0.2, 0.4)$.

Model 2 (DeAngelis Model)

The classical RM model can be modified to include the interference among individuals of predators. The Holling type II functional response term is replaced by one of the functional response terms proposed by Beddington–DeAngelis [43,78,79]. The modified RM model, which describes the dynamics of a specialist predator and its prey, is governed by the following system:

$$\frac{dX}{dt} = a_1 X - b_1 X^2 - \frac{wXY}{(\alpha + \beta Y + \gamma X)} = XF(X, Y), \qquad (2.71)$$

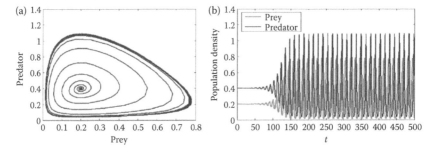

FIGURE 2.22
Limit-cycle solution: (a) Phase plot, (b) time series, for model (2.67) and (2.68).

$$\frac{dY}{dt} = -a_2 Y + \frac{w_1 XY}{(\alpha + \beta Y + \gamma X)} = YG(X, Y), \tag{2.72}$$

where

$$F(X, Y) = a_1 - b_1 X - \frac{wY}{(\alpha + \beta Y + \gamma X)},$$

$$G(X, Y) = -a_2 + \frac{w_1 X}{(\alpha + \beta Y + \gamma X)},$$

X: Prey for the specialist predator Y with Beddington–DeAngelis type functional response,

a_1: Per capita rate of self-reproduction for the prey,

b_1: Measures the intensity of competition among individuals of species X for space, food, and so on,

w: Maximum rate of per capita removal of prey species X due to predation by its predator Y,

α: Measures the protection provided to prey by its environment,

a_2: Measures how fast the predator Y will die when there is no prey to capture, kill, and eat,

β: Represents the intensity of interference between individuals of the specialist predator,

γ: Determines how fast the per capita feeding rate approaches its saturation value w,

w_1: Measures efficiency of biomass conversion from prey to predator.

The predator–prey system (2.71) and (2.72) with the Beddington–DeAngelis functional response was originally proposed by Beddington [16] and DeAngelis et al. [43], independently. The stability of the positive interior equilibrium point $E^*(X^*, Y^*)$ and the existence of limit cycles were studied by Cantrell and Cosner [26]. The global stability and the uniqueness of a limit cycle were recently investigated by Hwang [78,79]. The model with $\alpha = 0$ was analyzed by Ariditi and Ginzburg [11]. The Ariditi and Ginzburg model results from incorporation of a ratio-dependent functional response into the LV framework. The DeAngelis model with $\beta = 0$ is dynamically very similar to RM model and has similar dynamics. However, mutual interference between consumers reduces predator killing efficiency at high predator densities, therefore imposing a positive feedback on predator density.

The equilibrium points are $(0, 0)$, $(a_1/b_1, 0)$, and (X^*, Y^*), where

$$X^* = [p + \sqrt{p^2 + q}]/(2\beta w_1 b_1).$$

$$Y^* = [(w_1 - \gamma a_2)X^* - \alpha a_2]/(a_2 \beta),$$

where $p = \beta w_1 a_1 + w\gamma a_2 - ww_1$, $q = 4\beta w_1 b_1 w\alpha a_2$. (The chosen parameter values should satisfy $Y^* > 0$; see Problem 6, Exercise 2.2.) We find that $(0, 0)$ is a hyperbolic saddle point, which repels in the x-direction and gets attracted in the y-direction. The equilibrium point $(a_1/b_1, 0)$ is asymptotically stable when $[w_1 a_1/(b_1\alpha + a_1\gamma)] < a_2$. When $[w_1 a_1/(b_1\alpha + a_1\gamma)] > a_2$, $(a_1/b_1, 0)$ is a hyperbolic saddle point.

The sufficient conditions for the equilibrium point (X^*, Y^*) to be asymptotically stable are

$$(w_1 - \gamma a_2) > 0,$$

and

$$\frac{\gamma a_1 a_2}{(w_1 + \gamma a_2)} < b_1 X < a_1.$$

(see Problem 8, Exercise 2.2).

Applying the Kolmogorov theorem, we get the conditions on the parameters as (see Problem 7, Exercise 2.2)

$$w > a_1\beta, \quad w_1 > a_2\gamma$$

and

$$\frac{a_1}{b_1} > \frac{a_2\alpha}{(w_1 - a_2\gamma)}.$$

Model 3 Predator–Prey Model with Holling Type IV Functional Response

Holling type IV functional response

$$p(X) = \frac{cX}{(X^2/i) + X + a}$$

is relatively less studied in the population ecology. It was first introduced by Haldane [66] in enzymology. The parameter i is a measure of the predator's immunity from the prey. The parameters c and a can be interpreted as the maximum per capita predation rate and the half-saturation constant in the absence of any inhibitory effect. As the value of the parameter i decreases, the predator's foraging efficiency decreases. In the limit of large i, it reduces to a type II functional response [84]. This response function describes a situation in which the predator's per capita rate of predation decreases at sufficiently high prey densities.

We note that $p(0) = 0$. We assume that

$$\frac{dp}{dX} = c\left[\frac{a - (X^2/i)}{\{(X^2/i) + X + a\}^2}\right] > 0.$$

Van Gemerden [159] fitted a type IV functional response to the uptake of hydrogen sulfide by purple sulfur bacteria. Collings [33] used the response function in a mite predator–prey interaction model and called it a Holling type IV function.

Consider the following predator–prey model using the Holling type IV functional response system [157].

$$\frac{dP}{dt} = \frac{\xi NP}{(H_N + N)} - \phi P^2 - \frac{cPZ}{(P^2/i) + P + a} = PF(P, Z), \qquad (2.73)$$

$$\frac{dZ}{dt} = \frac{bcPZ}{(P^2/i) + P + a} - mZ = ZG(P, Z), \qquad (2.74)$$

where

$$F(P, Z) = \frac{\xi N}{(H_N + N)} - \phi P - \frac{cZ}{(P^2/i) + P + a},$$

and

$$G(P, Z) = \frac{bcP}{(P^2/i) + P + a} - m.$$

We explain the meaning of each variable or constant.

P, Z: Population densities of the prey and predator, respectively.

ξ: Maximum per capita growth rate of the prey.

ϕ: Intensity of competition among individuals of the prey population.

c: Rate at which the prey is grazed, which follows Holling type IV functional response.

b: Conversion coefficient from individuals of the prey into individuals of the predator.

m: Mortality rate of the predator.

i: A direct measure of the predator's immunity from the prey.

a: Half-saturation constant in the absence of any inhibitory effect.

N: Food for the prey.

H_N: Prey density at which specific growth rate becomes half of its saturation value.

The predator foraging efficiency decreases with decrease in the value of the parameter i. For example, at high prey densities, the predator's per capita grazing rate decreases with density because of inhibitory effects.

We introduce the following substitutions and notations to bring the system of equations into a nondimensional form

$$u = P/a, \quad v = (c/ar)Z, \quad T = r\,t,$$

$$\frac{\xi N}{(H_N + N)} = r, \quad \frac{r}{\phi} = 1, \quad K = \frac{l}{a}, \quad \alpha = \frac{i}{a}, \quad \beta = \frac{bc}{r}, \quad \gamma = \frac{m}{r}.$$

The model system, in dimensionless form becomes

$$\frac{du}{dT} = u\left(1 - \frac{u}{K}\right) - \frac{uv}{(u^2/\alpha) + u + 1} = uF(u,v), \tag{2.75}$$

$$\frac{dv}{dT} = \frac{\beta uv}{(u^2/\alpha) + u + 1} - \gamma v = vG(u,v), \tag{2.76}$$

where

$$F(u,v) = \left(1 - \frac{u}{K}\right) - \frac{v}{(u^2/\alpha) + u + 1},$$

and

$$G(u,v) = \frac{\beta u}{(u^2/\alpha) + u + 1} - \gamma.$$

The model system possesses four equilibrium points $E_0 = (0, 0)$, $E_1 = (K, 0)$, and $E_{2,3}^* = (u^*, v^*)$. The nontrivial points (u^*, v^*) are given by (see Problem 9, Exercise 2.2)

$$u^* = [-S \pm \sqrt{S^2 - 4\alpha}]/2, \quad v^* = (\beta/\gamma K)[K - u^*]u^*,$$

where $S = \alpha[1 - (\beta/\gamma)]$. The nontrivial positive points exist if $S < 0$, that is, $\beta > \gamma$; $[1 - (\beta/\gamma)]^2 > (4/\alpha)$, and $u^* < K$.

From the local stability analysis, we conclude the following:

i. The point of extinction of both prey and predator, that is, E_0, is always unstable.

ii. The equilibrium point E_1, which corresponds to the existence of the prey at its carrying capacity and extinction of the predator, is stable or unstable depending upon the parameter values. Denote $q = (\beta K)/[(K^2/\alpha) + K + 1]$. E_1 is locally asymptotically stable in the u–v plane if the inequality $q < \gamma$ is satisfied. If $q > \gamma$, then the equilibrium point E_1 is a saddle point with stable manifold locally in the u-direction and with unstable manifold locally in the v-direction. E_1 is a saddle node if $q = \gamma$.

Application of the Kolmogorov theorem to model system (2.75) and (2.76) yields the conditions (see Problem 10, Exercise 2.2)

$$\beta > \gamma, \quad \{(\beta/\gamma) - 1\}^2 > (4/\alpha), \tag{2.77}$$

$$u^2 < \alpha, \quad K > 0.5 \left[\{(\beta/\gamma) - 1\} + \alpha\sqrt{\{(\beta/\gamma) - 1\}^2 - (4/\alpha)}\right]. \tag{2.78}$$

Figure 2.23 shows the area in $(1/\alpha)$-(β/γ) plane (as given in Equation 2.78), which produces stable limit-cycle solutions [157]. The parameter values for simulation experiments are derived from the shaded area. An oscillatory predator–prey dynamics (time series) exhibited by model system (2.75) and (2.76) for a typical set of parameter values, $K = 1$, $\alpha = 0.3$, $\beta = 2.33$, and $\gamma = 0.3$ is presented in Figure 2.24.

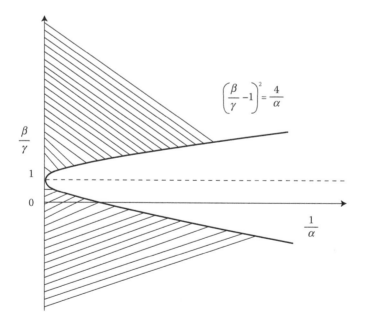

FIGURE 2.23
Shaded region which produces stable limit-cycle solutions. (From Upadhyay, R. K., Kumari, N., Rai, V. *Math. Model. Nat. Phenomena* 3, 71–95, 2008. With permission.)

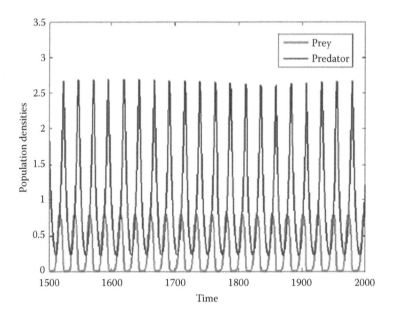

FIGURE 2.24
Time series displaying a limit cycle.

Example 2.8

Derive the local stability conditions for model system (2.75) and (2.76).

SOLUTION

The equilibrium points are (0, 0) and (K, 0) and (u^*, v^*), where

$$u^* = [-S \pm \sqrt{S^2 - 4\alpha}\,]/2, \quad \text{and} \quad v^* = (\beta/\gamma K)[K - u^*]u^*.$$

The nontrivial solutions exist if $S^2 > 4\alpha$, and $u^* < K$, that is, if $[1 - (\beta/\gamma)]^2 > (4/\alpha)$, and $u^* < K$. If $S < 0$, that is, $\beta > \gamma$, we obtain two positive equilibrium points.

The Jacobian matrix of the system evaluated at an equilibrium point (u^*, v^*) is given by Equation 2.59, where

$$a_{11} = \left(1 - \frac{2u}{K}\right) + \frac{v}{D^2}\left(\frac{u^2}{\alpha} - 1\right), \quad a_{12} = -\frac{u}{D}, \quad D = \frac{u^2}{\alpha} + u + 1,$$

$$a_{21} = \frac{\beta v}{D^2}\left(1 - \frac{u^2}{\alpha}\right), \quad a_{22} = \frac{\beta u}{D} - \gamma,$$

where * is dropped for convenience.

At the equilibrium point (0, 0): We obtain $a_{11} = 1$, $a_{12} = 0$, $a_{21} = 0$, $a_{22} = -\gamma$. The eigenvalues are $\lambda_1 = 1$, and $\lambda_2 = -\gamma < 0$. The equilibrium point (0, 0) is unstable. Since $\text{Re}(\lambda) \neq 0$ for both eigenvalues, the fixed point is hyperbolic. Since the eigenvalues are real and are of opposite signs, $(0, 0)$ is a hyperbolic saddle point which repels in the x-direction and attracts in the y-direction.

At the equilibrium point $(K, 0)$: We obtain

$$a_{11} = -1, \quad a_{12} = -K/D, \quad a_{21} = 0, \quad a_{22} = (\beta K/D) - \gamma, \quad D = (K^2/\alpha) + K + 1.$$

The eigenvalues are $\lambda_1 = -1$, and $\lambda_2 = a_{22} = (\beta K/D) - \gamma$. If $\lambda_2 < 0$, that is $(\beta K/D) < \gamma$, then the equilibrium point $(K, 0)$ is asymptotically stable. Otherwise, $(K, 0)$ is unstable. It depends on the values of the parameters α, β, γ. If $\lambda_2 > 0$, the eigenvalues are real and are of opposite signs and the fixed point $(K, 0)$ is a hyperbolic saddle point.

At the equilibrium point (u^*, v^*):

$$u^* = [-S \pm \sqrt{S^2 - 4\alpha}]/2, \quad S = \alpha[1 - (\beta/\gamma)], \quad \text{and} \quad v^* = (\beta/\gamma K)[K - u^*]u^*.$$

We obtain

$$a_{11} = 1 - \frac{2u^*}{K} + \frac{v^*}{D^2}\left(\frac{u^{*2}}{\alpha} - 1\right), \quad a_{12} = -\frac{u^*}{D} = -\frac{\gamma}{\beta},$$

$$a_{21} = \frac{\beta v}{D^2}\left(1 - \frac{u^{*2}}{\alpha}\right), \quad a_{22} = 0.$$

The eigenvalues of J are the roots of $\lambda^2 - a_{11}\lambda - a_{12}a_{21} = 0$. By the Routh–Hurwitz theorem, the necessary and sufficient conditions for local stability are $(-a_{11}) > 0$, $(-a_{12}\,a_{21}) > 0$.

$$-a_{11} = \frac{2u^*}{K} - 1 + \frac{v^*}{D^2}\left(1 - \frac{u^{*2}}{\alpha}\right) > 0,$$

and

$$-a_{12}a_{21} = \frac{\gamma v}{D^2}\left(1 - \frac{u^{*2}}{\alpha}\right) > 0.$$

We obtain a sufficient condition for the local stability as $(K/2) < u^* < \sqrt{\alpha}$.

Prey–Generalist Predator Model

Upadhyay [155] considered the case of an environment where there is a prey and a generalist predator. Following the Leslie–Gower formulation, the growth rate equations for the two populations are given by

$$\frac{dX}{dt} = AX\left(1 - \frac{X}{K}\right) - \frac{BXZ}{(D + dX + Z)} = XF(X, Z), \tag{2.79}$$

$$\frac{dZ}{dt} = cZ^2 - \frac{w_3 Z^2}{(X + D_3)} = ZG(X, Z), \tag{2.80}$$

where

X: prey, Z: generalist predator, A: rate of self-reproduction of the prey,
K: carrying capacity of the environment,
c: growth rate of the generalist predator Z due to sexual reproduction,
$A, B, D, K, c, d, D_3, w_3$ are positive constants and

$$F(X, Z) = A\left(1 - \frac{X}{K}\right) - \frac{BZ}{(D + dX + Z)}, \quad G(X, Z) = Z\left(c - \frac{w_3}{(X + D_3)}\right).$$

The term $BX/(D + dX + Z)$ on the right-hand side of Equation 2.79 gives the per capita rate of removal of species X by individuals of species Z whose searching efficiency decreases because of interference among them. On the right-hand side of Equation 2.80, c measures the rate of self-reproduction and the square term Z^2 signifies the fact that mating frequency is directly proportional to the number of male as well as female individuals. D_3 normalizes the residual reduction in the predator population because of severe scarcity of the favorite food. F and G are continuous functions of X and Z with continuous first partial derivatives in the domain $X \geq 0, Z \geq 0$.

Applying the Kolmogorov theorem, we obtain the following conditions (see Problem 12, Exercise 2.2):

i. $\dfrac{w_3 D_3}{(X + D_3)^2} < c < \dfrac{w_3}{(X + D_3)}$.

If X_m is the maximum value of X, we can choose $c < w_3/(X_m + D_3)$.

ii. $B > A$,

iii. $K > [(w_3 - cD_3)/c]$.

Linear stability of the system (2.79) *and* (2.80): The system has three equilibrium points. They are (0, 0), (K, 0), and (X*, Z*), where

$$X^* = \frac{1}{c}[w_3 - cD_3], \quad Z^* = \frac{A(K - X^*)(D + dX^*)}{BK - A(K - X^*)}.$$

The Jacobian matrix of the system evaluated at an equilibrium point is given by Equation 2.59, where

$$a_{11} = A\left(1 - \frac{2X^*}{K}\right) - \frac{BZ^*(D + Z^*)}{(D + dX^* + Z^*)^2}, \quad a_{12} = -\frac{BX^*(D + dX^*)}{(D + dX^* + Z^*)^2},$$

$$a_{21} = \frac{w_3(Z^*)^2}{(X^* + D_3)^2}, \quad a_{22} = 2Z^*\left[c - \frac{w_3}{(X^* + D_3)}\right].$$

At the equilibrium point $E_0(0,0)$: We obtain $a_{11} = A$, $a_{12} = 0$, $a_{21} = 0$, and $a_{22} = 0$. The eigenvalues are $\lambda_1 = A > 0$ and $\lambda_2 = 0$. The equilibrium point is unstable.

At the equilibrium point $E_1(K, 0)$: We obtain $a_{11} = -A$, $a_{12} = -BK/(D + dK)$, $a_{21} = 0$, and $a_{22} = 0$. The eigenvalues are $\lambda_1 = -A < 0$ and $\lambda_2 = 0$. Since one of the eigenvalues is zero, this equilibrium point is a nonhyperbolic point. The dynamical behavior near a nonhyperbolic point can be stable, periodic, or even chaotic. Nonhyperbolic fixed points are fragile.

At the equilibrium point $E^*(X^*, Z^*)$: We obtain $a_{22} = 0$. The eigenvalues of J are the roots of $\lambda^2 - a_{11}\lambda - a_{12}a_{21} = 0$. The system is locally stable if the eigenvalues are negative or have negative real parts. By the Routh–Hurwitz theorem, the necessary and sufficient conditions are given by $(-a_{11}) > 0$, $(-a_{12}a_{21}) > 0$.

$$A\left(\frac{2X^*}{K} - 1\right) + \frac{BZ^*(D + Z^*)}{(D + dX^* + Z^*)^2} > 0, \tag{2.81}$$

$$\frac{Bw_3 X^*(Z^*)^2(D + dX^*)}{(D + dX^* + Z^*)^2(X^* + D_3)^2} > 0. \tag{2.82}$$

Equation 2.82 is always satisfied.
Using $F(X^*, Z^*) = 0$ and simplifying Equation 2.81, we get

$$A\left(\frac{2X^*}{K} - 1\right) + A\left(1 - \frac{X^*}{K}\right)\left(\frac{(D + Z^*)}{(D + dX^* + Z^*)}\right) > 0.$$

By simplifying, we get

$$(D - Kd) + 2dX^* + Z^* > 0. \tag{2.83}$$

Suitable choices of the parameters satisfying Equation 2.83 will lead to a solution with a stable equilibrium point (a sufficient condition is $D > Kd$) and the choices violating Equation 2.83 will lead to limit cycles.

Numerical simulation of model system (2.79) and (2.80) is done using the initial condition $(X_0, Z_0) = (10, 10)$, and $A = 1.0$, $K = 100$, $B = 1.0$, and $d = 0.1$, $D = 1$, $c = 0.03$, $w_3 = 1.0$, $D_3 = 20$. The equilibrium point is obtained as $X^* = 13.3425$, $Z^* = 15.1690$. We obtain a stable focus as presented in Figure 2.25. The time series is also given in Figure 2.25.

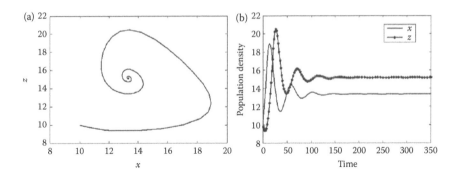

FIGURE 2.25
Stable focus: (a) Phase plot, (b) time series, for model (2.79) and (2.80).

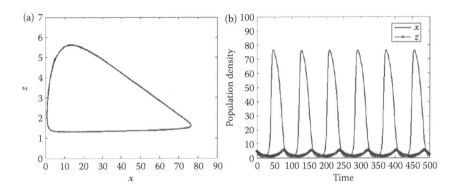

FIGURE 2.26
Limit-cycle solution. (a) Phase plot, (b) time series, for model (2.79) and (2.80).

The limit-cycle solution (phase plot and time series) of the model is plotted in Figure 2.26 for the following values of the parameters: $A = 1.0$, $K = 100$, $B = 1.5$, $d = 0.1$, $D = 1$, $c = 0.03$, $w_3 = 1.0$, $D_3 = 20$, and the initial values (X_0, Z_0) = (5, 5). We observe that the values of all the parameters are the same as for stable focus except for the control parameter $B = 1.5$ for which it shows the limit-cycle behavior.

Holling–Tanner Model

The Holling–Tanner (HT) model describes the dynamics of a generalist predator and its prey [99,128]. The ODEs that govern the dynamics are given by

$$\frac{dZ}{dt} = AZ\left(1 - \frac{Z}{K_1}\right) - \frac{w_3 UZ}{(Z + D_3)} = ZF(Z, U), \qquad (2.84)$$

$$\frac{dU}{dt} = cU - \frac{w_4 U^2}{Z} = UG(Z,U), \tag{2.85}$$

where

$$F(Z,U) = A\left(1 - \frac{Z}{K_1}\right) - \frac{w_3 U}{(Z + D_3)},$$

and

$$G(Z,U) = c - \frac{w_4 U}{Z}.$$

The meanings of the variables and constants are as follows:

U: Generalist predator.

Z: Prey. Most favorite food for the generalist predator U.

w_3: Maximum of the per capita rate of the removal of the prey population by its predator U.

D_3: Half-saturation constant for prey Z.

c: Per capita rate of self-reproduction of the generalist predator U.

w_4: Measures the severity of the limitation put to growth of predator population by per capita availability of its prey.

K_1: Carrying capacity for the prey population Z.

The prey population Z grows logistically with carrying capacity K_1 and intrinsic growth rate A in the absence of predation. The generalist predator U does not die out exponentially in the absence of its favorite prey Z, instead grows logistically to αZ, where $\alpha = (c/w_4)$ is the proportionality constant. The underlying assumption of this formulation is that the predator's carrying capacity is proportional to the population density of its most favorite prey. The positive aspect of this formulation of predator–prey interaction is that it takes care of our inability to write down growth equations for all the species on which a generalist predator feeds upon. Although this formulation is simple, it has some positive features, which are not found in other models of predator–prey interaction involving a generalist predator.

The Kolmogorov theorem cannot be applied to the system (2.84), (2.85) (see Problem 13, Exercise 2.2).

Linear stability of the system: Note that $(0, 0)$ is not an equilibrium point. The equilibrium points are the solutions of the equations

$$Z\left[A\left(1 - \frac{Z}{K_1}\right) - \frac{w_3 U}{(Z + D_3)}\right] = 0, \quad U\left[c - \frac{w_4 U}{Z}\right] = 0. \qquad (2.86)$$

The second equation gives $U = cZ/w_4$. Substituting in the first equation and simplifying, we get

$$Z^2 + Z(D_3 + qc - K_1) - K_1 D_3 = 0, \quad \text{where } q = (K_1 w_3)/(A w_4). \qquad (2.87)$$

The discriminant of the equation is positive (both the roots are real) and the product of the roots is negative. Therefore, we have one positive solution for the equation. The system has the equilibrium point (Z^*, U^*), where Z^* is the positive root of Equation 2.87 and $U^* = cZ^*/w_4$.

The Jacobian matrix of the system evaluated at an equilibrium point (Z^*, U^*) is given by Equation 2.59, where

$$a_{11} = A\left(1 - \frac{2Z^*}{K_1}\right) - \frac{w_3 U^* D_3}{(Z^* + D_3)^2}, \quad a_{12} = -\frac{w_3 Z^*}{Z^* + D_3},$$

$$a_{21} = w_4\left(\frac{U^*}{Z^*}\right)^2 = \frac{c^2}{w_4}, \quad a_{22} = c - 2w_4\left(\frac{U^*}{Z^*}\right) = -c.$$

The eigenvalues of J are the roots of $\lambda^2 - (a_{11} + a_{22})\lambda + (a_{11}a_{22} - a_{12}a_{21}) = 0$. By the Routh–Hurwitz theorem, the necessary and sufficient conditions for local stability are $-(a_{11} + a_{22}) > 0$, and $(a_{11}a_{22} - a_{12}a_{21}) > 0$.

After long computations and simplifications, we obtain the local stability criterion as (see [99])

$$\frac{c}{A} > \frac{2(\alpha - R)}{1 + \alpha + \beta + R}, \qquad (2.88)$$

where $\alpha = (w_3 c)/(w_4 A)$, $\beta = D_3/K_1$, and $R = [(1 - \alpha - \beta)^2 + 4\beta]^{1/2}$.

Parameter values are chosen satisfying the above criterion to produce a limit cycle. The condition divides the domain of dependence into two regions: (i) where the system has stable equilibrium points and (ii) where the system exhibits stable limit cycles. The unstable region signifies stable limit-cycle solutions which emanate from stable ones through supercritical Hopf bifurcations. It may be noted that Kolmogorov analysis of this system does not put any constraint on the parameter values of the system.

An oscillatory predator–prey dynamics exhibited by the model at a typical set of parameter values, $A = 2.5$, $K_1 = 150$, $w_3 = 0.75$, $D_3 = 10$, $c = 0.2$, and $w_4 = 0.0257$ (violates condition 2.88) is given in Figure 2.27. Graphical representation of the dynamical structure of the HT model in a 2D parameter space (time vs population density) was presented by May [99].

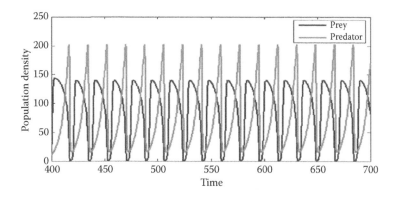

FIGURE 2.27
Oscillatory predator–prey dynamics exhibited by the HT model.

This model assumes interference among the individuals of the predator, but this interference does not affect the consumption rate in the prey equation. While this is possible, this is not general. The case of very high prey densities is the worst case for the model, but the decoupling of consumption and dynamics affects all densities.

Gasull et al. [57] have shown that the asymptotic stability of the positive equilibrium point in the HT model does not imply that it is also globally stable. They have given an example with two limit cycles.

For some values of the parameters in the HT model, the positive equilibrium is unstable and the model produces the interesting phenomenon of stable limit cycle. Studies of several pairs of interacting species ranging from house sparrow and European sparrow hawk, mule deer and mountain lion [147], and aphids and ladybugs *Cycloneda sanguinea* [122] show that the theoretical prediction of model system (2.84) and (2.85), based on the estimated parameter values, is broadly in agreement with reality. According to Morales and Buranr [111], the number of aphids captured per day by *C. sanguinea* (adults, male, and female) corresponds to Holling's type II functional response. Gakkhar and Naji [56] discussed the effect of seasonality on the HT model. Chaos was detected when seasonality is superimposed in the intrinsic growth rate of the system, which otherwise has a globally stable equilibrium state or global stable limit cycle.

Example 2.9

Derive the local stability criterion (2.88) for the HT model.

SOLUTION

The equilibrium point (Z^*, U^*) is the solution of the equations

$$A\left(1 - \frac{Z^*}{K_1}\right) - \frac{w_3 U^*}{(Z^* + D_3)} = 0, \quad c - \frac{w_4 U^*}{Z^*} = 0.$$

From the second equation, we obtain $U^* = cZ^*/w_4$. Substituting the value of U^* in the first equation, we obtain

$$A\left(1 - \frac{Z^*}{K_1}\right) - \frac{w_3 c Z^*}{w_4(Z^* + D_3)} = 0,$$

or

$$1 - \frac{Z^*}{K_1} = \frac{\alpha Z^*}{(Z^* + D_3)}, \tag{2.89}$$

where

$$\alpha = \frac{w_3 c}{w_4 A}.$$

Simplifying the equation, we obtain

$$(Z^*)^2 + (K_1 \alpha - K_1 + D_3)Z^* - K_1 D_3 = 0.$$

Solving, we get $Z^* = (1/2)[\{K_1(1 - \alpha) - D_3\} \pm \sqrt{\{K_1(1 - \alpha) - D_3\}^2 + 4K_1 D_3}]$

$$= \frac{K_1}{2}\left[(1 - \alpha - \beta) \pm \sqrt{(1 - \alpha - \beta)^2 + 4\beta}\,\right],$$

where $\beta = D_3/K_1$. Since $Z^* > 0$, we take the positive sign of the square root. We get

$$Z^* = \frac{K_1}{2}[(1 - \alpha - \beta) + R],$$

where

$$R = \sqrt{(1 - \alpha - \beta)^2 + 4\beta}.$$

The Jacobian matrix of the system evaluated at an equilibrium point (Z^*, U^*) is given by Equation 2.59, where

$$a_{11} = A\left(1 - \frac{2Z^*}{K_1}\right) - \frac{w_3 U^* D_3}{(Z^* + D_3)^2}, \quad a_{12} = -\frac{w_3 Z^*}{Z^* + D_3},$$

$$a_{21} = w_4\left(\frac{U^*}{Z^*}\right)^2 = \frac{c^2}{w_4}, \quad a_{22} = c - 2w_4\left(\frac{U^*}{Z^*}\right) = -c.$$

The eigenvalues of J are the roots of $\lambda^2 - (a_{11} + a_{22})\lambda + (a_{11}a_{22} - a_{12}a_{21}) = 0$. By the Routh–Hurwitz theorem, the necessary and sufficient conditions for local stability are given by $-(a_{11} + a_{22}) > 0$, and $(a_{11}a_{22} - a_{12}a_{21}) > 0$.

The first condition gives

$$A\left(\frac{2Z^*}{K_1} - 1\right) + \frac{w_3 U^* D_3}{(Z^* + D_3)^2} + c > 0,$$

$$A\left(\frac{2Z^*}{K_1} - 1\right) + \frac{w_3 c D_3 Z^*}{w_4 (Z^* + D_3)^2} + c > 0,$$

$$\left(\frac{2Z^*}{K_1} - 1\right) + \left(\frac{w_3 c}{A w_4}\right)\frac{D_3 Z^*}{(Z^* + D_3)^2} + \frac{c}{A} > 0,$$

$$\frac{c}{A} - \left(1 - \frac{Z^*}{K_1}\right) + \frac{Z^*}{K_1} + \frac{\alpha D_3 Z^*}{(Z^* + D_3)^2} > 0,$$

$$\frac{c}{A} - \frac{\alpha Z^*}{(Z^* + D_3)} + \frac{Z^*}{K_1} + \frac{\alpha D_3\, Z^*}{(Z^* + D_3)^2} > 0, \text{(using Equation 2.89)}$$

$$\frac{c}{A} - \frac{\alpha (Z^*)^2}{(Z^* + D_3)^2} + \frac{Z^*}{K_1} > 0,$$

or

$$\frac{c}{A} > Z^*\left[\frac{\alpha Z^*}{(Z^* + D_3)^2} - \frac{1}{K_1}\right].$$

Now,

$$Z^*\left[\frac{\alpha Z^*}{(Z^* + D_3)^2} - \frac{1}{K_1}\right] = Z^*\left[\frac{\alpha Z^*}{(Z^* + D_3)} \cdot \frac{1}{(Z^* + D_3)} - \frac{1}{K_1}\right]$$

$$= Z^*\left[\left(1 - \frac{Z^*}{K_1}\right)\frac{1}{(Z^* + D_3)} - \frac{1}{K_1}\right] = \frac{Z^*}{(Z^* + D_3)}\left[1 - \frac{2Z^* + D_3}{K_1}\right]$$

(using Equation 2.89).

Substituting the value of

$$Z^* = \frac{K_1}{2}[(1 - \alpha - \beta) + R] = \frac{D_3}{2\beta}[(1 - \alpha - \beta) + R],$$

we obtain

$$\frac{Z^*}{(Z^* + D_3)}\left[1 - \frac{2Z^* + D_3}{K_1}\right] = \frac{(1 - \alpha - \beta + R)}{(1 - \alpha + \beta + R)}\left[1 - \frac{1}{K_1}\left\{\frac{D_3}{\beta}(1 - \alpha - \beta + R) + D_3\right\}\right]$$

$$= \frac{(1 - \alpha - \beta + R)}{(1 - \alpha + \beta + R)}\left[1 - \frac{D_3}{K_1 \beta}\left\{(1 - \alpha - \beta + R) + \beta\right\}\right]$$

$$= \frac{(1 - \alpha - \beta + R)}{(1 - \alpha + \beta + R)}(\alpha - R),$$

since

$$\frac{D_3}{K_1\beta} = 1.$$

Now,

$$R^2 = (1 - \alpha - \beta)^2 + 4\beta = (1 - \alpha)^2 - {}^2(1 - \alpha)\beta + \beta^2 + 4\beta,$$

or

$$R^2 - (1 - \alpha)^2 = \beta^2 + 4\beta - 2(1 - \alpha)\beta.$$

$$\frac{(1 - \alpha - \beta + R)}{(1 - \alpha + \beta + R)} = \frac{(1 - \alpha - \beta + R)}{(1 - \alpha + \beta + R)} \cdot \frac{(1 - \alpha - \beta - R)}{(1 - \alpha - \beta - R)}$$

$$= \frac{(1 - \alpha - \beta)^2 - R^2}{(1 - \alpha)^2 - (\beta + R)^2} = -\frac{4\beta}{(1 - \alpha)^2 - \beta^2 - 2\beta R - R^2}$$

$$= \frac{4\beta}{\beta^2 + 4\beta - 2\beta(1 - \alpha) + \beta^2 + 2\beta R}$$

$$= \frac{4\beta}{2\beta^2 + 2\beta R + 4\beta - 2\beta(1 - \alpha)}$$

$$= \frac{2}{\beta + R + 2 - (1 - \alpha)} = \frac{2}{1 + \alpha + \beta + R}.$$

Hence, we obtain the stability condition $\dfrac{c}{A} > \dfrac{2(\alpha - R)}{1 + \alpha + \beta + R}$.

From the discussions in the previous subsections, we observe the following:

i. The LV scheme is suitable for describing the predator–prey interactions involving specialist predators.
ii. The Leslie–Gower scheme is better suited to model the rate of growth of the generalist predators. The criterion that any model predator–prey system must qualify as a K-system (Kolmogorov system) requires that an additional constant be added to the denominator of the second term in the growth equation for the generalist predator. It plays a legitimate functional role by reducing the per capita loss in the predator's growth by making an effective contribution denoted by this additional constant to the available prey at any point of time.
iii. The HT model assumes interference among the individuals of the predator, but this interference does not affect the consumption rate in the prey equation. Though this is possible, this is not general. The case of very high prey densities is the worst case for the model, but the decoupling of consumption and dynamics affects all densities.

It may be noted that while modeling the predator–prey interactions, the rate of growth equation of a prey population predated by both vertebrate and invertebrate predators must include a term with minus sign which describes per capita predation rate for the vertebrate predators [154]. The growth rate of the specialist predators with small body size (with less efficient biomass conversion) should be described using the Leslie–Gower scheme. One of the salient features of this scheme is that the per capita loss in predator's population is inversely proportional to the per capita availability of its potential prey.

Modified HT Model

One of the main demerits of the HT model describing the dynamics of a generalist predator and its prey (see Equations 2.84 and 2.85) is that there is no relationship between loss of prey and gain in predator population. To take care of a relationship (dynamics is coupled to consumption via the interference among individuals of the predator's population), the HT model is modified as follows [156]:

$$\frac{dZ}{dt} = AZ\left(1 - \frac{Z}{K}\right) - \frac{w_3 UZ}{(\alpha_1 + \beta_1 U + \gamma_1 Z)} = ZF(Z,U), \qquad (2.90)$$

$$\frac{dU}{dt} = cU - \frac{w_4 U^2}{Z} = UG(Z,U), \qquad (2.91)$$

where

$$F(Z,U) = A\left(1 - \frac{Z}{K}\right) - \frac{w_3 U}{(\alpha_1 + \beta_1 U + \gamma_1 Z)},$$

and

$$G(Z,U) = c - \frac{w_4 U}{Z}.$$

The modification has been implemented by replacing the Holling type II functional response term with the Beddington–DeAngelis response function. The variables and constants have the same meanings as given earlier. In this model, prey and predator both grow logistically. A and K are, respectively, the rate of self-reproduction and carrying capacity for the prey Z. The last term in Equation 2.91 describes how loss in species U depends on per capita availability of its prey Z.

The equilibrium points of the model system are $(K, 0)$ and (Z^*, U^*), where (see Problem 15, Exercise 2.2)

$$2Z^* = \left[-(p - K) + \sqrt{(p - K)^2 + 4q} \right], \text{ irrespective of the sign of } (p - K),$$

$$p = \frac{\alpha_1 w_4 + (w_3 Kc/A)}{(\beta_1 c + w_4 \gamma_1)}, \quad q = \frac{K\alpha_1 w_4}{(\beta_1 c + w_4 \gamma_1)},$$

and

$$U^* = cZ^*/w_4.$$

We find that Kolmogorov theorem cannot be applied for system (2.90) and (2.91) (see Problem 16, Exercise 2.2).

Example 2.10

Show that model system (2.90) and (2.91) is asymptotically stable when $c^2 > [A^2 \gamma_1 w_4/w_3]$.

SOLUTION

The equilibrium points of the system are $(K, 0)$ and (Z^*, U^*), where (see Problem 15, Exercise 2.2)

$$2Z^* = \left[-(p - K) + \sqrt{(p - K)^2 + 4q} \right], \quad U^* = cZ^*/w_4,$$

$$p = \frac{\alpha_1 w_4 + (w_3 Kc/A)}{(\beta_1 c + w_4 \gamma_1)}, \quad q = \frac{K\alpha_1 w_4}{(\beta_1 c + w_4 \gamma_1)}.$$

The Jacobian matrix of the system evaluated at an equilibrium point (Z^*, U^*) is given by Equation 2.59, where

$$a_{11} = A\left(1 - \frac{2Z}{K}\right) - \frac{w_3 U(\alpha_1 + \beta_1 U)}{D^2}, \quad a_{12} = -\frac{w_3 Z(\alpha_1 + \gamma_1 Z)}{D^2},$$

$$D = \alpha_1 + \beta_1 U + \gamma_1 Z, \quad a_{21} = \frac{w_4 U^2}{Z^2}, \quad a_{22} = c - \frac{2 w_4 U}{Z}.$$

At the equilibrium point $(K, 0)$: We obtain

$$a_{11} = -A, \quad a_{12} = -(w_3 K)/(\alpha_1 + \gamma_1 K), \quad a_{21} = 0, \quad \text{and } a_{22} = c.$$

The eigenvalues are $\lambda_1 = -A$, and $\lambda_2 = c$. Therefore, $(K, 0)$ is unstable.
At the equilibrium point (Z^*, U^*): The eigenvalues of J are the roots of $\lambda^2 - (a_{11} + a_{22})\lambda + (a_{11}a_{22} - a_{12}a_{21}) = 0$. By the Routh–Hurwitz theorem, the necessary and sufficient conditions for local stability are given by $-(a_{11} + a_{22}) > 0, (a_{11}a_{22} - a_{12}a_{21}) > 0$. Now, (dropping the * for convenience)

$$a_{11} = A\left(1 - \frac{2Z}{K}\right) - \left(\frac{w_3 U}{D}\right)\frac{(\alpha_1 + \beta_1 U)}{D}$$

$$= A\left(1 - \frac{Z}{K}\right) - \left(\frac{w_3 U}{D}\right)\frac{(\alpha_1 + \beta_1 U)}{D} - \frac{AZ}{K}$$

$$= A\left(1 - \frac{Z}{K}\right) - A\left(1 - \frac{Z}{K}\right)\frac{(\alpha_1 + \beta_1 U)}{D} - \frac{AZ}{K}$$

$$= A\left(1 - \frac{Z}{K}\right)\left(\frac{\gamma_1 Z}{D}\right) - \frac{AZ}{K}$$

$$= A^2\left(1 - \frac{Z}{K}\right)^2\left(\frac{\gamma_1 w_4}{cw_3}\right) - \frac{AZ}{K}$$

(using both the equilibrium equations)

$$a_{12} = -\frac{w_3 Z(\alpha_1 + \gamma_1 Z)}{D^2}, \quad a_{21} = \frac{w_4 U^2}{Z^2}, \quad a_{22} = c - \frac{2w_4 U}{Z} = -c.$$

Now,

$$-(a_{11} + a_{22}) > 0$$

gives

$$c + \frac{AZ}{K} - \left(1 - \frac{Z}{K}\right)^2\left(\frac{A^2\gamma_1 w_4}{cw_3}\right) > 0.$$

Since $0 < Z \le K$, a sufficient condition is $c^2 > [A^2\gamma_1 w_4/w_3]$.
We have,

$$a_{12}a_{21} = -\frac{w_3 Z(\alpha_1 + \gamma_1 Z)}{D^2}\left(\frac{c^2}{w_4}\right) = -\left(\frac{c^2 w_3}{w_4}\right)Z(\alpha_1 + \gamma_1 Z)\left(\frac{A}{w_3 U}\left\{1 - \frac{Z}{K}\right\}\right)^2$$

$$= -\left(\frac{c^2 A^2}{w_3 w_4}\right)Z(\alpha_1 + \gamma_1 Z)\left\{1 - \frac{Z}{K}\right\}^2\left(\frac{w_4}{cZ}\right)^2$$

$$= -\left(\frac{A^2 w_4}{w_3}\right)\left(\frac{\alpha_1 + \gamma_1 Z}{Z}\right)\left(1 - \frac{Z}{K}\right)^2.$$

Now,

$$a_{11}a_{22} - a_{12}a_{21} = \left[A^2\left(1 - \frac{Z}{K}\right)^2 \left(\frac{\gamma_1 w_4}{c w_3}\right) - \frac{AZ}{K}\right](-c)$$

$$+ \left(\frac{A^2 w_4}{w_3}\right)\left(\frac{\alpha_1 + \gamma_1 Z}{Z}\right)\left(1 - \frac{Z}{K}\right)^2$$

$$= \frac{AcZ}{K} + \left(\frac{A^2 w_4}{w_3}\right)\left(1 - \frac{Z}{K}\right)^2 \left(\frac{\alpha_1 + \gamma_1 Z}{Z} - \gamma_1\right)$$

$$= \frac{AcZ}{K} + \left(\frac{A^2 w_4}{w_3}\right)\left(1 - \frac{Z}{K}\right)^2 \left(\frac{\alpha_1}{Z}\right) > 0$$

is always satisfied.

Therefore, the equilibrium point (Z^*, U^*) is asymptotically stable when $c^2 > [A^2 \gamma_1 w_4/w_3]$.

Example 2.11

Draw the phase plot and time series for model system (2.90) and (2.91) assuming the parameter values as $A = 2$, $K = 100$, $w_3 = 2.1$, $w_4 = 1.0$, $\alpha_1 = 0.5$, $\beta_1 = 0.65$, $\gamma_1 = 0.25$, and $c = 0.55$.

SOLUTION

The phase plot and time series are given in Figure 2.28.

Competition Model

Competitive interaction occurs in nature in the same community or in different communities when two or more members of the same community

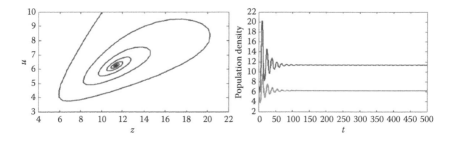

FIGURE 2.28
Phase plot and time series for the model system (2.90) and (2.91) for the given set of parameter values.

or species, respectively, compete for limited resources (such as food, water, light and territory, etc.) in an ecosystem. The utilization of these resources by one species inhibits the ability of another species to survive and grow.

There are two aspects of competition: (i) exploitation, when the competitor uses the resources itself like sharing the common pasture, and (ii) interference, where the population behaves in such a manner as to prevent the competitor from utilizing the resources. An example of exploitation of the same trophic niches was given in Malchow et al. [97]. The exploitation is between goats and the wild chamois (*Rupicapra rupicapra*) and ibex (*Capra ibex*), in the Gran Paradiso National Park in Italy. An example of direct competition for the pastures was given in Malchow et al. [97]. The competition arises between the red deer and roe deer, the red deer suffering the most as the roe deer is more generalist in the habitat selection as well as in feeding habits. Review articles by Pianka [127] and Waltman [166] deal with some practical aspects of competition. Flores' [51] proposed a model after analyzing the palaeontological data of 5000–10,000 years for competition between Neanderthal man and early modern man with slightly different mortality rates and determines the conditions under which the Neanderthal man became extinct. The LV competition model is an interference competition model: two species are assumed to diminish each other's per capita growth rate by direct interference.

Model 1: Gause Model

Consider an ideal case of two species that occupy the identical trophic niche. Gause [58] studied two species that both fed on the same bacterial cells. When the two species were placed in a single culture, one always drove the other to extinction. Many later experiments have supported *Gause's Law*, now called the *Principle of Competitive Exclusion*. It states that any two species that utilize identical resources cannot coexist indefinitely or complete competitors cannot coexist [68].

Many experiments have demonstrated that if the use of resources by the two species differ, then they are more likely to coexist [85]. Over time, members of each species may come to specialize in a subdivision of some category of similar resources. For example, if both feed upon apples, one may feed upon small green fruits and the other upon larger ripe ones.

To model the interactions of two competing species, let the population densities (or biomass) of competing species be denoted by X_1 and X_2. The model describing the above situation is given by [4]

$$\frac{dX_1}{dt} = X_1(a_1 - b_1 X_2) = F(X_1, X_2), \tag{2.92}$$

$$\frac{dX_2}{dt} = X_2(a_2 - b_2 X_1) = G(X_1, X_2), \qquad (2.93)$$

where all the parameters are positive, a_1, a_2 are the per capita growth rates and b_1, b_2 are the interaction parameters or competition coefficients of X_1 and X_2, respectively.

Setting $F(X_1, X_2) = 0$, $G(X_1, X_2) = 0$, we find that the system has two equilibrium points E_0 $(0,0)$ and $E^*(X_1^*, X_2^*)$ where $X_1^* = (a_2/b_2)$ and $X_2^* = (a_1/b_1)$.

Linear stability analysis: Linear stability is determined by computing the eigenvalues of the Jacobian matrix about each equilibrium point. The Jacobian matrix of the system is given by Equation 2.59, where $a_{11} = a_1 - b_1 X_2$, $a_{12} = -b_1 X_1$, $a_{21} = -b_2 X_2$, and $a_{22} = a_2 - b_2 X_1$.

At the equilibrium point E_0 $(0,0)$: We obtain $a_{11} = a_1$, $a_{12} = 0$, $a_{21} = 0$, $a_{22} = a_2$. The eigenvalues are $\lambda_1 = a_1 > 0$ and $\lambda_2 = a_2 > 0$. Since the eigenvalues are real and positive, we find that E_0 is an unstable, hyperbolic point and repels in both the directions.

At the equilibrium point $E^*(X_1^*, X_2^*)$: We obtain

$$a_{11} = 0, \quad a_{12} = -(a_2 b_1)/b_2, \quad a_{21} = -(a_1 b_2)/b_1, \quad \text{and} \quad a_{22} = 0.$$

The eigenvalues are $\lambda_{1,2} = \pm\sqrt{a_1 a_2}$. The system is not stable. Using the concept of separatrix (an orbit separating two different types of orbits in the phase space), Freedman [55] suggested a graphical approach to determine which species, X_1 or X_2, approaches to zero as $t \to \infty$.

Model 2

The competition model (Gause model) can be generalized to take into account the logistic growth in both species, in the absence of other species [4,84]. The governing equations are taken as

$$\frac{dX_1}{dt} = X_1\left[a_1\left(1 - \frac{X_1}{K_1}\right) - b_1 X_2\right], \quad \frac{dX_2}{dt} = X_2\left[a_2\left(1 - \frac{X_2}{K_2}\right) - b_2 X_1\right]. \qquad (2.94)$$

The terms $[-(a_1/K_1)X_1^2]$ and $[-(a_2/K_2)X_2^2]$ arise due to intraspecific competition between the species X_1 and X_2. Here, K_1, K_2 are the carrying capacities for the species X_1 and X_2, respectively.

We find that the system has four equilibrium points $E_0(0, 0)$, $E_1(K_1, 0)$, $E_2(0, K_2)$, and a positive equilibrium point $E^*(X_1^*, X_2^*)$. The positive equilibrium point $E^*(X_1^*, X_2^*)$ is the solution of the equations $p_1 X_1 + b_1 X_2 = a_1$, $b_2 X_1 + p_2 X_2 = a_2$, where $p_1 = a_1/K_1$, $p_2 = a_2/K_2$. The solution is given by

$$X_1^* = \frac{a_1 p_2 - a_2 b_1}{p_1 p_2 - b_1 b_2}, \quad X_2^* = \frac{p_1 a_2 - b_2 a_1}{p_1 p_2 - b_1 b_2}.$$

Since, $X_1^* > 0$ and $X_2^* > 0$, the necessary conditions are either

$$a_1 p_2 - a_2 b_1 > 0, \quad p_1 a_2 - b_2 a_1 > 0, \quad \text{and} \quad p_1 p_2 - b_1 b_2 > 0 \tag{2.95}$$

or

$$a_1 p_2 - a_2 b_1 < 0, \quad p_1 a_2 - b_2 a_1 < 0, \quad \text{and} \quad p_1 p_2 - b_1 b_2 < 0. \tag{2.96}$$

The first set of inequalities gives the condition

$$\frac{b_1}{p_2} < \frac{a_1}{a_2} < \frac{p_1}{b_2}. \tag{2.97}$$

We can write the condition as

$$\frac{K_2 b_1}{a_2} < \frac{a_1}{a_2} < \frac{a_1}{K_1 b_2}.$$

The condition simplifies to

$$K_2 b_1 < a_1 \quad \text{and} \quad K_1 b_2 < a_2. \tag{2.98}$$

The second set of inequalities (2.96) gives the condition with reversed inequalities in Equation 2.97, which we find is not useful.

Linear stability analysis: The Jacobian matrix of the system is given by Equation 2.59, where

$$a_{11} = a_1 - 2 p_1 X_1 - b_1 X_2, \; a_{12} = -b_1 X_1, \; a_{21} = -b_2 X_2, \quad \text{and} \quad a_{22} = a_2 - 2 p_2 X_2 - b_2 X_1.$$

At the equilibrium point E_0 (0, 0): We obtain $a_{11} = a_1$, $a_{12} = 0$, $a_{21} = 0$, $a_{22} = a_2$. The eigenvalues are $\lambda_1 = a_1 > 0$ and $\lambda_2 = a_2 > 0$. Since the eigenvalues are real and positive, we find that E_0 is an unstable hyperbolic point and repels in both the directions.

At the equilibrium point $E_1(K_1, 0)$: We obtain $a_{11} = a_1 - 2 p_1 K_1 = -a_1$, $a_{12} = -b_1 K_1$, $a_{21} = 0$, and $a_{22} = a_2 - b_2 K_1$. The eigenvalues are $\lambda_1 = -a_1$, and $\lambda_2 = a_2 - b_2 K_1 < 0$, for $(a_2/b_2) < K_1$. The equilibrium point $E_1(K_1, 0)$ is asymptotically stable for $(a_2/b_2) < K_1$. The system leads to the extinction of the species X_2.

At the equilibrium point $E_2(0, K_2)$: We obtain $a_{11} = a_1 - b_1 K_2$, $a_{12} = 0$, $a_{21} = -b_2 K_2$, and $a_{22} = a_2 - 2 p_2 K_2 = -a_2$. The eigenvalues are $\lambda_1 = -a_2$ and $\lambda_2 = a_1 - b_1 K_2 < 0$, for $(a_1/b_1) < K_2$. The equilibrium point $E_2(0, K_2)$ is asymptotically stable for $(a_1/b_1) < K_2$. The system leads to the extinction of the species X_1.

At the equilibrium point $E^*(X_1^*, X_2^*)$: We obtain

$$a_{11} = a_1 - 2p_1X_1^* - b_1X_2^* = -p_1X_1^*, \quad a_{12} = -b_1X_1^*,$$

$$a_{21} = -b_2X_2^*, \quad \text{and} \quad a_{22} = a_2 - 2p_2X_2^* - b_2X_1^* = -p_2X_2^*.$$

The characteristic equation of the Jacobian matrix is

$$\lambda^2 + \lambda(p_1X_1^* + p_2X_2^*) + (p_1p_2 - b_1b_2)X_1^*X_2^* = 0.$$

The roots are negative or have negative real parts if the Routh–Hurwitz criterion is satisfied. The necessary and sufficient conditions are

$$p_1X_1^* + p_2X_2^* > 0, \quad \text{and} \quad p_1p_2 - b_1b_2 > 0, \quad \text{since } X_1^* > 0, \text{ and } X_2^* > 0.$$

Both the conditions are satisfied (see Equation 2.95). The equilibrium point $E^*(X_1^*, X_2^*)$ is asymptotically stable if conditions (2.97) or (2.98) are satisfied. The species coexist.

2D Discrete Models

Predator–prey models can also be formulated as discrete-time models. These models apply most readily to groups such as insect population where there is rather natural division of time into discrete generations. This is appropriate, when organisms have discrete, nonoverlapping generations. By introducing density-dependent coefficients such as logistic growth, the models become nonlinear. A model which has received considerable attention from theoretical and experimental biologists is the *host–parasitoid system*. A simple model for this system has the following set of assumptions [14]:

i. Hosts that have been parasitized will give rise to the next generation of parasitoids.

ii. Hosts that have not been parasitized will give rise to their own prodigy.

iii. The fraction of hosts that are parasitized depends on the rate of encounter of the two species. In general, this fraction may depend on the densities of one or both species.

Define N_n: Density (number) of host species in nth generation,

P_n: Density (number) of parasitoid in nth generation,

$f = f(N_n, P_n)$: Fraction of hosts not parasitized,

r: Host reproduction rate, and

c: Average number of viable eggs laid by a parasitoid in a single host.

Then, the assumptions (i) through (iii) led to the model equations

$$N_{n+1} = rN_n f(N_n, P_n), \quad P_{n+1} = cN_n[1 - f(N_n, P_n)]. \tag{2.99}$$

Nicholson–Bailey Model

The Nicholson–Bailey (NB) model [116] considers the host–parasitoid inter-action, which is inherently unstable. It generates increasingly large oscil-lations. These oscillations, when coupled with stochasticity, drive the host and/or the parasitoid to become extinct [84]. The model assumes that the interaction between host and parasitoid is intrinsically unstable and that the effect of increasing oscillations is to break up the parent population into local populations which grow/decay, go extinct, and are again colonized. The model also assumes that the spatial heterogeneity is a factor that keeps host–parasitoid systems intact. This view point was supported by the laboratory experiments of Huffaker [77] on a six-spotted mite *Eoterranychus sexmacula-tus*, which feeds on oranges and the predatory mite *Typhlodromus occidentalis*.

Nicholson and Bailey [116] added two more assumptions to the above assumptions (i) through (iii).

iv. The number of encounters N_e of the host with the parasitoid is propor-tional to the product of their densities, $N_e = \alpha N_n P_n$, where α is a constant. The equation represents the searching efficiency of the parasitoids.

v. Only the first encounter between a host and parasitoid is significant (once the host has been parasitized it gives rise exactly c parasitoid progeny); a second encounter with an egg-laying parasitoid will not increase or decrease this number.

The assumption that the parasitoids search independently and their searching efficiency is a constant a leads to the NB model

$$N_{n+1} = rN_n e^{-aP_n} = F(N_n, P_n), \quad P_{n+1} = cN_n[1 - e^{-aP_n}] = G(N_n, P_n)]. \tag{2.100}$$

The equilibrium points are obtained by solving $N = rNe^{-aP}; P = cN[1 - e^{-aP}]$.

One equilibrium point is $(N_0, P_0) = (0, 0)$. The second equilibrium point is obtained from

$$r = e^{aP}, \quad \text{or} \quad P = \frac{\ln r}{a}; \quad N = \frac{r\ln r}{(r-1)ac}. \text{ Hence, } (N^*, P^*) = \left(\frac{r\ln r}{(r-1)ac}, \frac{\ln r}{a}\right).$$

We have earlier noted in the discussion of the stability of continuous growth models that the system is stable if and only if all the eigenvalues of the community matrix are negative or have negative real parts. This implies that all the eigenvalues lie in the left half of the complex plane. If one or more eigenvalues are positive or have positive real parts, then the equilibrium point is unstable. However, for discrete systems, stability requires a stronger condition. The eigenvalues should lie inside a circle in the left half of the complex plane with the center at $-1/T$ and radius $1/T$, where T is the time interval or gestation time. If $T = 1$, as is often the case in the discrete systems, then the stability region is given by $|\lambda + 1| < 1$. If $\lambda = x + iy$, then we get the stability region as the circular region $(x + 1)^2 + y^2 < 1$. This requirement shows that the stability of the difference system implies the stability of the differential system and the converse may not be true.

The stability requirement is given in the following theorem (Kot [84], May [99]).

Theorem 2.4

For the system $X_{n+1} = F(X_n)$, define the community matrix as

$$J(X) = \left[\frac{\partial F(X)}{\partial X} \right]. \tag{2.101}$$

The equilibrium is asymptotically stable if the eigenvalues λ of J satisfy the condition $|\lambda| < 1$. Otherwise, equilibrium is unstable.

We now require the conditions to test the above stability requirement which are analogous to the Routh–Hurwitz conditions for continuous systems. We use the following conditions given by Miller [107].

Theorem 2.5

For a 2×2 system given in Equation 2.100, let the characteristic equation of the matrix in Equation 2.101, be $A\lambda^2 + B\lambda + C = 0$. The necessary and sufficient conditions for $|\lambda| < 1$ are

$$A + B + C > 0, \quad A - B + C > 0, \quad \text{and} \quad A - C > 0. \tag{2.102}$$

These conditions are the same as the *Jury conditions* or the *Jury test* [84].

Example 2.12

Derive the local stability conditions for model system (2.100).

SOLUTION
The equilibrium points are $(0, 0)$, and

$$(N^*, P^*) = \left(\frac{r \ln r}{(r-1)ac}, \frac{\ln r}{a} \right). \ N^* \text{ is positive when } r > 1.$$

The Jacobian matrix of the system evaluated at an equilibrium point (N^*, P^*) is given by Equation 2.59, where $a_{11} = re^{-aP}$, $a_{12} = -arNe^{-aP}$, $a_{21} = c(1 - e^{-aP})$, and $a_{22} = acNe^{-aP}$.

At the equilibrium point (0, 0): We obtain $a_{11} = r$, $a_{12} = 0$, $a_{21} = c$, and $a_{22} = 0$. The eigenvalues are $\lambda_1 = r > 1$ and $\lambda_2 = 0$. Therefore, the origin is an unstable saddle point.

At the equilibrium point (N^*, P^*): We obtain

$$a_{11} = 1, \quad a_{12} = -\frac{r \ln r}{c(r-1)}, \quad a_{21} = \frac{c(r-1)}{r}, \quad \text{and} \quad a_{22} = \frac{\ln r}{(r-1)}.$$

The eigenvalues of J are roots of the equation $\lambda^2 - (a_{11} + a_{22})\lambda + (a_{11}a_{22} - a_{12}a_{21}) = 0$. We have

$$a_{11} + a_{22} = 1 + \frac{\ln r}{r-1}, \quad a_{11}a_{22} - a_{12}a_{21} = \frac{r \ln r}{r-1}, \quad r > 1.$$

For asymptotic stability, we require that both the eigenvalues have magnitude < 1.

Applying the Miller's conditions or Jury conditions, we get

$$A + B + C = 1 - 1 - \frac{\ln r}{r-1} + \frac{r \ln r}{r-1} = \ln r > 0, \quad \text{for } r > 1.$$

$$A - B + C = 1 + 1 + \frac{\ln r}{r-1} + \frac{r \ln r}{r-1} = 2 + \frac{(r+1)}{(r-1)} \ln r > 0,$$

$$A - C = 1 - \frac{r \ln r}{r-1} = \frac{r(1 - \ln r) - 1}{r-1}.$$

The condition $A - C > 0$ is not satisfied for any value of r. The equilibrium of coexistence (N^*, P^*) is unstable. The only equilibrium point is $(0, 0)$. This leads to the conclusion that both prey and predator become extinct. This indicates that the model is not a satisfactory representation of real systems.

Modified NB Model

Most natural host–parasitoid systems in nature are more stable than the NB model. In the absence of parasitoids, the host population grows to some limited density determined by the carrying capacity K of the environment. Incorporating this observation, the modified model can be written as [17]

$$N_{n+1} = N_n \exp\left[r\left(1 - \frac{N_n}{K}\right) - aP_n\right] = F(N_n, P_n), \qquad (2.103)$$

$$P_{n+1} = cN_n[1 - \exp(-aP_n)] = G(N_n, P_n). \qquad (2.104)$$

Scale the variables as $u_n = N_n/K$, $v_n = aP_n$. Then, model systems (2.103) and (2.104) simplify to

$$u_{n+1} = u_n \exp[r(1 - u_n) - v_n], \qquad (2.105)$$

$$v_{n+1} = \alpha u_n[1 - \exp(-v_n)], \quad \text{where } \alpha = Kca. \qquad (2.106)$$

The equilibrium points are the solutions of the equations

$$u = u \exp[r(1 - u) - v], \quad \text{and} \quad v = \alpha u[1 - \exp(-v)].$$

One of the equilibrium points is $(0, 0)$, that is, the extinction of both prey and predator. We find that $(1, 0)$ is also an equilibrium point, which suggests that the predator becomes extinct.

The positive equilibrium point is the solution of the equations

$$1 = \exp[r(1 - u) - v], \quad \text{and} \quad v = \alpha u[1 - \exp(-v)].$$

From the first equation, we get

$$[r(1 - u) - v] = 0, \quad \text{or} \quad r(1 - u) = v. \qquad (2.107)$$

From the second equation, we get

$$r(1 - u) = \alpha u[1 - e^{-v}],$$

or

$$e^{-v} = 1 + \frac{r(u - 1)}{\alpha u},$$

or

$$-v = \ln\left[1 + \frac{r(u - 1)}{\alpha u}\right]. \qquad (2.108)$$

Since positive solutions are required, we require $v > 0$. The left-hand side of Equation 2.108 is negative. Hence, we require

$$0 < 1 + \frac{r(u - 1)}{\alpha u} < 1.$$

The right inequality gives $u < 1$. The left inequality gives

$$0 < \alpha u + ur - r, \quad \text{or} \quad u > [r/(r + \alpha)]. \tag{2.109}$$

Therefore,

$$\frac{r}{r + \alpha} < u < 1. \tag{2.110}$$

From Equations 2.107 and 2.108, we get

$$r(u - 1) = \ln\left[1 + \frac{r(u - 1)}{\alpha u}\right]. \tag{2.111}$$

Hence, u^* is a solution of Equation 2.111 satisfying Equation 2.110. v^* is given by $v^* = r(1 - u^*)$.

The existence of a nonzero positive equilibrium point is proved.

For example, let $r = 1$, and $\alpha = 2$. Then, $(1/3) < u^* < 1$. u^* is a solution of

$$u^* - 1 = \ln\left[1 + \frac{u^* - 1}{2u^*}\right].$$

We obtain the solution using Newton–Raphson method. Define

$$f(u^*) = 1 + \frac{u^* - 1}{2u^*} - e^{u^* - 1} = \frac{3}{2} - \frac{1}{2u^*} - e^{u^* - 1}.$$

Applying the Newton–Raphson method to find a root of $f(u^*) = 0$, with initial approximation taken as $u_0 = 0.5$, we obtain the sequence of iterates as $0.576449948, 0.602446606, \ldots, 0.60518257$. With $u^* = 0.60518257$, we get $v^* = 0.39481743$.

Example 2.13

Derive the local stability conditions for the nonzero equilibrium point of model system (2.105) and (2.106). Is the equilibrium point asymptotically stable for the parameter values $r = 1$ and $\alpha = 2$?

SOLUTION

Define $F(u,v) = u \exp[r(1 - u) - v]$, $G(u,v) = \alpha u[1 - e^{-v}]$, where $\alpha = Kca$.

The nonzero equilibrium point (u^*, v^*) is the solution of the equation

$$r(u - 1) = \ln\left[1 + \frac{r(u - 1)}{\alpha u}\right],$$

where

$$\frac{r}{r + \alpha} < u < 1. \quad v^* = r(1 - u^*).$$

The Jacobian matrix of the system evaluated at an equilibrium point (u^*, v^*) is given by Equation 2.59, where

$$a_{11} = (1 - ru) \exp[r(1 - u) - v], \quad a_{12} = -u \exp[r(1 - u) - v],$$

$$a_{21} = \alpha[1 - e^{-v}], \quad a_{22} = \alpha u e^{-v}.$$

At the equilibrium point (u^*, v^*): We obtain

$$a_{11} = (1 - ru^*), \quad a_{12} = -u^*, \quad a_{21} = \frac{r(1 - u^*)}{u^*}, \quad \text{and} \quad a_{22} = \alpha u^* - r(1 - u^*).$$

The eigenvalues of J are roots of the equation $\lambda^2 - (a_{11} + a_{22})\lambda + (a_{11}a_{22} - a_{12}a_{21}) = 0$.
We have

$$a_{11} + a_{22} = 1 + \alpha u^* - r,$$

$$a_{11}a_{22} - a_{12}a_{21} = (1 - ru^*)(\alpha u^* - r + ru^*) + r(1 - u^*) = u^*[\alpha - rp],$$

where $p = (\alpha u^* + ru^* - r) > 0$, from Equation 2.109.
For asymptotic stability, we require that both the eigenvalues have magnitude < 1.
Applying the Miller's conditions or Jury conditions, we require

$$A + B + C = 1 - (1 + \alpha u^* - r) + u^*[\alpha - rp] = r[1 - u^*p] > 0,$$

$$A - B + C = 1 + (1 + \alpha u^* - r) + u^*[\alpha - rp] = 2 - r + u^*[2\alpha - rp] > 0,$$

$$A - C = 1 - u^*[\alpha - rp] > 0.$$

When $r = 1$ and $\alpha = 2$, we obtain

$$u^* = 0.60518257, \quad v^* = 0.39481743, \quad p = 3u^* - 1 = 0.8155477,$$

$$A + B + C = 1 - pu^* = 0.50644474 > 0,$$

$$A - B + C = 1 + u^*[4 - p] = 2.927175 > 0,$$

$$A - C = 1 - u^*[2 - p] = 0.28319 > 0.$$

Hence, the equilibrium point is asymptotically stable.

Aihara Model

A brilliant practical application of chaos is the fashion dress conceptualized by Professor Kazuyuki Aihara (University of Tokyo, Japan) (see Kazuyuki A., private communication [82]) for a fashion show. The dress is based on a 2D chaotic map. The dress was designed and made by Eri Matsui, Keiko Kimoto, and Kazuyuki Aihara (Eri Matsui is a famous fashion designer in Japan). The dress was presented at a famous fashion show "Tokyo Collection" held in Japan in March 2010 (see Figure 2.29). The development of the dresses was partially supported by the Aihara Project, the FIRST program from JSPS, initiated by CSTP. The picture in the figure was actually that on the stage of Tokyo Collection.

The model is described by the equations

$$x_{n+1} = x_n - (x_n \cos\theta - y_n \sin\theta)/(x_n^2 + y_n^2) = F(x_n, y_n), \qquad (2.112)$$

$$y_{n+1} = y_n - (x_n \sin\theta + y_n \cos\theta)/(x_n^2 + y_n^2) = G(x_n, y_n). \qquad (2.113)$$

FIGURE 2.29
Fashion dress designed and created using model (2.112) and (2.113). (Courtesy Eri Matsui, Keiko Kimoto, and Kazuyuki Aihara. Fashion Show in Japan, 2010; Kazuyuki Aihara, private communication.)

We can write the model in matrix form as

$$\begin{bmatrix} x_{n+1} \\ y_{n+1} \end{bmatrix} = \begin{bmatrix} x_n \\ y_n \end{bmatrix} - \frac{1}{(x_n^2 + y_n^2)} \begin{bmatrix} \cos\theta & -\sin\theta \\ \sin\theta & \cos\theta \end{bmatrix} \begin{bmatrix} x_n \\ y_n \end{bmatrix}.$$

We can conceptualize this mapping as an orthogonal rotation by an angle θ, followed by magnification (contraction) using the factor $[1/(x_n^2 + y_n^2)]$ and subtraction (shifting) from the previous values. Note that $(0, 0)$ is not a point in the domain and not an equilibrium point. Let $F(x, y) = x - (x \cos\theta - y \sin\theta)/(x^2 + y^2)$, $G(x,y) = y - (x \sin\theta + y \cos\theta)/(x^2 + y^2)$.

The elements of the Jacobian matrix J are

$$a_{11} = 1 - \frac{1}{D}[(y^2 - x^2)C + 2xyS], \quad a_{21} = a_{12} = \frac{1}{D}[(x^2 - y^2)S + 2xyC],$$

$$a_{22} = 1 + \frac{1}{D}[(y^2 - x^2)C + 2xyS], \quad C = \cos\theta, \quad S = \sin\theta, \quad D = (x^2 + y^2)^2.$$

Now, $\left|\det(J(x, y))\right| = \left|a_{11}a_{22} - a_{12}a_{21}\right| = \left|1 - \frac{1}{D}\right|$. (Independent of θ).

The map is *conservative* if $\left|\det (J(x,y))\right| = 1$. The map is *dissipative* if $\left|\det (J(x, y))\right| < 1$. The map is *area expanding* if $\left|\det (J(x, y))\right| > 1$. For $D < 1$, or $x^2 + y^2 < 1$, $\left|\det(J(x,y))\right| > 1$. The map is area expanding. For $D > 1$, or $x^2 + y^2 > 1$, $\left|\det(J(x, y))\right| < 1$. The map is dissipative.

EXERCISE 2.2

1. Find the equilibrium solutions and discuss the local stability analysis of the following system:

$$\frac{dX}{dt} = X\left(1 - \frac{X}{100}\right) - \frac{1.6XY}{X + 5Y}, \quad \frac{dY}{dt} = \frac{0.6XY}{X + 5Y} - 0.05Y.$$

2. Show that Kolmogorov theorem is not applicable for model system (2.61).

3. Derive the conditions so that the model described by Equations 2.62 and 2.63 satisfies the Kolmogorov theorem.

4. Perform the linear stability analysis of the equilibrium points of the RM model (2.62) and (2.63).

5. Discuss the linear stability of the equilibrium points of the system (2.67) and (2.68).

6. Find the nonzero equilibrium points of the model system (2.71) and (2.72).

7. Check the applicability of the Kolmogorov theorem for the model system (2.71) and (2.72). Draw the time series for the following set of parameter values, $a_1 = 2.5$, $b_1 = 0.05$, $w = 0.85$, $\alpha = 0.45$, $\beta = 0.2$, $\gamma = 0.6$, $a_2 = 0.95$, and $w_1 = 1.65$.

8. Derive the local stability conditions for the model system (2.71) and (2.72).

9. Find the equilibrium points of the model system (2.75) and (2.76).

10. Derive the conditions for the model system (2.75) and (2.76) to satisfy the Kolmogorov theorem.

11. Draw the phase plot and time series for the model system (2.75) and (2.76) assuming the parameter values as $K = 1$, $\alpha = 3$, $\beta = 2.3$, and $\gamma = 0.3$.

12. Derive the conditions for the model described by (2.79) and (2.80) to satisfy the Kolmogorov theorem.

13. Show that the Kolmogorov theorem cannot be applied for the model system (2.84) and (2.85).

14. Draw the oscillatory predator–prey dynamics exhibited by the HT model (2.84) and (2.85) for the parameter values $A = 1.5$, $K_1 = 125$, $w_3 = 0.45$, $D_3 = 20$, $c = 0.1$, and $w_4 = 0.03$.

15. Find the equilibrium points of the model system (2.90) and (2.91). Draw the time series for the model system assuming the parameter values as $A = 2$, $K = 100$, $w_3 = 2.1$, $\alpha_1 = 0.45$, $\beta_1 = 0.2$, $\gamma_1 = 0.6$, $c = 0.95$, and $w_4 = 1.65$.

16. Test whether the Kolmogorov theorem can be applied for the model system (2.90) and (2.91).

17. Analyze the LV competition model for the asymptotic stability of positive equilibrium [4].

$$\frac{dX_1}{dt} = (r_1/K_1)X_1(K_1 - X_1 - b_1X_2), \quad \frac{dX_2}{dt} = (r_2/K_2)X_2(K_2 - b_2X_1 - X_2).$$

18. Find the conditions for the asymptotic stability of the positive equilibrium point for the logarithmic competition model [32]

$$\frac{dX_1}{dt} = X_1(a_1 - b_1 \ln X_1 - c_1 \ln X_2), \quad \frac{dX_2}{dt} = X_2(a_2 - b_2 \ln X_1 - c_2 \ln X_2),$$

where all the parameters are positive. Check for asymptotic stability when $a_1 = 11$, $b_1 = 3$, $c_1 = 2$, $a_2 = 13$, $b_2 = 4$, $c_2 = 3$.

19. For model system (2.105) and (2.106)

$$u_{n+1} = u_n \exp[r(1 - u_n) - v_n], \quad v_{n+1} = \alpha u_n[1 - \exp(-v_n)],$$

where $\alpha = Kca$, find whether the system is asymptotically stable for $r = 1.5$, $\alpha = 3$.

20. Find the equilibrium solutions and discuss the local stability analysis of the following model system:

$$N_{n+1} = N_n [1 + 1.5(1 - N_n)] - 0.5N_n P_n, \quad P_{n+1} = 0.2P_n + 1.8N_n P_n.$$

References

1. Abrams, P. A., Ginzburg, L. R. 2000. The nature of predation: Prey dependent, ratio-dependent or neither? *TREE* 15, 337–341.
2. Aguirre, P., Gonzlez-Olivares, E., Sez, E. 2009a. Three limit cycles in a Leslie–Gower predator–prey model with additive Allee effect. *SIAM J. Appl. Math.* 69(5), 1244–1262.
3. Aguirre, P., Gonzlez-Olivares, E., Sez, E. 2009b. Two limit cycles in a Leslie–Gower predator–prey model with additive Allee effect. *Nonlinear Anal.: Real World Appl.* 10(3), 1401–1416.
4. Ahmad, S., Rao, M. R. M. 1999. *Theory of Ordinary Differential Equations.* Affiliated East-West Press Private Limited, New Delhi.
5. Alicki, R., Messer, J. 1983. Nonlinear quantum dynamical semigroups for many-body open systems. *J. Stat. Phys.* 32, 299–312.
6. Allee, W. C. 1931. *Animal Aggregations.* University of Chicago Press, Chicage, IL.
7. Allee, W. C. 1951. *Cooperation among Animals.* Henry Schuman, New York.
8. Amarasekare, P. 1998. Allee effects in metapopulation dynamics. *Am. Naturalist* 152, 298–302.
9. Andrews, J. F. 1968. A mathematical model for the continuous culture of microorganisms utilizing inhibitory substrates. *Biotechnol. Bioeng.* 10, 707–723.
10. Arditi, R., Akcakaya, H. R. 1990. Underestimation of the mutual interference of predators. *Oecologia* 83, 358–361.
11. Arditi, R., Ginzburg, L. R. 1989. Coupling in predator–prey dynamics: Ratio dependence. *J. Theor. Biol.* 139, 311–329.
12. Ardito, A., Ricciardi, P. 1995. Lyapunov functions for a generalized Gause-type model. *J. Math. Biol.* 33, 816–828.
13. Aziz-Alaoui, M. A. 2002. Study of a Leslie–Gower type tri-trophic population model. *Chaos, Solitons Fractals* 14(8), 1275–1293.
14. Banasiak, J. 2005. *Notes on Mathematical Models in Biology.* University of KwaZulu-Natal, Durban. http://users.aims.ac.za/~geofrey/MBlectnotes.pdf
15. Barnes, B., Fulford, G. R. 2002. *Mathematcal Modeling with Case Studies.* Taylor & Francis, Boca Raton, FL.
16. Beddington, J. R. 1975. Mutual interference between parasites or predators and its effect on searching efficiency. *J. Anim. Ecol.* 44, 331–340.
17. Beddington, J. R., Free, C. A., Lawton, J. H. 1975. Dynamic complexity in predator–prey models framed in difference equations. *Nature* 255, 58–60.

18. Berec, L. 2008. Models of Allee effects and their implications for population and community dynamics. *Biophys. Rev. Lett.* 3, 157–181.
19. Berec, L., Angulo, E., Counchamp, F. 2006. Multiple Allee effects and population management. *TREE* 22(4), 185–191.
20. Berryman, A. A., Gutierrez, A. P., Arditi, R. 1995. Credible, parsimonious and useful predator–prey models: A reply to Abrams, Gleeson, and Sarnelle. *Ecology* 76(6), 1980–1985.
21. Beth, F. T. B., Hassall, M. 2005. The existence of an Allee effect in populations of *Porcellio scaber* (Isopoda: Oniscidea). *Eur. J. Soil Biol.* 41, 123–127.
22. Beverton, R. J. H., Holt, S. J. 1957. On the dynamics of exploited fish populations. *Fishery Invest. Ser. II* 19, 1–533.
23. Blythe, S. P. 1982. Instability and complex dynamic behaviour in population models with long time delays. *Theor. Popul. Biol.* 22, 147–176.
24. Brauer, F., Sanchez, D. A. 2003. Periodic environments and periodic harvesting. *Nat. Res. Model.* 16, 233–244.
25. Campbell, S. A., Edwards, R., van den Driessche, P. 2004. Delayed coupling between two neural network loops. *SIAM J. Appl. Math.* 65(1), 316–335.
26. Cantrell, R. S., Cosner, C. 2001. On the dynamics of predator–prey models with the Beddington–DeAngelis functional response. *J. Math. Anal. Appl.*, 257, 206–222.
27. Causton, D. R., Venus, J. C. 1981. *The Biometry of Plant Growth*. Edward Arnold, London.
28. Cheng, K. S. 1981. Uniqueness of a limit cycle for predator–prey system. *SIAM J. Math. Anal.* 12(4), 541–548.
29. Ciupe, M. S., Bivort, B. L., Bortz, D. M., Nelson, P. W. 2006. Estimating kinetic parameters from HIV primary infection data through the eyes of three different mathematical models. *Math. Biosci.* 200(1), 1–27.
30. Clark, C. W. 1976. *Mathematical Bioeconomics—The Optimal Control of Renewable Resources*. John Wiley, New York.
31. Clark, C. W. 2005. *Mathematical Bioeconomics—The Optimal Management of Renewable Resources*, 2nd edition. Wiley-Interscience, John Wiley & Sons Inc., New York.
32. Coleman, T. P., Gomatam, J. 1972. Application of new model of species competition to *Drosophila*. *Nature (New Biol.)* 239, 251.
33. Collings, J. B. 1997. The effects of the functional response on the bifurcation behavior of a mite predator–prey interaction model. *J. Math. Biol.* 36, 149–168.
34. Cooke, K., Kuang, Y., Li, B. 1998. Analyses of an antiviral immune response model with time delays. *Canad. Appl. Math. Quart.* 6(4), 321–354.
35. Cooke, K. L., van den Driessche, P., Zou, X. 1999. Interaction of maturation delay and nonlinear birth in population and epidemic models. *J. Math. Biol.* 39, 332–352.
36. Correigh, M. G. 2003. Habitat selection reduces extinction of populations subject to Allee effects. *Theor. Popul. Biol.* 64, 1–10.
37. Cosner, C., DeAngelis, D. L., Ault, J. S., Olson, D. B. 1999. Effect of spatial grouping on the functional response of predators. *Theor. Popul. Biol.* 56, 65–75.
38. Courchamp, F., Clutton-Brock, T., Grenfell, B. 1999. Inverse density dependence and the Allee effect. *TREE* 14(10), 405–410.
39. Courchamp, F., Berec, L., Gascoigne, J. 2008. *Allee Effects in Ecology and Conservation*. Oxford University Press, New York.

40. Crick, F. H. C. 1954. The structure of the hereditary material. *Sci. Am.* 191(4), 54–61. (A wonderful account of the discovery of the double-helix. Reprinted in *Scientific American: Science of the 20th Century*, 1991.)

41. David, S. B., Berec, L. 2002. Single-species models of the Allee effect: Extinction boundaries, sex ratios and mate encounters. *J. Theor. Biol.* 218, 375–394.

42. David, S. B., Maurice, W. S., Berec, L. 2007. How predator functional responses and Allee effects in prey affect the paradox of enrichment and population collapses. *Theor. Popul. Biol.* 72, 136–147.

43. DeAngelis, D. L., Goldstein, R. A., O'Neill, R. V. 1975. A model for trophic interaction. *Ecology* 56, 881–892.

44. Dennis, B. 1989. Allee effects: Population growth, critical density, and the chance of extinction. *Nat. Res. Model.* 3, 481–538.

45. Dong, L., Chen, L., Sun, L. 2007. Optimal harvesting policies for periodic Gompertz systems. *Nonlinear Anal.: Real World Appl.* 8(2), 572–578.

45a. Dubey, B., Kumari, N., Upadhyay, R. K. 2009. Spatiotemporal pattern formation in a diffusive predator–prey system: An analytical approach. *J. Appl. Math. Comput.* 31, 413–432.

46. Easton, D. M. 1995. Gompertz survival kinetics: Fall in number alive or growth in number dead? *Theor. Popul. Biol.* 48, 1–6.

47. Edelstein-Keshet, L. 1987. *Mathematical Models in Biology*. McGraw-Hill, Inc., New York.

48. Eifert, H.-J., Held, S., Messer, J. A. 2009. A one-parameter model for the spread of avian influenza A/H5N1. *Chaos, Solitons Fractals* 41, 2271–2276.

49. Feigenbaum, M. J. 1978. Quantitative universality for a class of nonlinear transformations. *J. Stat. Phys.* 19, 25–52.

50. Feigenbaum, M. J. 1980. Universal behavior in nonlinear systems. *Los Alamos Sci.* 1, 4.

51. Flores, J. C. 1998. A mathematical model for Neanderthal extinction. *J. Theor. Biol.* 191, 295–298.

52. Forde, J. E. 2005. *Delay Differential Equation Models in Mathematical Biology*. PhD thesis, University of Michigan.

53. Fox, W. W. 1970. An exponential surplus yield model for optimizing in exploited fish populations. *Trans. Am. Fish. Soc.* 99, 80–88.

54. Frank, M. H., Langlais, M., Sergei, V. P., Malchow, H. 2007. A diffusive SI model with Allee effect and application to FIV. *Math. Biosci.* 206, 61–80.

55. Freedman, H. I. 1980. *Deterministic Mathematical Models in Population Ecology*. Marcel Dekker, New York.

56. Gakkhar, S., Naji, R. K. 2003. Chaos in seasonally perturbed ratio-dependent prey–predator system. *Chaos, Solitons Fractals* 15, 107–118.

57. Gasull, A., Koolj, R. E., Torregrosa, J. 1997. Limit cycles in the Holling–Tanner model. *Publ. Mat.* 41, 149–167.

58. Gause, G. F. 1934. *The Struggle for Existence*. Williams & Wilkins, Baltimore, New York.

59. Ginzburg, L. R. 1998. Assessing reproduction to be a function of consumption raises doubts about some popular predator–prey models. *J. Anim. Ecol.* 67, 325–327.

60. Gompertz, B. 1825. On the nature of the function expressive of the law of mortality, and on a new method of determining the value of life contingencies. *Philos. Trans. Roy. Soc.* 27, 513–585.

61. Gorini, V., Frigerio, A., Verri, M., Kossakowski, A., Sudarshan, E. C. G. 1978. Properties of quantum Markovian master equations. *Rep. Math. Phys.* 13, 149–173.

62. Gotelli, N. J. 1995. *A Primer of Ecology.* Sinauer Associates, Sunderland, MA.

63. Gurney, W. S. C., Blythe, S. P., Nisbet, R. M. 1980. Nicholson's blowfly revisited. *Nature* (London) 287, 17–21.

64. Gurney, W. S. C., Nisbet, R. M. 1998. *Ecological Dynamics.* Oxford University Press, New York.

65. Gyllenberg, M., Hemminki, J., Tammaru, T. 1999. Allee effects can both conserve and create spatial heterogeneity in population densities. *Theor. Popul. Biol.* 56, 231–242.

66. Haldane, J. B. S. 1930. *Enzymes.* Longman, London.

67. Hanski, I., Hansson, L., Henttonen, H. 1991. Specialist predators, generalist predators, and the microtine rodent cycle. *J. Anim. Ecol.* 60, 353–367.

68. Harden, G. 1960. The competitive exclusion principle. *Science* 131, 1292–1297.

69. Hasik, K. 2000. Uniqueness of limit cycle in the predator–prey system with symmetric prey isocline. *Math. Biosci.* 164, 203–215.

70. Hassell, M. P. 1975. Density-dependence in single-species populations. *J. Anim. Ecol.* 44, 283–295.

71. Hassell, M. P., Lawton, J. N., May, R. M. 1976. Patterns of dynamical behavior in single-species populations. *J. Anim. Ecol.* 45, 471–486.

72. Hassell, M. P., Varley, C. C. 1969. New inductive population model for insect parasites and its bearing on biological control. *Nature* 223, 1133–1137.

73. Hesaaraki, M., Moghadas, S. M. 1999. Nonexistence of limit cycles in a predator–prey system with a sigmoid functional response. *Can. Appl. Math. Quart.* 7(4), 401–408.

74. Hesaaraki, M., Moghadas, S. M. 2001. Existence of limit cycles for predator–prey systems with a class of functional responses. *Ecol Modell.* 142, 1–9.

75. Hirsch, M. W., Smale, S. 1974. *Differential Equations, Dynamical Systems, and Linear Algebra.* Academic Press, San Diego.

76. Hopper, K. R., Roush, R. T. 1993. Mate finding, dispersal, number released, and the success of biological-control introductions. *Ecol. Entomol.* 18, 321–331.

77. Huffaker, C. B. 1958. Experimental studies on predation: Dispersion factors and predator–prey oscillations. *Hilgardia* 27, 343–383.

78. Hwang, T. W. 2003. Global analysis of the predator–prey system with Beddington–DeAngelis functional response. *J. Math. Anal. Appl.* 281, 395–401.

79. Hwang, T. W. 2004. Uniqueness of limit cycles of the predator–prey system with Beddington–DeAngelis functional response. *J. Math. Anal. Appl.* 290, 113–122.

80. Ivlev, V. S. 1961. *Experimental Ecology of the Feeding of Fishes.* Yale University Press, New Haven, CT.

81. Jost, C., Ellner, S. P. 2000. Testing for predator dependence in predator–prey dynamics: A non-parametric approach. *Proc. R. Soc. Lond.* B, 267, 1611–1620.

82. Kazuyuki Aihara, Private communication. See also Preface.

82a. Kolmogorov, A. N. 1936. Sulla Teoria di Voltera della Lotta per l'Esisttenza, Giorn. *Instituto Ital. Attuari* 7, 74–80.

83. Koromondy, E. J. 1969. *Concepts of Ecology.* Prentice-Hall, Englewood Cliffs, NJ.

84. Kot, M. 2001. *Elements of Mathematical Ecology.* Cambridge University Press, Cambridge, UK.

85. Krebs, C. J. 1994. Ecology. *The Experimental Analysis of Distribution and Abundance*, 4th edition. Harper Collins, New York.
86. Kumar, N. 1996. *Deterministic Chaos: Complex Chance out of Simple Necessity*. University Press, India.
87. Laird, A. K. 1965. Dynamics of tumour growth: Comparison of growth rates and extrapolation of growth curve to one cell. *Br. J. Cancer* 19, 278–291.
88. Leslie, P. H., Gower, J. C. 1960. The properties of a stochastic model for the predator–prey type of interaction between two species. *Biometrika* 47, 219–234.
89. Lewis, M. A., Kareiva, P. 1993. Allee dynamics and the spread of invading organisms. *Theor. Popul. Biol.* 43, 141–158.
90. Lin, Z.-S., Li, B.-L. 2002. The maximum sustainable yield of Allee dynamical system. *Ecol. Model.* 154, 1–7.
91. Lindblad, G. 1976. On the generators of quantum dynamical semigroups. *Commun. Math. Phys.* 48, 119–130.
92. Liu, S., Edoardo, B. 2006. A stage-structured predator–prey model of Beddington–DeAngelis type. *SIAM J. Appl. Math.* 66, 1101–1129.
93. Lotka, A. J. 1925. *Elements of Physical Biology*. Williams & Wilkins, New York.
94. Ludwig, D., Jones, D., Holling, C. 1978. Qualitative analysis of an insect outbreak system: The spruce budworm and forest. *J. Anim. Ecol.* 47, 315–332.
95. Malchow, H. 2000. Motional instabilities in predator–prey systems. *J. Theor. Biol.* 204, 639–647.
96. Malchow, H., Hilker, F. M., Petrovskii, S. V. 2004. Noise and productivity dependence of spatiotemporal pattern formation in a prey–predator system. *Discrete Continuous Dynam. Systems* B 4(3), 705–711.
97. Malchow, H., Petrovskii, S. V., Venturino, E. 2008. *Spatiotemporal Patterns in Ecology and Epidemiology: Theory, Models and Simulation*. Chapman & Hall/CRC, Taylor & Francis Group, New York.
98. Matsui, E., Kimoto, K., Aihara, K. 2010. *Fashion Show in Japan*. [Kazuyuki Aihara, private communication].
99. May, R. M. 1973. *Stability and Complexity in Model Ecosystems*. Princeton University Press, Princeton, NJ. (Reprinted in 2001.)
100. May, R. M. 1975. Biological populations obeying difference equations: Stable points, stable cycles and chaos. *J. Theor. Biol.* 51, 511–524.
101. May, R. M. 1976. Simple mathematical models with very complicated dynamics. *Nature* 261, 459–467.
102. May, R. M. 1981. *Models for Single Populations. Theoretical Ecology: Principles and Applications*, 2nd edition. Sinauer Associates, Sunderland, MA.
103. May, R. M. 1981. *Theoretical Ecology: Principles and Applications*, 2nd edition. Blackwell Scientific Publications, Oxford.
104. May, R. M., Oster, G. F. 1976. Bifurcations and dynamic complexity in simple ecological models. *Am. Naturalist* 110, 573–599.
105. Medvinsky, A. B., Petrovskii, S. V., Tikhonova, I. A., Malchow, H., Li, B.-L. 2002. Spatiotemporal complexity of plankton and fish dynamics. *SIAM Rev.* 44(3), 311–370.
106. Messer, J. A. 2009. A non-equilibrium phase transition in a dissipative forest model. *Chaos, Solitons Fractals* 41(5), 2456–2462.
107. Miller, J. J. H. 1971. On location of zeros of certain classes of polynomials with applications to numerical analysis. *J. Inst. Math. Appl.* 8(2), 397–406.

108. Moghadas, S. M. 2002. Some conditions for the nonexistence of limit cycles in a predator–prey system. *Appl. Anal.* 81, 51–67.
109. Moghadas, S. M., Alexander, M. E., Corbett, B. D. 2004. A non-standard numerical scheme for a generalized Gause-type predator–prey model. *Phys. D.* 188, 134–151.
110. Moghadas, S. M., Corbett, B. D. 2008. Limit cycles in a generalized Gause-type predator–prey model. *Chaos, Solitons Fractals* 37, 1343–1355.
111. Morales, J., Buranr, Jr. V. 1985. Interactions between *Cycloneda sanguinea* and the brown citrus aphid: Adult feeding and larval mortality. *Environ. Entomol.* 14(4), 520–522.
112. Morozov, A., Petrovskii, S., Li, B.-L. 2004. Bifurcations and chaos in a predator–prey system with the Allee effect. *Proc. Roy. Soc. Lond. Ser. B* 271, 1407–1414.
113. Murray, J. D. 1989. *Mathematical Biology*. Springer-Verlag, New York.
114. Nedorezov, L. V. 2011. Analysis of some experimental time series by Gause: Application of simple mathematical models. *Comput. Ecol. Software* 1(1), 25–36.
115. Nelson, P. W., Murray, J. D., Perelson, A. S. 2000. A model of HIV-1 pathogenesis that includes an intracellular delay. *Math. Biosci.* 163, 201–215.
116. Nicholson, A. J., Bailey, V. A. 1935. The balance of animal populations. Part I. *Proc. Z. Soc. Lond.* 3, 551–598.
117. Okubo, A. 1980. *Diffusion and Ecological Problems: Mathematical Models*. Springer-Verlag, Berlin.
118. Oster, G. F., Guckenheimer, J. 1976. Bifurcation phenomena in population modes. In: Marsden, ed. *The Hopf-Bifurcation*. Springer-Verlag, Lecture Notes in Mathematics, New York.
119. Owen, M. R., Lewis, M. A. 2001. How predation can slow, stop or reverse a prey invasion? *Bull. Math. Biol.* 63, 655–684.
120. Pascual, M. 1993. Diffusion-induced chaos in a spatial predator–prey system. *Proc. R. Soc. Lond. Ser. B* 251, 1–7.
121. Pearl, R. 1927. The growth of populations. *Q. Rev. Biol.* 2, 532–548.
122. Peixoto, M. S., Barros, L. C., Bassanezi, R. C. 2008. Predator–prey fuzzy model. *Ecol. Model.* 214, 39–44.
123. Petrovskii, S. V., Malchow, H. 2001. Wave of chaos: New mechanism of pattern formation in spatio-temporal population dynamics. *Theor. Popul. Biol.* 59, 157–174.
124. Petrovskii, S. V., Malchow, H. 1999. A minimal model of pattern formation in a prey–predator system. *Math. Comput. Model.* 29, 49–63.
125. Petrovskii, S. V., Morozov, A. Y., Venturino, E. 2002a. Allee effect makes possible patchy invasion in a predator–prey system. *Ecol. Lett.* 5, 345–352.
126. Petrovskii, S. V., Vinogradov, M. E., Morozov, A. Y. 2002b. Formation of the patchiness in the plankton horizontal distribution due to biological invasion in a two-species model with account for the Allee effect. *Oceanology* 42(3), 363–372.
127. Pianka, E. R. 1981. Competition and niche theory. In: *Theoretical Ecology, Principles and Applications*, (Ed. R. M. May), pp. 167–196. Blackwell Scientific, Oxford.
128. Pielou, E. C. 1977. *Mathematical Ecology: An Introduction*. Wiley Interscience, New York.
129. Pradhan, T., Chaudhuri, K. S. 1998. Bioeconomic modeling of a single species fishery with Gompertz law of growth. *J. Biol. Systems* 6(4), 393–409.
130. Rai, V., Upadhyay, R. K. 2004. Chaotic population dynamics and biology of the top-predator. *Chaos, Solitons Fractals* 21, 1195–1204.
131. Ricker, W. E. 1954. Stock and recruitment. *J. Fishery Res. Board of Canada* 11, 559–623.

132. Rosenzweig, M. L. 1971. Paradox of enrichment: Destabilization of exploitation ecosystems in ecological time. *Science* 171, 385–387.
133. Rosenzweig, M. L., MacArthur, R. H. 1963. Graphical representation and stability conditions of predator–prey interactions. *Am. Naturalist* 97, 209–223.
134. Ruan, S., Xiao, D. 2001. Global analysis in a predator–prey system with non-monotonic functional response. *SIAM J. Appl. Math.* 61(4), 1445–1472.
135. Scheffer, M. 1991a. Fish and nutrients interplay determines algal biomass: A minimal model. *Oikos* 62, 271–282.
136. Scheffer, M. 1991b. Should we expect strange attractors behind plankton dynamics—And if so, should we bother? *J. Plankton Res.* 13, 1291–1305.
137. Schaffer, W. M., Kot, M. 1986. Chaos in ecological systems: The coals that Newcastle forgot. *Trends Ecol. Evol.* 1(3), 58–63.
138. Sharov, A. A Quantitative Population Ecology-on-linelectures, http://home. comcast.net/~sharov/PopEcol/lec9/stabil.html
139. Sherratt, J. A. 2001. Periodic travelling waves in cyclic predator–prey systems. *Ecol. Lett.* 4, 30–37.
140. Skalski, G. T., Gilliam, J. F. 2001. Functional responses with predator interference: Viable alternatives to the Holling type II model. *Ecology* 82, 3083–3092.
141. Smolen, P., Baxter, D., Byrne, J. 2002. A reduced model clarifies the role of feedback loops and time delays in the *Drosophila* circadian oscillator. *Biophys. J.* 83, 2349–2359.
142. Sokol, W., Howell, J. A. 1980. Kinetics of phenol oxidation by washed cells. *Biotechnol. Bioeng.* 23, 2039–2049.
143. Stephens, P. A., Sutherland, W. J. 1999. Consequences of Allee effect for behavior, ecology and conservation. *TREE* 14(10), 401–405.
144. Stephens, P. A., Sutherland, W. J., Freckleton, R. P. 1999. What is Allee effect? *Oikos* 87, 185–190,
145. Strogatz, S. H. 1994. *Nonlinear Dynamics and Chaos.* Addison-Wesley Publishing House, New York.
146. Svirezhev, Y. M., Logofet, D. O. 1983. *Stability of Biological Communities.* Mir, Moscow, Russia.
147. Tanner, J. T. 1975. The stability and intrinsic growth rates of prey and predator populations. *Ecology* 56, 855–867.
148. Tikhonova, I., Li, B.-L., Malchow, H., Medvinsky, A. 2003. The impact of the phytoplankton growth rate on spatial and temporal dynamics of plankton communities in a heterogeneous environment. *Biofizika* 48, 891–899.
149. Truscott, J. E., Brindley, J. 1994. Ocean plankton populations as excitable media. *Bull. Math. Biol.* 56, 981–998.
150. Turchin, P. 2001. Does population ecology have general law? *Oikos* 94, 17–26.
151. Turchin, P. 2003a. *Complex Population Dynamics: A Theoretical/Empirical Synthesis.* Princeton University Press, Princeton, NJ.
152. Turchin, P. 2003b. *Historical Dynamics: Why States Rise and Fall.* Princeton University Press, Princeton.
153. Turchin, P., Ellner, S. P. 2000. Living on the edge of chaos: Population dynamics of Fennoscandian voles. *Ecology* 81(11), 3099–3116.
154. Turchin, P., Hanski, I. 1997. An empirically based model for the latitudinal gradient in vole population dynamics. *Am. Naturalist* 149, 842–874.
155. Upadhyay, R. K. 1999. *Dynamical System Studies in Model Ecosystems.* PhD thesis, IIT Delhi.

156. Upadhyay, R. K. 2009. Observability of chaos and cycles in ecological systems: Lessons from predator–prey models. *Int. J. Bif. Chaos* 19(10), 3169–3234.
157. Upadhyay, R. K., Kumari, N., Rai, V. 2008. Wave of chaos and pattern formation in a spatial predator–prey system with Holling type IV functional response. *Math. Model. Nat. Phenomena* 3(4), 71–95.
158. Upadhyay, R. K., Kumari, N., Rai, V. 2009. Wave of chaos in a diffusive system: Generating realistic patterns of patchiness in plankton-fish dynamics. *Chaos, Solitons Fractals* 40, 262–276.
159. Van Gemerden, H. 1974. Coexistence of organisms competing for the same substrate: An example among the purple sulfur bacteria. *Microb. Ecol.* 1, 104–119.
160. Verhulst, P.-F. 1838. Notice sur la loi que la population suit dans son accroissement. *Corr. Math. Phys.* 10, 113–121.
161. Vielle, B., Chauvet, G. 1998. Delay equation analysis of human respiratory stability. *Math. Biosci.* 152(2), 105–122.
162. Villasana, M., Radunskaya, A. 2003. A delay differential equation model for tumor growth. *J. Math. Biol.* 47(3), 270–294.
163. Volterra, V. 1927. Varia zioni e fluttu a zioni del numero d'indidui in specie animali conviventi. *Memoria R. Comita to talassografica Italiano.* 131, 1–142.
164. Volterra, V. 1931. *Lecons sur la theorie mathematique de la lute pour la vie.* Gauthiers-Vilars, Paris.
165. Wang, J., Wang, Ke 2004. Optimal control of harvesting for single population. *Appl. Math. Comput.* 156, 235–247.
166. Waltman, P. 1984. *Competition Models in Population Biology.* CMBS Lectures 45, SIAM Publications, Philadelphia.
167. Wheldon, T. E. 1988. *Mathematical Models in Cancer Research.* Adam Hilger, Bristol.
168. Winsor, C. P. 1932. The Gompertz curve as a growth curve. *Proc. Natl. Acad. Sci. USA* 18, 1–8.
169. Yodzis, P. 1989. *Introduction to Theoretical Ecology.* Harper & Row, New York.
170. Zeng, Z. 2011. Asymptotically periodic solution and optimal harvesting policy for Gompertz system. *Nonlinear Anal. Real World Appl.* 12(3), 1401–1409.
171. Zhao, T. 1995. Global periodic solutions for a differential delay system modeling a microbial population in the chemostat. *J. Math. Anal. Appl.* 193, 329–352.
172. Zhou, S.-R., Liu, C.-Z., Wang, G. 2004. The competitive dynamics of metapopulations subject to the Allee-like effect. *Theor. Popul. Biol.* 65, 29–37.
173. Zhou, S.-R., Liu, Y.-F., Wang, G. 2005. The stability of predator–prey systems subjected to the Allee effects. *Theor. Popul. Biol.* 67, 23–31.
174. Zhou, S.-R., Wang, G. 2004. Allee like effects in metapopulation dynamics. *Math. Biosci.* 189, 103–113.

3

Introduction to Chaotic Dynamics

Introduction

The concept of chaos is one of the major discoveries of recent times. Chaos is observed in day-to-day life, in the areas of metrology, fluid flow (turbulence), cardiology, population biology, the stock market, economic modeling, and so on. Predictions are possible in many systems. For example, eclipses can be predicted thousands of years in advance. The motion of a simple pendulum (under suitable assumptions), which is governed by a differential equation, is predictable and a closed-form solution can be written. However, there are many natural phenomena which are not predictable, such as weather predictions, roll of dice, and so forth, even though they obey the same laws of physics.

It is believed until recently that predictability can be achieved by having more information and processing it. Laplace (1776) was quoted of having said that "if we can conceive of an intelligence which at a given instant comprehends all the relations of the entities of this universe, it could state the respective positions, motions and general affects of all these entities at any time in the past or future" [47a]. Poincaré (1903) cautioned against this approach. He was quoted of having said that "even if it were the case that the natural laws are no longer any secret for us, we could still only know the initial situation approximately. It enabled us to predict the succeeding situation with the same approximation. But, it is not always so. It may happen that a small difference in the initial conditions produces very great difference in the final phenomena. A small error in the former may produce an enormous error in the latter" [75a]. Simple deterministic systems can exhibit random behavior—it will not go by collecting more information.

Chaos and Chaotic Dynamics

A map, $x_{n+1} = g(\alpha, x_n)$, is a system that is discrete in time.

A flow is a dynamical system that is continuous in time. Consider an autonomous vector field defined on $C^r(r \geq 1)$

$$x' = f(\mu, x), \quad \text{where } \mu \text{ is a parameter.} \tag{3.1}$$

Denote the flow generated by the vector field (3.1) by $\phi(x, t)$. The flow is said to have sensitive dependence on initial conditions if two trajectories that are starting very close together rapidly diverge from each other, and thereafter have totally different features. Strogatz [103] suggested that the rate of divergence of nearby trajectories should be exponential. The exponential divergence of adjacent phase points has the implication that long-term prediction becomes impossible in a system where small uncertainties are amplified enormously. Even if there are careful measurements and calculations, the system is not predictable in the long term. The nonperiodic solutions of Equation 3.1 correspond to strange chaotic attractors (SCAs) of complex geometric structure. They have at least one positive Lyapunov exponent.

In principle, a 2D map can represent a 3D flow.

Wiggins [116] defined an attractor and a chaotic compact set of a map as follows:

Definition 3.1

Let f be a map from n-dimensional space to itself. A compact set Ω in the n-dimensional space is an *attractor* for f if it satisfies the following properties [116]:

  a. *Invariance:* $f(\Omega) = \Omega$, (Ω is an invariant).
  b. *Density:* Ω contains an initial point from which trajectories travel throughout Ω (the flow $\phi(x, t)$ is dense in Ω).
  c. *Stability and attraction:* Initial points starting close to Ω have trajectories that stay close to Ω and tend asymptotically to Ω.

Thus, an attractor is a topologically transitive attracting set [116], that is, the system ultimately settles down or gets attracted to the attractor.

The presence of a stable focus in the phase space of a dynamical system suggests that the system trajectories would tend to an equilibrium state as $t \to \infty$. In the case of limit-cycle solutions, this equilibrium point is not a point, but oscillates persistently between upper and lower limits.

For dissipative dynamical systems, the state of the system evolves on an invariant geometrical object known as *attractor* whose dimension is less than the dimension of the original state space. Attractors are geometric forms in the phase space to which the phase trajectories of the dynamical system converge or are attracted and on which they eventually settle down. This happens independently of the initial conditions. For example, for a 3D dissipative dynamical system involving three first-order ODEs, only three distinct possibilities of attractors are of physical interest. They are represented by a stable focus, a stable limit cycle, and an SCA. In nonlinear systems, these possibilities very often coexist. The simplest attractor is a stable fixed point. The strange attractor was discovered and introduced by David Ruelle and Floris Takens [83] as the mathematical image of chaos in a dynamical system. They connected turbulent behavior with the chaotic dynamics. Thus, attractors having noninteger dimensions are called strange attractors [4].

Definition 3.2

A compact set Ω is said to be *chaotic* if [116]

    a. $\phi(t, x)$ has SIC (sensitive dependence on initial conditions) on Ω.

    b. $\phi(t, x)$ is topologically transitive on Ω.

    c. The periodic orbits of $\phi(t, x)$ are dense in Ω.

A dynamical system displaying sensitive dependence on initial conditions on a closed invariant set (which consists of more than one orbit) is called chaotic. Suppose Ω is an attractor. It is called a strange attractor if it is chaotic. Thus, we say that if the system trajectory meanders in a bounded phase space of finite volume, then the corresponding attractor is known as a strange chaotic attractor. The dynamics of an SCA are complex and the system wanders off among different states on a surface of complex geometry, but enclosing finite volume of the phase space. In other words, the future state of the system is not solely decided by its initial state and by evolution law in a deterministic manner, but is very sensitive to any error which invariably occurs in specifying the initial state. There is complete loss of predictability of the future state of the system after a finite amount of time has elapsed. This time is inversely proportional to the maximum Lyapunov exponent that quantifies the sensitivity of the system's asymptotic dynamics with respect to the initial state.

Chaotic attractors have trajectories that never repeat and exhibit erratic fluctuations despite being deterministic. The attractors are confined to a finite volume of the phase space, and there is a fine-scale structure that is fractal and is visible at all magnifications. Thus, attractors do not fill the volume in the phase space, but have a noninteger (fractal) dimension. The system's trajectories move on an SCA in the phase space of the system. The geometry of this attractor guarantees exponential divergence of nearby trajectories [36]. A dynamical behavior is robust if it exists in a large region of a 2D parameter space. Nonchaotic modes of dynamical behavior always exists preceding chaos like stable focus, stable limit cycle, or quasiperiodicity (found in 4D dissipative dynamical systems or in a 3D-driven dissipative system). It is possible that one of these dynamical behaviors changes into any other when a system parameter is varied. The inertia of a system for such a change in temporal dynamics measures its stability. A model system is said to be stable if it displays a particular type of attractor in the phase space in a wide range of parametric regimes. If the dynamics do not truncate and go to newer and newer places, but never go beyond the region specified by the vector field, then we say that the system is chaotic. If the two time histories (generated by integrating the system of differential equations and recorded after the transients have died out) corresponding to two different, but very close, initial conditions overlap each other completely, the absence of dynamical chaos at that particular set of parameter values is inferred and it is taken to be the case of cyclic dynamics (stable limit cycle in phase space). In the case when two time histories appear to be quite different and intersect each other, then it confirms that the dynamics of the system is chaotic for that particular set of values of parameters. This may be a simple and efficient way of detecting chaos in dynamical systems. This approach enables us to study the two forms of chaos in dissipative dynamical systems—dynamical chaos and intermittent chaos [114]. However, the method is not helpful in distinguishing the long-lived chaotic transients and chaotic dynamics on a chaotic attractor [78].

Definition 3.3

A closed invariant set A is called *topologically transitive* if, for any two open sets $U, V \subset A$, $\exists\, t \in \mathbb{R} \ni \phi(t, U) \cap V \neq \varnothing$.

Remark 3.1

For a flow, chaotic regimes may exist if there are three degrees of freedom. The corresponding Poincaré section has two dimensions and therefore the 2D invertible map may also exhibit chaos.

An algorithm for testing the existence of a strange attractor for a dynamical system was given by Wiggins [116]. Strogatz [103] defines chaos as an aperiodic long-term behavior in a deterministic system that exhibits SIC. Yorke [120] describes chaos as a mathematical concept in nonlinear dynamics for systems that vary according to precise deterministic laws, but appear to behave in random fashion.

Using topology and differential geometry, Ruelle and Taken [83] showed that the Landau scenario to turbulence can actually be truncated to only three stages of instability, which is sufficient to drive the system into turbulence.

Chaotic dynamics provides an introduction to chaotic phenomena, based on geometrical interpretation and simple arguments, without in-depth scientific and mathematical knowledge [104]. This phenomenon has been successfully used to device a better pacemaker for the heart and it may also find a place in defense as a means of secure communication.

Chaos theory studies the behavior of dynamical systems that are highly sensitive to initial conditions, an effect that is popularly referred to as the *butterfly effect,* the notion that a butterfly flapping its wings at any particular place may set off a tornado in a faraway place a few days later [51]. The deterministic nature of these systems does not make them predictable [8].

Basin of Attraction

A system may have several attractors, sometimes infinitely many. Different initial conditions may evolve into different attractors. The set of points that evolves into an attractor is called the basin of the attractor. It is a common occurrence that two or more dynamical possibilities coexist at the same set of parameter values. Every attractor has a bowl of initial conditions such that the trajectories land up on the attractor if we start from any of the initial conditions from the bowl. This bowl is known as its basin of attraction. The boundary between two basins of attraction is called a *separatrix*. Basins of attraction are separated by simple curves. This curve goes through an unstable fixed point. The initial conditions on the basin boundary generate trajectories that eventually approach to an unstable fixed point, that is, the basin boundary is the stable manifold of an unstable fixed point. The basin can stretch to infinity. For example, the pendulum clock has two basins of attractors. If it is given small displacements, then the pendulum comes to rest after some time, that is, the attractor is a fixed point. The set of all these points is the first basin of attractor. If it is given sufficiently large displacements, then the pendulum clock sets in motion and finally reaches stable oscillations. The attractor in this case is a limit cycle. The set of all these points is the second basin of attractor.

The boundary of a basin of attraction may be fractal. There can be hundreds of coexisting attractors with fractal basin boundaries. The existence of fractal basin boundaries shows that even a system with periodic response can show a sensitive dependence on the initial conditions in the sense that trajectories of nearby initial conditions converge to different periodic attractors [40]. Systems with multiple attractors are said to be *multistable*. Since chaos usually requires a 3D state space, the basin of a strange attractor is the volume inside which the attractor resides.

Primary Routes to Study Chaos

The following are the five primary routes to study chaos in dissipative dynamical systems.

a. Period-doubling route [68]
b. Ruelle–Takens–Newhouse route or quasiperiodicity route [19,67]

c. Intermittency transition route [76]
d. Crisis route [28,29,102]
e. The route to chaos in quasiperiodically driven systems [30]

We discuss briefly the five routes leading to chaos.

a. The period-doubling route is most commonly used to study the existence of chaos in dissipative dynamical systems. An SCA presents itself in the phase space of the system after an accumulation point in the crucial parameter is crossed. Successive bifurcation points are characterized by the presence of new oscillations with frequencies half of the fundamental frequency in the power spectrum. The numerical noise does not generally allow us to observe period doublings beyond $p8$ or $p16$. This route assumes a one-parametric analysis.

Let us consider the logistic map $x_{n+1} = f(x,\mu) = 1 - \mu\, x_n^2$, $x \in (-1, 1)$, $\mu \in (0, 2)$. The two fixed points of the map are $x_1^* = (-1 + \sqrt{1 + 4\mu})/(2\mu)$, and $x_2^* = (-1 - \sqrt{1 + 4\mu})/(2\mu)$. We find, $|f'(\mu, x_1^*)| = |-2\mu\, x_1^*| = |1 - \sqrt{1 + 4\mu}| < 1$, for $\mu < 0.75$. Hence, x_1^* is stable for $\mu < 0.75$. Now, $|f'(\mu, x_2^*)| = |-2\mu\, x_2^*| = |1 + \sqrt{1 + 4\mu}| > 1$, for $\mu \in (0, 2)$. Hence, x_2^* is unstable for $\mu \in (0, 2)$. For $\mu = 0.75$, we have $(\partial f/\partial x) = -1$, at $x = x_1^*$. We find that the map enters into a 2-cycle map, that is, the period has doubled from 1 to 2 fixed points, where $x_2^* = 1 - \mu\, x_1^*$, $x_1^* = 1 - \mu\, x_2^*$. The bifurcation from a fixed point to a 2^1-*cycle* occurs at $\mu_1 = 0.75$, the bifurcation from the 2^1-*cycle* to a 2^2-*cycle* takes place at $\mu_2 = 1.25$, and so on. This process leads to a sequence of periods 2^n, $n = 0, 1, 2, \ldots, \infty$. The stability interval of each period diminishes quickly with increasing n. Infinitely many periods accumulate at the limiting value $\mu_\infty = 1.401155189$. Beyond μ_∞, the system enters a chaotic regime. A chaotic attractor appears in a parameter region immediately following the accumulation of an infinite number of period doublings [32]. The period-doubling phenomena have been observed in systems from many scientific disciplines ranging from the simplest maps to distributed systems. Chaotic behavior is interrupted by a *window of periodicity* in the chaotic regime. Periodic windows can be eliminated by stretching the map everywhere. However, in such models, periodic windows disappear together with a period-doubling cascade. The theory of the period-doubling route to chaos was developed by Feigenbaum on the basis of 1D maps [21,22] and the bifurcation mechanism is known as the Feigenbaum scenario. The Feigenbaum scenario is universal and its universality manifests itself in the behavior of spectral amplitudes of subharmonics that appear with each period doubling [3]. In the context of ecology, the first such effort was made by Gardini et al. [25]. Rai and Sreenivasan [80] observed the period-doubling bifurcation leading to chaos in a model food chain.

b. The second route to chaos is not applicable for 3D dissipative dynamical systems. It is applicable to 4D dissipative systems or 3D-driven dissipative systems. An SCA is observed after three successive Hopf bifurcations, which give rise to a stable limit cycle and tori T^2 and T^3 with two and three incommensurate frequencies. In this route to chaos, the following sequential behavior is observed when the bifurcation parameter is changed: A stationary point bifurcates to a periodic orbit, which then bifurcates to a doubly periodic orbit formed by the surface of a torus, which then bifurcates to a system with chaotic behavior. According to this route, a 2D torus T^2 in the phase space is destroyed, and the trajectories fall in a set with fractal dimension $2 + d$, $d < 1$. This set is created in the vicinity of T^2 and thus called a *torus-chaos* [92]. This route may be thought of as a special case of the quasiperiodic transition to chaos and it requires a two-parameter analysis. The transition from quasiperiodicity to chaos occurs after the birth of a third frequency when the unstable and chaotic trajectories appear on a 3D torus [65,83]. This scenario is ruled out for the Lorenz model since doubly periodic orbits are not possible in an autonomous system of three first-order ODEs with phase space contraction, that is, *div* $\mathbf{V} < 0$. Numerical investigations using this route to chaos were carried out by Curry [13] for an autonomous system of 14 first-order differential equations and by Curry and Yorke [14] for a 2D map.

c. The intermittency route is characterized by the presence of chaotic motion interrupted by regular (periodic) motion after equal intervals. A famous example for this route to chaos is the Lorenz system [68]. In the intermittency route, as a parameter passes through a critical value, a simple periodic orbit is replaced by a chaotic attractor in such a way that the chaotic behavior is interspersed with a cyclic-like behavior in an intermittent fashion. The abrupt transition to chaos and the intermittency phenomenon was first considered by Pomeau and Manneville [76]. The corresponding bifurcation mechanism of the onset of chaos is called the *Pomeau–Manneville scenario* [92]. Intermittency is a well-known phenomenon both in experimental data and in the numerical solution of dynamical systems. It is related to the tangent bifurcation of cycles and is most typical for a variety of dynamical systems [76]. This phenomenon was discovered and studied much earlier than other types of intermittency and is called *type-I intermittency*. Other types of intermittency are associated with a subcritical Andronov–Hopf bifurcation in the Poincaré section and to a subcritical period doubling. They are called *type-II* and *type-III* intermittency, respectively [7]. Long phases of regular behavior are interrupted at seemingly random times by short regular bursts. Laminar (regular) periods occur in the flow as trajectories intersect the return plane several times near to where the periodic orbits used to exist. Intermittency brings in unexpected richness to the logistic map in the so-called chaotic regime.

d. In the fourth route to chaos, a chaotic attractor is suddenly created to replace a nonattracting chaotic saddle as the parameter passes through the crisis value [29]. In a crisis (a phenomenon of sudden destruction or expansion of a chaotic attractor that is defined as a sudden discontinuous change in a chaotic attractor as a system parameter is varied), a chaotic transient gets converted into a chaotic attractor. For example, a typical way in which this happens can be found in the case of the map $x_{n+1} = \mu - x_n^2$. A periodic-3 window exists in the range $1.72 < \mu < 1.82$. The window begins with a saddle-node bifurcation at $\mu = \mu_0 = 1.75$, from which the stable and unstable period-3 orbits are born. A stable period-3 orbit goes through a period-doubling cascade and becomes chaotic as μ is increased. The end of the window occurs at the crisis point, $\mu = \mu_0 = 1.79$, when the unstable period-3 orbit created at the original saddle-node bifurcation at $\mu = \mu_0$ collides with the three-piece chaotic attractor. A good review of the crisis route was given by Grebogi [26]. The third and fourth routes to chaos have not been investigated to any appreciable extent in ecology.

e. In quasiperiodically driven systems, the route to chaos can be schematically represented as (three frequency quasiperiodicity) → (strange nonchaotic behavior) → (chaos). This route to chaos differs from the one in periodically forced systems or in 4D dissipative systems in the way that the two-frequency torus is broken before the creation of a three-frequency torus. The three-frequency torus appears after the strange nonchaotic attractor (SNA) and before the chaotic attractor appears. SNAs appear to be an integral part of the creation of chaotic dynamics in quasiperiodically forced systems. An SNA is defined as an attracting limit set of a dynamical system that is not a manifold and for which there is no exponential divergence of phase trajectories. Numerous studies of maps and flows with quasiperiodic forcing have shown the appearance of SNAs [27]. This route is relevant to many physical and biological situations.

Types of Chaos, Transients, and Attractors

Li and Yorke [48] had demonstrated that the dynamics of a map as simple as a difference map (which maps an interval on to itself) could shuttle between and through three points. The authors proved that it implied chaos. Later, May [59] discovered that simple rules can generate complex dynamical patterns including chaos.

Mathematical models of ecological systems based on difference and differential equations have been designed and studied since the 1970s. Iterations and simulation experiments of these mathematical models have suggested that deterministic chaos could be a representation of temporal dynamics of these ecological systems. However, unambiguous evidence of deterministic chaos from a natural ecological system was not available. The detection of chaotic dynamics in any of the mathematical models does not necessarily imply the existence of the same in the real-world ecological models. One of the reasons why chaos is not observed in natural ecological systems may be that the observed system is not an isolated one and it constantly interacts with its environment. It is not known how the environment influences the dynamics of the observed system, which is always nonlinear. Another reason why mathematical and ecological chaos may be two different and distinct entities is that in chaotic dynamical (mathematical) systems, SIC exists, which means exponential amplification of the error occurs either due to the limitations of the equipment or insufficient expertise of solution procedures. If we model the environmental fluctuations as "noise," then this amplification would not follow a uniform distribution. The trajectory of the model system does not represent the trajectory of the actual system it represents.

The pseudo-prey method (see Example 3.1) helps one to prepare sets of biologically realistic parameter values to perform simulation experiments. These experiments provide an idea about the dominant mode of the dynamics of the model system. It is necessary to find suitable values of the parameters for these experiments. For example, logistic and Ricker maps yield chaotic solutions for biologically unrealistic values [79].

There are two types of dynamics: nonchaotic and chaotic. Nonchaotic dynamics consists of stable focus, stable limit cycles, and quasiperiodicity. Observable chaos can be divided into two categories: robust chaos and short-term recurrent chaos (STRC).

Robust chaos: If chaotic dynamics exists in a sufficiently large region in a 2D parameter space spanned by two crucial parameters of a model system and also if the basin of attraction for the associated attractor is large, then we say that the chaotic dynamics displayed by the system is robust. Further, the attractor's basin should have only smooth boundaries. In this case, chaos would be a dominant dynamical mode of the corresponding system. Unless the chaotic behavior is robust, there is no possibility that the same may be observed in a laboratory experiment or it can be captured in a natural setting. An ecological model displaying chaotic dynamics at an isolated point in a parameter space has no meaning.

Short-term recurrent chaos: STRC is a form of chaotic behavior, wherein it is interrupted by other kinds of dynamics at irregular and unpredictable intervals. It can also result because of the coexistence of regular and chaotic attractors at the same set of parametric values. If chaotic dynamics is displayed in a region in a 2D parameter space of a model system, it can support STRC provided the basins of attraction of different attractors is intermixed. STRC is characterized by chaotic bursts repeated at unpredictable intervals. A possible reason for the termination of chaotic behavior is the existence of chaos at discrete parameter values or with narrow parameter ranges. It can be caused either by deterministic changes in the system's parameters or by exogenous stochastic influences (abrupt changes in initial conditions). As chaos (exponential sensitivity of system's trajectories on initial conditions) takes time (inversely proportional to the maximum Lyapunov exponent of the system) to develop, the system gets time to lock itself on to a nonchaotic attractor before it settles into a fully developed chaotic state. It may be noted that chaotic attractors without extinction-sized densities could only be used to explain STRC. It was shown by Medvinsky et al. [62], that chaotic behavior might occur in a wide range of parameter values in diffusively coupled predator–prey systems. They have also shown that the basin boundaries of these regular and chaotic attractors can be fractal. In such a case, the dynamical system under the influence of environmental fluctuations can display STRC. This phenomenon was observed in the vole population in Northern Fennoscandia. Such behavior was also found in Dungeness crab [37], larval fish [18], laboratory population of flour beetles [12], and blowflies [95].

Edge of chaos: The phrase "Edge of chaos" was first used in 1990 by Chris Langton [47], which has been a defining concept in the study of complex systems. This phrase was introduced in biological systems by Kaufmann [41]. Later, it was studied in ecology by Rai [77] and Rai and Upadhyay [81]. Edge of chaos can be defined as the phenomenon when "complexity is at its maximum just where order transitions into chaos" [47]. The classical examples of complex systems have some mixture of characteristics of order (recognizable patterns, coherence over time) and characteristics of chaos (sensitive dependence on initial conditions, seemingly infinite variety of behavior). As mentioned earlier, as chaos takes time to develop, the system gets time to lock itself onto a nonchaotic attractor before it settles into a fully developed chaotic state. These dynamical transitions are caused by changes in system parameters. The chaotic dynamics are frequently interrupted by a purely deterministic cause, which is intrinsic to the system.

Wave of chaos (WOC): It can be defined as the dynamics in which for each moment of time there exist distinct boundaries or interfaces separating the regions of chaotic and regular patterns [73,110,111]. The phenomenon is essentially spatio-temporal. Chaos prevails over the regular pattern by displacing it. In short, WOC can be defined as the moving interface between the two regions, chaotic and regular. This phenomenon has been observed in models of aquatic systems with Holling type II [110] and Holling type IV [111] functional response functions and in a community of competing species [57,70].

Chaos in natural populations: The first attempt to observe chaos in natural populations was made by Schaffer and Kot [88]. There has been evidence of chaotic dynamics in an experimental flour beetle population *Tribolium castaneum* [11,12]. The observation of chaos in this cannibalistic population was based on visual inspection. Cannibalism is an extremely potent feedback mechanism, which artificially brings in elements of chaotic dynamics to the system. The comprehensive search for ecological chaos was carried out by Ellner and Turchin [20] who used three different Lyapunov exponent estimators (the number of Lyapunov exponents is equal to the dimensionality of the phase space) to analyze a large collection of empirical data. The authors concluded that ecological chaos is not to be expected in the wild. It was suggested that evolution tends to preserve populations from such chaotic behavior. Ferriere and Gatto [23] suggested that chaos could be an optimal behavior for various ecological systems.

The concept of chaos originally came into existence in the context of temporal dynamics of a spatially homogeneous system. The consideration of space in the model system makes the dynamics more complex and creates the possibility of getting chaos even in those cases where it is almost impossible otherwise. This phenomenon is termed as spatiotemporal chaos, to distinguish it from the purely temporal chaos of a homogeneous system. In spatiotemporal chaos, dynamics is disordered in both time and space for a large homogeneous system. To study the effects of space and time on the interacting species, Jansen [38] had extended the scope of the simple Lotka–Volterra (LV) system and Rosenzweig–MacArthur (RM) model to a patchy environment. The spatial interaction can bind the fluctuations of a predator–prey system and regulate the abundance of the populations [38]. Pascual [69] observed that spatial distribution of nutrients along a linear gradient can induce temporal chaos. Spatiotemporal chaos in population dynamics was observed by many authors [5,11,12,15,16,31,50,105,106,115,119]. Medvinsky et al. [61] extensively reviewed the spatiotemporal pattern formation in plankton dynamics and found that modeling by reaction–diffusion equations was the most appropriate tool for investigation of complex spatiotemporal plankton dynamics. Conceptual prey–predator models have successfully been used to (i) model phytoplankton–zooplankton interactions, (ii) study the mechanisms of spatiotemporal pattern formation like patchiness and blooming [54,69,86,90,99], and (iii) study different routes to local and spatiotemporal chaos [45,55,56,71 –73,82,87,89,91,100,108].

Chaos was generated in a mathematical model by some authors by using the following approaches: (a) coupling two nonlinear oscillators [77,114], (b) incorporating the seasonal variation as a model parameter [82,85,109], and (c) incorporating the effect of space in an oscillatory predator–prey system [61,63,69,71]. Ecological systems are classified as weakly dissipative systems. The factors that determine the parameter region in the 2D space is the strength of this dissipation, which, in turn, depends upon values of the chosen system parameters. It would also depend on which of the three aforementioned mechanisms operates to generate chaos in a given case.

Transients and attractors: Long-lived chaotic transients have been discovered in many ecological models [33,34,93]. The lifespan of these transients increases many fold in spatially extended systems. These have been termed as super transients [35,79]. In STRC, the interruptions in chaotic dynamics are caused either by changes in system parameters or changes in initial conditions [77]. The terms, transient chaos and chaotic transients [24], are used as synonyms in this book. In dissipative systems, transient chaos appears primarily in the dynamics of approaching the attractor(s). It is therefore also called the chaotic transient. Tél and Gruiz [104] describe it as the motion starting near a fractal boundary, which remains irregular for a while exhibiting transient chaos (lasting for a finite period of time), but ultimately ends up on one of the attractors. Existence of transient chaos in an ecological system may present an opportunity for humans to intervene and help in reducing the risk of species extinction. The control scheme suggested by Shulenburger et al. [93] involves the application of feedback perturbations to the populations of an endangered species at appropriate, but rare times. An advantage of this scheme over other such schemes is that it does not require accurate information about the dynamics, such as the Jacobian matrices, the stable and unstable eigenvalues associated with target states. The control technique is robust to environmental noise. The phase space can be reconstructed even if data on one of the interacting species are available [106]. Edward Spiegel refers to transient chaos as "pandemonium" [96]. Transient chaos can be viewed as a situation in which an attractor touches its basin of attraction, but only at places that are rarely visited by the trajectory. The trajectory is initially drawn to the attractor and wanders around on it for a long time before eventually coming to a place outside the basin of attraction, from where it escapes [98].

An attractor is a bounded region of phase space of a dynamical system towards which the nearby trajectories approach asymptotically. The initial part of the trajectory approaching the attractor is called a transient. A trajectory which starts directly on the attractor does not have any transient. Attractors are of the following types: regular, hyperbolic, quasihyperbolic, and nonhyperbolic attractors. The point attractor, limit-cycle attractor, and a torus are regular attractors. A system is hyperbolic if all of its phase trajectories are saddle ones. A point as the image of a trajectory in the Poincaré section is always a saddle. Hyperbolic attractors must satisfy three conditions [1], which are not fulfilled by real dynamical systems. The Smale–Williams attractor [117] and Plykin attractor [74] are hyperbolic attractors. Quasihyperbolic attractors violate at least one of the three hyperbolicity conditions and are not structurally stable. Lozi, Belykh, and Lorenz attractors are quasihyperbolic attractors. Systems with nonhyperbolic attractors exhibit regimes of deterministic chaos, which are characterized by exponential instability of phase trajectories and a fractal structure of the attractor.

Further, nonhyperbolic attractors can coexist with a set (countable) of different chaotic and regular attracting subsets in a bounded phase volume space [3]. Strange attractors are always robust hyperbolic limit sets. SCAs are characterized by exponential instability of the phase trajectory on the attractor.

Methods of Investigation for Detecting Chaos

There are a number of ways to detect chaos in dynamical systems. In this section, we present a method for finding possible values of the parameters for numerical simulations and for the dynamical study of three and four species ecosystems. The parameter regimes displaying different dynamical possibilities can be identified. Time-series analysis, phase-space representations, 2D scans, bifurcation diagrams, and basin boundary calculations are used in our studies.

Method for Selection of Parameter Values

The methodology to select parameter values for simulation experiments is based on the dynamical representation of a given model. For example, consider the 4D model system given below, obtained by coupling the RM model with the Holling–Tanner (HT) model, which is schematically given in Figure 3.1.

$$\frac{dX}{dt} = a_1 X - b_1 X^2 - \frac{wXY}{X + D}, \tag{3.2}$$

$$\frac{dY}{dt} = -a_2 Y + \frac{w_1 XY}{X + D_1} - \frac{w_2 Y^2 U}{Y^2 + D_2^2}, \tag{3.3}$$

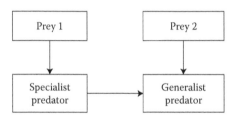

FIGURE 3.1
Coupling of RM and HT models.

$$\frac{dZ}{dt} = AZ\left(1 - \frac{Z}{K_1}\right) - \frac{w_3 UZ}{Z + D_3}, \tag{3.4}$$

$$\frac{dU}{dt} = cU - \frac{w_4 U^2}{Y + Z}. \tag{3.5}$$

The methodology is based on the idea of decoupling the model system into two 2D subsystems. The first subsystem is obtained by dropping the last term in Equation 3.3, that is, we assume that the generalist predator (U) is absent from the system.

Subsystem I is, therefore, defined by the equations

$$\frac{dX}{dt} = a_1 X - b_1 X^2 - \frac{wXY}{X + D}, \quad \frac{dY}{dt} = -a_2 Y + \frac{w_1 XY}{X + D_1}. \tag{3.6}$$

Subsystem II is obtained by breaking the link between the specialist predator and generalist predator. The resulting system is given in Figure 3.2.

The corresponding evolution equations for subsystem II are given by

$$\frac{dZ}{dt} = AZ\left(1 - \frac{Z}{K_1}\right) - \frac{w_3 UZ}{Z + D_3}, \quad \frac{dU}{dt} = cU - \frac{w_4 U^2}{Z}. \tag{3.7}$$

Each of the two subsystems can be written in the form

$$\frac{dX}{dt} = XF(X, Y), \quad \frac{dY}{dt} = YG(X, Y), \tag{3.8}$$

where F and G are analytic functions in the domain $X \geq 0$, $Y \geq 0$.

We now require that both the subsystems should qualify as a Kolmogorov system. According to the Kolmogorov theorem, if a 2D system satisfies all the nine conditions, then it admits either a stable limit cycle or a stable

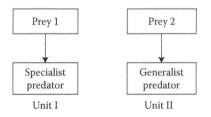

FIGURE 3.2
Breaking of link between the specialist and generalist predator.

equilibrium point solution. From the linear stability theory, we know that the conditions on the parametric values can be obtained in a manner such that the stable equilibrium point solutions are observed in the phase space of the system. If the parametric values are chosen in such a way that the conditions of Kolmogorov theorem are satisfied and the conditions of linear stability are violated, then the system yields stable limit-cycle solutions in the phase space.

For subsystem I, the Kolmogorov constraints give rise to the following conditions (see Problem 3, Exercise 2.2)

$$w_1 > a_2, \quad \frac{a_1}{b_1} > \frac{D_1 a_2}{(w_1 - a_2)}. \tag{3.9}$$

The local stability condition is given by

$$2b_1 \left(\frac{D_1 a_2}{w_1 - a_2} \right) + b_1 D - a_1 > 0 \quad \text{(see Problem 4, Exercise 2.2).} \tag{3.10}$$

Violation of this condition produces a stable limit cycle in the phase space. This stable limit cycle signifies persistent periodic oscillations whose amplitude becomes independent of initial conditions asymptotically. For example, a set of parametric values satisfying the Kolmogorov conditions and violating the local stability condition is $a_1 = 2$, $b_1 = 0.05$, $w = 1$, $D = 10$, $a_2 = 1$, $w_1 = 2$, and $D_1 = 10$.

For subsystem II, Kolmogorov conditions are satisfied automatically.

Linear Stability Analysis of Subsystem II: The equilibrium points of the system are $(K_1, 0)$ and (Z^*, U^*), where (see Holling–Tanner Model)

$$Z^* = \frac{1}{2Aw_4} \left[q \pm \sqrt{q^2 + 4A^2 K_1 D_3 w_4^2} \right], \ q = A(K_1 - D_3)w_4 - K_1 c w_3, \ \text{and} \ U^* = \frac{cZ^*}{w_4}.$$

The local stability conditions are given by (see also Example 2.9)

$$c + q_2 > q_3, \quad \left(\frac{Z^* - D_3}{Z^* + D_3} \right)(q_3 - q_2) > \frac{c w_3 Z^*}{w_4(Z^* + D_3)}, \tag{3.11}$$

where $q_2 = D_3 w_3 c Z^* / [w_4(Z^* + D_3)^2]$, $q_3 = A(1 - (2Z^*/K_1))$.

Subsystem II given by Equation 3.7 would admit stable limit-cycle solutions when one or both of the above constraints are violated. For example, for the set of parametric values, $A = 1.5$, $K_1 = 100$, $w_3 = 0.74$, $D_3 = 10$, $c = 0.5$, $w_4 = 0.2$, subsystem II has limit-cycle solutions. For this set of values, we obtain the equilibrium point as $Z^* \approx 19.08$, $U^* \approx 47.7$ In this case, both the local stability conditions are violated.

Now, couple the two units. Essentially, this amounts to finding out the values of w_2 taking into consideration their biological significance. The choice of w_2 is guided by the values of w, w_1, and w_3 which were obtained from the application of the Kolmogorov theorem and the linear stability theory. w_2 can be varied from 0.01 to 1.

In the case of the 3D systems, we assume that there exists a second prey (other than the specialist predator), called as *pseudo-prey* [109], for the generalist predator. It seems to be a realistic assumption as it is in conformity with the biology of all the other species in the ecosystem. Subsystem I corresponding to Equation 3.6 is obtained by dropping the last term of the second equation. It turns out that this subsystem is identical to the first subsystem that we have obtained in the case of the 4D model system. The second subsystem is designed in the following way. Since Z is a generalist predator, there may not be any contradiction if we assume that there exists a prey $\tilde{X}$, in addition to Y whose growth rate is governed by

$$\frac{d\tilde{X}}{dt} = a_1 \tilde{X} - b_1 \tilde{X}^2 - \frac{w \tilde{X} Z}{\tilde{X} + E}. \tag{3.12}$$

The growth rate of Z is governed by

$$\frac{dZ}{dt} = cZ - \frac{w_3 Z^2}{\tilde{X}}. \tag{3.13}$$

Equations 3.12 and 3.13 define subsystem II. The application of the Kolmogorov theorem and the local stability analysis yield sets of parametric values for the simulation experiments. It may be mentioned that in the final selection of the parametric values, pseudo-prey terms are omitted. A similar analysis provides us meaningful sets of parametric values to perform simulation experiments on the Allee variant of the 3D system, and on the ecological system with learning behavior in the generalist predator Z. The ranges for the variations of the values of the parameters for the study of the dynamical behavior can be chosen on the basis of values reported in Jorgensen [39].

Example 3.1

Apply the *pseudo-prey* method for the following 3D system to find biologically realistic parameter values for simulation experiments [109].

$$\frac{dx}{dt} = a_1 x - b_1 x^2 - \frac{wxy}{x + D}, \tag{3.14}$$

$$\frac{dy}{dt} = -a_2 y + \frac{w_1 xy}{(x + D_1)} - \frac{w_2 yz}{y + D_2}, \tag{3.15}$$

$$\frac{dz}{dt} = cz^2 - \frac{w_3 z^2}{y + D_3}. \tag{3.16}$$

SOLUTION

The top prey and the middle predator give a biologically meaningful subsystem. The subsystem should qualify as a Kolmogorov (K) system. The last term in Equation 3.15 is omitted to get subsystem I, which is the same as the equations in system (3.6). Kolmogorov theorem gives the conditions as given in Equation 3.9 and the local stability conditions are given in Equation 3.10. As mentioned earlier, a set of parametric values satisfying the Kolmogorov conditions and violating the local stability condition are $a_1 = 2$, $b_1 = 0.05$, $D = 10$, $a_2 = 1$, $w_1 = 2$, and $D_1 = 10$.

Since z is a generalist predator, we assume that there exists a prey $\tilde{x}$, in addition to Y whose growth rate is governed by

$$\frac{d\tilde{x}}{dt} = A\tilde{x}\left(1 - \frac{\tilde{x}}{K}\right) - \frac{B\tilde{x}z}{\tilde{x} + E},$$

where A is the rate of self-reproduction for this prey and K is the carrying capacity of its environment. In the Leslie–Gower scheme, the growth rate equations for the two populations will give the subsystem II

$$\frac{d\tilde{x}}{dt} = A\tilde{x}\left(1 - \frac{\tilde{x}}{K}\right) - \frac{B\tilde{x}z}{\tilde{x} + E}, \quad \frac{dz}{dt} = cz^2 - \frac{w_3 z^2}{\tilde{x} + D_3}. \tag{3.17}$$

For this subsystem to be a K-system, the following conditions are to be satisfied

$$\frac{B\tilde{x}z}{(\tilde{x} + E)^2} < \frac{Bz}{\tilde{x} + E} + \frac{A\tilde{x}}{K}, \quad c(\tilde{x} + D_3) < w_3. \tag{3.18}$$

Subsystem II exhibits a stable equilibrium when the following conditions are satisfied:

$$A + 2cz^* < \frac{2A\tilde{x}^*}{K} + \frac{2w_3 z^*}{\tilde{x}^* + D_3} + \frac{BEz^*}{(\tilde{x}^* + E)^2},$$

$$\left\{A - \frac{2A\tilde{x}^*}{K} + \frac{BEz^*}{(\tilde{x}^* + E)^2}\right\}\left\{2cz^* - \frac{2w_3 z^*}{\tilde{x}^* + D_3}\right\} + \frac{Bw_3 \tilde{x}^* z^{*2}}{(\tilde{x}^* + E)(\tilde{x}^* + D_3)^2} > 0, \tag{3.19}$$

where $\tilde{x}^* = (w_3/c) - D_3, z^* = (A/B)(1 - (\tilde{x}^*/K))(\tilde{x}^* + E)$ are the equilibrium populations.

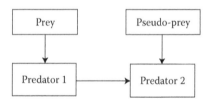

FIGURE 3.3
Relationship between food-chain species and pseudo prey.

When the values of the parameters of subsystem II are chosen in such a way that constraints (3.18) are satisfied and inequality (3.19) is violated, subsystem II admits a limit-cycle solution. For example, for the set of parameter values, $A = 1.0$, $K = 50$, $B = 1$, $E = 20$, $c = 0.0062$, $w_3 = 0.2$, $D_3 = 20$, the subsystem II has limit-cycle solutions. We choose the maximum value of $w = 1$ as it is the maximum per capita removal rate of prey population.

Now, link the two subsystems. One possible way is given in Figure 3.3.

The linking scheme would depend on how the individual populations of the two subsystems are related with each other. The link scheme can be mathematically represented by adding a term $\{-[w_2yz/(y + D_2)]\}$ to the second equation of subsystem I. This gives Equation 3.15. Since the meanings of D_1 and D_2 are the same and D_2 is not an important parameter as far as asymptotic dynamics of the complete system is concerned, D_2 can also be assigned numerical values in the range of D_1. The selection of values for w_2 is based on the fact that it plays a role similar to that of w. w_2 can be varied from 0.1 to 1.

The parameter values for the original system are selected by omitting those appearing in the growth equation of pseudo prey ($\tilde{x}$). A set of parameter values for which the system admits limit-cycle solution is $a_1 = 2$, $b_1 = 0.05$, $w = 1$, $D = 10$, $a_2 = 1$, $w_1 = 2$, $D_1 = 10$, $w_2 = 0.3$, $D_2 = 10$, $c = 0.0062$, $w_2 = 1$, $D_3 = 20$. There exist other sets of parameter values, which satisfy the above criteria.

Calculation of the Basin Boundary Structures

It is a common occurrence that two or more of the dynamical possibilities coexist for a particular set of parameter values. Every attractor has a bowl of initial conditions called the basin of attraction, such that we land up on the attractor if we start from any one of the initial conditions from the bowl. In the case of coexisting attractors, we have boundaries formed by the two basins. These boundaries may have very simple or very complicated geometries. The geometry of the basin boundary and the nature and strength of the external perturbations together decide how frequent would be the jumps, in the system dynamics. The frequency of these jumps can sometimes be unpredictable, for example, when the basin boundary has a fractal structure. In such a case, even the final outcome of these jumps may be unpredictable, that is, to which attractor a system eventually goes. This property of a system is known as *final state sensitivity* [60]. A chaotic attractor may be suddenly destroyed in a collision with the basin boundary. Such events are known as *crisis* [29,60]. Beyond a crisis point, there exist long chaotic transients. Detailed discussions on the basin boundaries and the related concepts were given by McDonald et al. [60].

Sommerer and Ott [97] have discovered a basin boundary structure (BBS) known as a *riddled basin*. It was shown (for a class of dynamical systems having invariant subspace) that (i) if there is a chaotic attractor in the invariant subspaces; (ii) if there is another attractor in the phase space, and (iii) if the Lyapunov exponent transverse to the subspace is negative, then the basin of the chaotic attractor in the invariant subspace can be riddled with the holes belonging to the basin of the other attractor. This means that for every initial condition that belongs to the chaotic attractor in the invariant subspace, there are arbitrarily nearby initial conditions which tend to the asymptotes of the other attractor. Invariant subspaces are commonly found in systems with symmetry.

To obtain information about how a given system would behave when acted upon by external disturbances/perturbations, it is necessary to know the structure of the basin boundaries of coexisting attractors. For a dynamical behavior to be of any practical interest, it is essential that it should exist in wide parameter ranges and must fulfill the requirement that it exists for a set of initial conditions whose natural measure (area or volume) is nonzero. When these two conditions are met by a dynamical system for a particular dynamical behavior, then the same is understood to be robust and is considered to have some significance.

We now mention briefly how basin boundary calculations are performed. First, we define the *basin of infinity*. Let SD denote the diameter of the computer screen. It may be possible that the point at infinity is an attractor. Since we cannot examine rigorously whether the trajectory of a point goes to infinity, we conclude that a trajectory diverges or is diverging if it leaves the computer screen area, that is, it goes to the left or to the right of the screen by more than one SD width of the screen, or goes above or below the screen area by more than one SD screen height. The basin of infinity is the set of initial points whose trajectories are diverging. The Maryland Chaos group has done pioneering work in this area and has developed a tool to calculate BBS. We have used the research version of the software which accompanied the book titled *Dynamics: Numerical Explorations* authored by Nusse and Yorke [66]. We have used the BAS (basins and attractors structure) method for all computations. This method divides the basin into the following two groups: (i) the basin of attraction A whose points will be plotted, and (ii) the basin B whose points will not be plotted. A generalized attractor is the union of finitely many attractors, and a generalized basin is the basin of a generalized attractor. The BAS routine does not plot the bowl lying outside. It considers a 100×100 grid of boxes covering the screen. The strategy is to test each grid point, which is the center of the grid box. In the event that the center of a grid box is in basin A, the same is plotted (colored). In the default case, basin A is the set of points whose trajectories are diverging, while basin B is empty. Therefore, the BAS routine will plot a grid box if the trajectory of its center is diverging. The important aspects of the basin boundary calculations are to specify the basins A, B and to find the radius RA, where RA stands for the radius of attraction for storage vectors which help to specify the basins A and B. The value of RA will be different for different dynamical systems. It must be set appropriately to avoid any misleading basin picture.

The meanings of the various colors are as follows:

Dark blue: Color of the points that diverges from the screen area

Green: Color of the first attractor, sky blue: basin of the first attractor

Red: Color of the second attractor, maroon: basin of the second attractor

Brown: Color of the third attractor, white: basin of the third attractor

Example 3.2

Draw the BBS for the following model, for the parameter values: $a_1 = 2.0$, $a_2 = 1.0$, $w = 1$, $w_1 = 2.0$, $w_2 = 0.55$, $w_3 = 1$, $b_1 = 0.05$, $\theta = 0.009$, $D_3 = 20$, $D = D_1 = D_2 = 10$, $c = 0.0257$ [113].

$$\frac{dx}{dt} = a_1 x - b_1 x^2 - \frac{wxy}{(x + D)},$$

$$\frac{dy}{dt} = -a_2 y + \frac{w_1 xy}{(x + D_1)} - \frac{w_2 yz}{(y + D_2)} - \theta f(x)y, \quad \text{where } f(0) = 0,$$

$$\frac{dz}{dt} = cz^2 - \frac{w_3 z^2}{y + D_3}.$$

Explain the significance of the diagrams.

SOLUTION

We use the BAS routine to compute the BBS. The BBS are shown in Figure 3.4a–c.

From the plots, we observe that basin boundaries of the chaotic attractor are fractal. It is also seen that basins of attraction of different attractors are intermixed. The encroachment into the basin of chaotic attractor by basin of attractor at infinity (shown in green color) can also be observed. It appears between the first attractor (shown in dark blue color) and its basin (shown in sky blue color). The interesting feature is that the riddled basin with fractal boundary lies in the basin of the repeller, which has many rectangular and square holes created by stable focus, limit cycles, and chaotic attractors. This complicated BBS suggests that the system

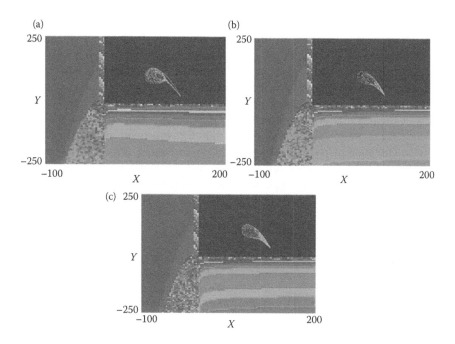

FIGURE 3.4
BBS for chaotic attractor for (a) $f(x) = x$, $\theta = 0.009$. (b) $f(x) = x/(x + 10)$, $\theta = 0.2$, (c) $f(x) = x^2/(x^2 + 100)$, $\theta = 0.15$. Example 3.2. (Reprinted from *Chaos, Solitons Fractals*, 39, Upadhyay, R. K., Rao, V. S. H., Short term recurrent chaos and role of toxin producing phytoplankton on chaotic dynamics in aquatic systems, 1550–1564, Copyright 2009, with permission from Elsevier.)

dynamics may have loss of even qualitative predictability in the case of external disturbances. This establishes that STRC in this model system is generated by deterministic changes in the crucial parameters of the system. However, the negligible amounts of intermixing of basins of different attractors suggest that exogenous stochastic influences like drastic changes in environmental conditions have no appreciable influence over the dynamics of the system.

2D Parameter Scans

For a dynamical behavior to be of any practical interest, it is essential that it should exist in wide parameter ranges and the corresponding natural measure (area) in 2D parameter scans should be nonzero. We can perform 2D parameter scans to identify the parameter regimes in which different dynamics (stable focus, stable limit cycle, chaos, and torus) exist. The parameters chosen for the 2D scans are the parameters which control the dynamics of the systems. The basis for performing the 2D scans is the belief that the changes in physical conditions may bring corresponding changes in at least two parameters at a time. The change in the nature of dynamics is monitored. In the 2D scans, phase-space studies are carried out in two dimensions by fixing a control parameter on the X-axis and exploring the whole range of another control parameter on the Y-axis. The analysis is repeated by giving an increment to the value of the control parameter plotted along the X-axis. Various combinations of such pairs of parameter combinations can be studied. 2D parameter spaces comprising critical parameters of the systems are prepared and a single parameter of the system is varied at a time on both sides of the parameter set corresponding to a stable limit cycle in the phase space of the system. The rest of the parameters are fixed at their limit cycle values. The system parameters are chosen in such a way that asymptotic dynamics of both the subsystems confine to a stable limit cycle. Chaos in the models can be detected using one of the distinguishing properties of chaos, for example, *sensitive dependence on initial conditions*.

Example 3.3

Draw the 2D scan diagrams using the pairs of the parameters (a_1, D), (a_1, D_2), for the following model in the parameter ranges $1.75 \leq a_1 \leq 2$, $5 \leq D \leq 15$, $5 \leq D_2 \leq 25$, and for the parameter values $K = 31$, $c = 0.1$, $w = 1$, $a_2 = 1$, $w_1 = 2$, $w_2 = 0.55$, $D_1 = 10$, $D_3 = 20$, $w_3 = 0.25$. Explain the significance of the diagrams.

$$\frac{dX}{dt} = a_1 X\left(1 - \frac{X}{K}\right) - \frac{wYX}{X + D}, \quad \frac{dY}{dt} = -a_2 Y + \frac{w_1 YX}{X + D_1} - \frac{w_2 YZ}{Y + D_2},$$

$$\frac{dZ}{dt} = -cZ + \frac{w_3 YZ}{Y + D_3}.$$

SOLUTION

The system is solved using MATLAB® 7.0. The 2D scan diagrams given in Figure 3.5 are plotted with Microsoft Excel.

The points shown in Figure 3.5 represent "edge of chaos" as chaos exists at discrete isolated points in these parameter spaces. The chosen parameter combinations are (a_1, D) and (a_1, D_2) as these parameters change in

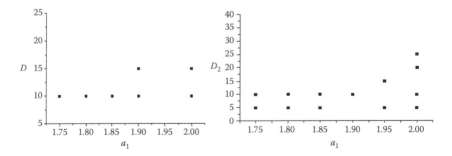

FIGURE 3.5
2D scan diagrams.

response to the changes in environmental conditions and, therefore, are suitable for the dynamical study of the system. The unpredictability of transitions between dynamical states (cyclic and chaotic) of the model is caused by a joint action of smooth or deterministic changes in system parameters and sudden exogenous stochastic factors (drastic changes in environmental conditions) which operate on initial conditions. This unpredictability in dynamical transitions would be present no matter at what scale the system's dynamics is measured. The phenomena of edge of chaos in the parameter spaces reveal something fundamental about the dynamics of real-world ecological systems.

Bifurcation Diagrams

Bifurcation diagrams are constructed by plotting parameter values as abscissas and dynamical variables or related quantities as ordinates. Bifurcation diagrams in parameter space or phase-parametric diagrams are often used for descriptive representation. Bifurcation can be defined as a sudden qualitative change in the nature of the dynamics that occur at critical parameter values. Typically, one type of motion loses stability at a critical value of the parameter as it varies smoothly giving rise to a new type of stable motion. The parameter values at which bifurcations occur are called bifurcation points or bifurcation values. In a multidimensional parameter space of a system, bifurcation points may correspond to certain sets representing points or lines or surfaces, and so on. If several limit sets are realized in the same parameter range, the bifurcation diagram appears to be multisheeted. Bifurcation points may appear to be dense everywhere in the parameter space when a large (even infinite) number of limit sets coexist. In this case, construction of a complete bifurcation diagram becomes impossible and only its separate leaves and parts can be considered.

Bifurcation is characterized by conditions that impose certain requirements on system parameters. The number of such conditions is called the *codimension* of bifurcation. For example, codimension 1 means that there is only one bifurcation condition. There are basically four types of bifurcations that may occur in simple low-dimensional nonlinear systems. They are: (i) saddle node, (ii) pitchfork, (iii) transcritical, and (iv) Hopf bifurcations.

In pitchfork and transcritical bifurcations, the real eigenvalues of the least stable equilibrium point increases through zero as a control parameter is varied through a critical value. One or two new stable points may arise after bifurcation. Local bifurcations are associated with the local neighborhood of a trajectory on a limit set. For example, when one of the Lyapunov exponents of a trajectory on a limit set changes its sign, this testifies to a local bifurcation of the limit set. Nonlocal bifurcations are related to the behavior of the manifolds of saddle limit sets like the formation of separatrix loops, and so forth. Bifurcations can happen with any limit sets, but bifurcations of attractors are more important because they are experimentally observed regimes of change. Attractor bifurcations are classified into soft (internal) bifurcation and hard (crises) bifurcation [2,28]. Internal bifurcations result in topological changes in attracting limit sets, but do not affect their basin of attraction. Attractor crises are accompanied by a qualitative modification of the basin boundaries [3].

One of the useful ways of studying a dynamical system is to monitor the amplitude (maxima) of the subsequent oscillations as the control parameter of the system is varied. When the maxima of one of the system's phase variables are plotted as the ordinate against the control parameter of the system, then the resulting figure is known as the bifurcation diagram. Small changes in parameter values may lead to a bifurcation: an abrupt, qualitative change in the dynamics. These bifurcation diagrams may show a number of qualitative changes, that is, many bifurcations, in the chaotic attractor as well as in the periodic orbits. These changes include shifts among qualitative attractor types, doubling the period of a limit cycle, and so on. In systems with multiple attractors, complete destabilization of one of the attractors may take place [42]. Algorithms are available for plotting the bifurcation diagrams. Some of these algorithms are given in a book by Baker and Gollub [4]. Another kind of bifurcation diagram is known as the Lyapunov bifurcation diagram, which plots the maximum Lyapunov exponent against a system parameter. It gives direct information about how the system dynamics change when the control parameter is varied. A stable focus in the phase space is characterized by a negative maximum Lyapunov exponent. The Lyapunov exponent of a limit cycle is zero, and that of an SCA is positive. A positive Lyapunov exponent means that the nearby system trajectories would diverge from each other exponentially fast with time, generating loss of predictability observed in chaotic dynamical systems. It is important to understand the mechanisms that bring in the above-mentioned changes in the dynamical behavior when a system parameter is varied.

Example 3.4

Draw the bifurcation diagrams for the model given in Example 3.2 taking θ as a control parameter and for (a) $f(x) = x$, $0 \le \theta \le 0.008$; (b) $f(x) = x/(x + D_2)$, $0 \le \theta \le 0.2$; (c) $f(x) = x^2/(x^2 + D_2^2)$, $0 \le \theta \le 0.15$. Take the parameter values as $a_1 = 1.93$, $a_2 = 1$, $b_1 = 0.06$, $c = 0.03$, $w = 1$, $w_1 = 2$, $w_2 = 0.405$, $w_3 = 1$, $D_3 = 20$, $D = D_1 = D_2 = 10$.

SOLUTION

The bifurcation diagrams are generated using MATLAB 7.0. In the example, (a) Holling type I, (b) Holling type II, and (c) Holling type III functional responses are considered. The plots show the transition from chaos to order through a sequence of period halving bifurcations. In a typical model system, if $f(x)$ denotes a form of a toxic substance liberation process by toxin-producing phytoplankton (TPP), then it is observed that an increase in the value of toxic substance released (θ) has a stabilizing effect. The bifurcation diagrams show that the model system possesses

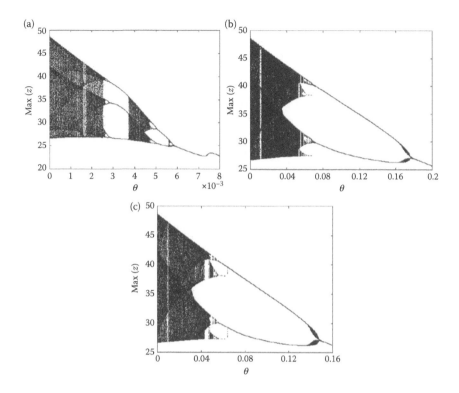

FIGURE 3.6
(a) Bifurcation diagram for $f(x) = x$, $0 \le \theta \le 0.008$. (b) Bifurcation diagram for $f(x) = x/(x + D_2)$, $0 \le \theta \le 0.2$. (c) Bifurcation diagram for $f(x) = x^2/(x^2 + D_2^2)$, $0 \le \theta \le 0.15$. (From Upadhyay, R. K., Naji, R. K., Nitu Kumari. Dynamical complicity in some ecological models: Effect of toxin production by phytoplankton. *Nonlinear Anal.: Model. Control* 12(1), 123–138. Copyright 2007, Lithuanian Association of Nonlinear Analysts (LANA). Reprinted with permission.)

a rich variety of dynamical behavior for the bifurcation parameter θ only in the ranges [0, 0.0029] for Holling type I (Figure 3.6a), [0, 0.07] for Holling type II (Figure 3.6b) and [0, 0.06] for Holling type III (Figure 3.6c) functional response [111a]. In all the three cases, a period-doubling cascade is observed.

Hopf Bifurcation Analysis

In Hopf bifurcation, the real parts of a pair of complex conjugate eigenvalues of the least-stable equilibrium point increase through zero as a control parameter is varied through a critical value. A time-periodic solution may arise after bifurcation. For example, as a control parameter is varied, a stable equilibrium point may lose its stability at a critical value and give rise to a limit cycle. Hopf bifurcation can occur only in nonlinear systems with dimension >1. Hopf bifurcation can be supercritical or subcritical. In a supercritical Hopf bifurcation, the limit cycle is stable above the bifurcation point. The bifurcation is supercritical if a small attracting limit cycle appears immediately after the fixed point goes unstable and its amplitude shrinks back to zero as the parameter is reversed. In a subcritical Hopf bifurcation, the bifurcated limit cycle will be unstable. After the bifurcation, the trajectory jumps to a distant attractor (a fixed point, another limit cycle, etc.). In this case, the nearest attractor might be far from the fixed point. For example, consider the dynamics of the damped pendulum $\ddot{x} + \mu\dot{x} + \sin x = 0$. As the damping parameter μ is varied from positive to negative, the fixed point at the origin changes from a stable to an unstable spiral. However, Hopf bifurcation does not occur at $\mu = 0$ as there is no limit cycle on either side of the bifurcation and exhibits only a continuous band of closed orbits surrounding the origin at $\mu = 0$. This is a case of degenerate Hopf bifurcation, which typically arises when a nonconservative system suddenly becomes conservative at the bifurcation point. It would also require that the origin be a nonlinear center when $\mu = 0$ [103].

Consider a 2D dynamical system of the form

$$\frac{dx}{dt} = P(x, y), \quad \frac{dy}{dt} = Q(x, y), \tag{3.20}$$

where P and Q are well-defined functions of x and y. Let the characteristic equation of the Jacobian matrix

$$J(\mu) = \begin{bmatrix} \partial P/\partial x & \partial P/\partial y \\ \partial Q/\partial x & \partial Q/\partial y \end{bmatrix}$$

be

$$p(\lambda; \mu) = \det(\lambda I_2 - J(\mu)) = p_0(\mu) + p_1(\mu)\lambda + p_2(\mu)\lambda^2 = 0 \tag{3.21}$$

with $p_2(\mu) = 1$. Then, Hopf bifurcation occurs at $\mu = \mu_0$, if the following conditions are satisfied [49]

$$p_0(\mu_0) > 0, \, p_1(\mu_0) = 0, \quad \text{and} \quad dp_1(\mu_0)/d\mu \neq 0. \tag{3.22}$$

The condition $p_0(\mu_0) > 0$ implies that $\lambda = 0$ is not a root when $\mu = \mu_0$. For a 3D system, the characteristic equation is a cubic

$$p(\lambda; \mu) = \det(\lambda I_3 - J(\mu)) = p_0(\mu) + p_1(\mu)\lambda + p_2(\mu)\lambda^2 + p_3(\mu)\lambda^3 = 0,$$

with $p_3(\mu) = 1$. Hopf bifurcation occurs at $\mu = \mu_0$, if

$$p_0(\mu_0) > 0, D_1(\mu_0) = p_1(\mu_0) > 0, D_2(\mu_0) = p_1(\mu_0)p_2(\mu_0) - p_0(\mu_0) = 0,$$

and

$$[dD_2(\mu_0)/d\mu] \neq 0. \tag{3.23}$$

In general, we define $D_n(\mu) = \det(L_n(\mu)) > 0$, and

$$
L_n(\mu) = \begin{bmatrix}
p_1(\mu) & p_0(\mu) & \cdots & 0 \\
p_3(\mu) & p_2(\mu) & \cdots & 0 \\
\cdots & \cdots & \cdots & \cdots \\
p_{2n-1}(\mu) & p_{2n-2}(\mu) & \cdots & p_n(\mu)
\end{bmatrix},
$$

where $p_i(\mu) = 0$, for $i < 0$, or $i > n$.

Example 3.5

Perform the Hopf bifurcation analysis of the positive equilibrium point for the model system in Example 3.2.

SOLUTION

We follow the method of Liu [49]. The model system given in Example 3.2 has the following equilibrium points: $E_0 = (0,0,0)$, $E_1 = (a_1/b_1, 0, 0)$, $E_2 = (\bar{x}, \bar{y}, 0)$, where $\bar{x}$ is the positive root of the equation

$$
\left(w_1 - a_2 - \theta f(\bar{x})\right)\bar{x} - D_1(a_2 + \theta f(\bar{x})) = 0,
$$

and

$$
\bar{y} = \frac{1}{w}\left(a_1 - b_1 \bar{x}\right)\left(\bar{x} + D\right).
$$

E_2 exists when $0 < \bar{x} < a_1/b_1$. The nontrivial equilibrium $E_3 = (x^*, y^*, z^*)$ is given by $y^* = (w_3 - cD_3)/c$, x^* is the positive root of the equation

$$
b_1 x^{*2} + (b_1 D - a_1)x^* + wy^* - Da_1 = 0,
$$

and

$$
z^* = \frac{(y^* + D_2)}{w_2}\left[\frac{w_1 x^*}{x^* + D_1} - (a_2 + \theta f(x^*))\right].
$$

E_3 exists when $w_3 - cD_3 > 0$, and $w_1 x^* - (x^* + D_1)(a_2 + \theta f(x^*)) > 0$. The variational matrix V_0 at E_0 is given by

$$
V_0 = \begin{bmatrix}
a_1 & 0 & 0 \\
0 & -a_2 & 0 \\
0 & 0 & 0
\end{bmatrix}.
$$

The eigenvalues of V_0 are 0, a_1 and $(-a_2)$. There is an unstable manifold along the x-direction and a stable manifold along the y-direction. The equilibrium point E_0 is a saddle point.

The variational matrix V_1 at $E_1 = (a_1/b_1, 0, 0)$ is given by

$$V_1 = \begin{bmatrix} -a_1 & -c_1 & 0 \\ 0 & -c_2 & 0 \\ 0 & 0 & 0 \end{bmatrix},$$

where

$$c_1 = \frac{wa_1}{a_1 + b_1D}, \quad c_2 = a_2 - \frac{w_1a_1}{a_1 + b_1D_1} + \theta f\left(\frac{a_1}{b_1}\right).$$

The eigenvalues of V_1 are $0, -a_1, (-c_2)$. There are stable manifolds along both the x- and y-directions if $c_2 > 0$.

The variational matrix about the equilibrium point $E_2 = (\bar{x}, \bar{y}, 0)$ is given by

$$V_2 = \begin{bmatrix} a_{11} & -a_{12} & 0 \\ a_{21} & 0 & -a_{23} \\ 0 & 0 & 0 \end{bmatrix},$$

where

$$a_{11} = -\bar{x}\,[b_1 - (w\bar{y})/(\bar{x} + D)^2], \quad a_{12} = (w\bar{x})/(\bar{x} + D),$$

$$a_{21} = \bar{y}\,[\{(w_1D_1)/(\bar{x} + D_1)^2\} - \theta f'(\bar{x})], \quad a_{23} = (w_2\bar{y})/(\bar{y} + D_2).$$

One eigenvalue is 0. The other eigenvalues are the roots of the equation $\lambda^2 - a_{11}\lambda + a_{21}a_{12} = 0$. By the Routh–Hurwitz criterion, the roots are negative or have negative real parts if $a_{21}a_{12} > 0$, and $a_{11} < 0$. This gives the conditions $w_1D_1 > [(\bar{x} + D_1)^2\theta f'(\bar{x})]$; and $b_1(\bar{x} + D)^2 > w\bar{y}$ or $a_1 < b_1(D + 2\bar{x})$. There are stable manifolds along both the x- and y-directions if these conditions are satisfied. Since one of the eigenvalues is zero in all the above cases, the equilibrium points are nonhyperbolic. The dynamical behavior near them can be stable, periodic, or even chaotic.

The variational matrix about the equilibrium point $E_3 = (x^*, y^*, z^*)$ is given by

$$V_3 = \begin{bmatrix} a_{11} & -a_{12} & 0 \\ a_{21} & a_{22} & -a_{23} \\ 0 & a_{32} & 0 \end{bmatrix},$$

where

$$a_{11} = -x^*[b_1 - (wy^*)/(x^* + D)^2], \quad a_{12} = (wx^*)/(x^* + D),$$
$$a_{21} = y^*[\{(w_1D_1)/(x^* + D_1)^2\} - \theta f'(x^*)], \quad a_{22} = w_2y^*z^*/(y^* + D_2)^2,$$
$$a_{23} = (w_2y^*)/(y^* + D_2), \quad a_{32} = c^2z^{*2}/w_3.$$

The characteristic equation of V_3 is given by

$$p(\lambda; \mu) = \lambda^3 + p_2(\mu)\lambda^2 + p_1(\mu)\lambda + p_0(\mu) = 0,$$

where $p_2 = -(a_{11} + a_{22})$, $p_1 = a_{11}a_{22} + a_{23}a_{32} + a_{21}a_{12}$, $p_0 = -a_{11}a_{23}a_{32}$.

Now, $a_{12} > 0$, $a_{22} > 0$, $a_{23} > 0$, and $a_{32} > 0$. According to the Routh–Hurwitz criterion, E_3 is locally asymptotically stable if $p_0 > 0$, $p_1 > 0$, $p_2 > 0$, and $p_1 p_2 - p_0 > 0$.

$p_0 > 0$, if $a_{11} < 0$. This gives the condition $b_1(x^* + D)^2 > wy^*$ or $a_1 < b_1(D + 2x^*)$.

$p_2 > 0$, if $(a_{11} + a_{22}) < 0$, that is $y^*[wx^*/(x^* + D)^2 + w_2 z^*/(y^* + D_2)^2] < b_1 x^*$.

$p_1 > 0$, if $a_{11}a_{22} + a_{23}a_{32} + a_{12}a_{21} > 0$. Let $a_{21} > 0$, that is $(x^* + D)^2 \theta f'(x^*) < w_1 D_1$. Also, $a_{22} = [(w_2 y^* z^*)/(y^* + D_2)^2] = [(z^* a_{23})/(y^* + D_2)]$. A sufficient condition is $a_{32} + [z^*/(y^* + D_2)]a_{11} > 0$, since $a_{23} > 0$. We obtain the condition

$$\frac{c^2 z^*}{w_3} > \frac{x^*}{y^* + D_2}\left[b_1 - \frac{wy^*}{(x^* + D)^2}\right].$$

Simplifying $p_1 p_2 - p_0$, we obtain the requirement

$$[-(a_{11} + a_{22})a_{12}a_{21} - a_{11}a_{22}^2] - a_{22}(a_{11}^2 + a_{23}a_{32}) > 0.$$

The first term is positive and the second term is negative.

Now, let c, the growth rate of the generalist predator, be the bifurcation parameter. The parameter c appears only in a_{32}. Hopf bifurcation occurs at $c = c_0$, if the conditions given in Equation 3.23 are satisfied. The conditions for $p_0(c_0) > 0$, $p_1(c_0) > 0$, are derived earlier. The requirement D_2 $(c_0) = p_1(c_0)p_2(c_0) - p_0(c_0) = 0$ gives

$$D_2(c_0) = p_2(a_{11}a_{22} + a_{12}a_{21}) - a_{22}a_{23}(c_0^2 z^{*2}/w_3) = 0,$$

or

$$c_0^2 = \frac{w_3 p_2(a_{11}a_{22} + a_{12}a_{21})}{z^{*2} a_{22}a_{23}}.$$

We also require the condition $a_{11}a_{22} + a_{12}a_{21} > 0$. Now,

$$\frac{dD_2}{dc} = -\frac{2cz^{*2}a_{22}a_{23}}{w_3} \neq 0, \quad \text{for } c = c_0.$$

Hopf bifurcation occurs if all the above conditions are satisfied.

Time-Series Analysis and Phase-Space Diagram

Time series, which displays fluctuations as $t \to \infty$, is a familiar representation of dynamical systems. The model systems are numerically integrated to get the time series corresponding to the variables of the model systems. Since every nonlinear system has a finite amount of transience, the data points representing transient behavior are discarded. Phase portraits are drawn using these data to obtain the geometry of the attractors. The geometrical object (phase portrait) with zero phase volume and represented by an isolated point in the phase space is stable focus. In the case of limit-cycle solutions, this equilibrium state is not a point, but oscillates persistently between top and bottom limits. The trajectory follows a closed loop in the phase space. Chaotic attractors have trajectories that never repeat, and exhibit erratic fluctuations despite being deterministic.

Time series enables us to study the two forms of chaos in dissipative dynamical systems: dynamical chaos and intermittent chaos [114]. However, the method is not helpful in distinguishing the long-lived chaotic transients and chaotic dynamics on a chaotic attractor [78].

The following typical computational procedure is used to find the existence of chaotic attractors:

a. Fix the initial conditions arbitrarily and plot the time series. Then, change the first initial condition by a small amount, say, 0.01. If the two time histories corresponding to two different, but very close, initial conditions overlap each other completely, then it is taken to be the case of stable limit cycle in phase space. If the two time histories are quite different and intersect each other, then it confirms that the system's dynamics are chaotic for that particular set of values of the parameters.

b. We draw the phase diagram in the 2D and 3D planes. If the dynamics do not truncate and go to newer and newer places, but never go beyond the region specified by the vector field, then we say that the system is chaotic. If the dynamics/trajectory follow a closed loop in the phase space, then we say that the system is periodic or cyclic.

c. A powerful method to characterize a chaotic attractor is through Lyapunov exponents (see the section "Lyapunov Exponents"). The sensitivity to initial conditions can be quantified by calculating the dominant Lyapunov exponent λ, which measures the long-term average rate of divergence of nearby trajectories. Positive values of Lyapunov exponent λ demonstrate the sensitive dependence of initial conditions, whereas $\lambda = 0$ for cyclic motion. A chaotic system has one or more positive Lyapunov exponents, but the total sum of the exponents is negative. One zero and two negative exponents indicate limit cycles in a three species systems.

Example 3.6

Draw the phase plane diagram and time series for the following model, for the parameter values $a_1 = 1.93$, $b_1 = 0.06$, $w = 1$, $D = D_1 = D_2 = 10$, $a_2 = 1$, $w_1 = 2$,

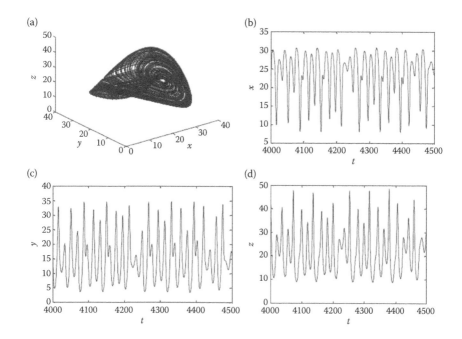

FIGURE 3.7

(a) Chaotic attractor. (b) Temporal evolution (t vs. x). (c) Temporal evolution (t vs. y). (d) Temporal evolution (t vs. z), in Example 3.6. (Reprinted from *Chaos, Solitons Fractals*, 40, Upadhyay, R. K., Rai, V. Complex dynamics and synchronization in two non-identical chaotic ecological systems, 2233–2241, Copyright 2009, with permission from Elsevier.)

$$w_2 = 0.405,\ c = 0.03,\ w_3 = 1,\ D_3 = 20\ [112].$$

$$\frac{dx}{dt} = a_1 x - b_1 x^2 - \frac{wxy}{(x + D)},\quad \frac{dy}{dt} = -a_2 y + \frac{w_1 xy}{(x + D_1)} - \frac{w_2 yz}{(y + D_2)},$$

$$\frac{dz}{dt} = cz^2 - \frac{w_3 z^2}{y + D_3}.$$

SOLUTION

The phase plane diagram and the time series are drawn using MATLAB 7.0. Chaotic attractor and the time series for the model system are given in Figure 3.7a–d.

Types of Bifurcations

Saddle-Node Bifurcation or Tangent Bifurcation

The saddle-node bifurcation is the basic mechanism by which fixed points are created and destroyed. As a parameter is varied, two fixed points move toward each other, collide, and mutually annihilate. A one-parameter family of 1D vector fields $(dx/dt) = f(x, \mu)$, $x, \mu \in$ in R^1, undergoes a saddle-node bifurcation if [116]

$$f(0,0) = 0, \quad \frac{\partial f}{\partial x}(0,0) = 0, \quad \frac{\partial f}{\partial \mu}(0,0) \neq 0, \quad \text{and} \quad \frac{\partial^2 f}{\partial x^2}(0,0) \neq 0. \quad (3.24)$$

The fixed point is a nonhyperbolic fixed point. In saddle-node bifurcation, on one side of a parameter value, there are no fixed points, and on the other side there are two fixed points.

Example 3.7

Discuss the saddle node bifurcation of the system

$$\frac{dx}{dt} = \frac{\mu}{4} - x^2, \quad \frac{dy}{dt} = -2y, \quad x, y, \mu \in R^1.$$

SOLUTION

We have a 2D system with one parameter μ in the first equation. Strogatz [103] had shown that all the action is confined to a 1D subspace along which the bifurcations occur; while in the extra dimensions, the flow is either simple attraction or repulsion from that subspace. Equilibrium points of the system are $E_1 = (\sqrt{\mu}/2, 0)$ and $E_2 = (-\sqrt{\mu}/2, 0)$. The eigenvalues of the Jacobian matrix are $\lambda_1 = -2$, $\lambda_2 = -2x^*$. For $\mu < 0$, the system has no real equilibrium points. For $\mu = 0$, origin $(0, 0)$ becomes a unique equilibrium point and the eigenvalues are $\lambda_1 = -2$, $\lambda_2 = 0$. In this case, the solutions of the equations are

$$x(t) = \frac{x(0)}{1 + x(0)t} \quad \text{and} \quad y(t) = y(0)e^{-2t}.$$

For $x(0) > 0$, $x(t) \to 0$, $y(t) \to 0$, as $t \to \infty$. All the trajectories approach the origin. For $x(0) < 0$, $y(t) \to 0$, as $t \to \infty$, while $x(t) \to -\infty$, as $t \to [1/|x(0)|]$. All the trajectories diverge from the origin. That is, in the region $x < 0$, trajectories are in the vicinity of a saddle; while in the region $x > 0$, they appear to be related to a stable node. This type of an equilibrium point is called a saddle-node. As μ is increased beyond the zero value, the equilibrium point $(0, 0)$ bifurcates into two new equilibrium points E_1 and E_2. The eigenvalues corresponding to these equilibrium points are $-2, -\sqrt{\mu}$, and $-2, +\sqrt{\mu}$, respectively. The equilibrium point E_2 is a saddle point, which is unstable, while E_1 is a stable node. As $\mu \to 0$ both the equilibrium points tend to $(0, 0)$ and the eigenvalues tend to 0 and -2. Therefore, at the critical value $\mu = 0$, a bifurcation occurs or a saddle node is born. This then bifurcates into a stable node and a saddle for $\mu > 0$. For $\mu < 0$, the equilibrium point disappears.

Example 3.8

Show that the system $(dx/dt) = \mu - (x^2/2) = f(\mu, x)$ undergoes a saddle-node bifurcation as μ is varied.

SOLUTION

We have

$$f(0,0) = 0, \quad \frac{\partial f}{\partial x}(0, 0) = 0, \quad \frac{\partial f}{\partial \mu}(0, 0) = 1, \quad \text{and} \quad \frac{\partial^2 f}{\partial x^2}(0, 0) = -1.$$

For, $\mu < 0$, the model has no fixed points and the vector field is decreasing in x.

For, $\mu > 0$, the model has two fixed points $x = \pm\sqrt{2\mu}$. The fixed point $x = \sqrt{2\mu}$ is stable, and the second fixed point $x = -\sqrt{2\mu}$ is unstable. We refer $(\mu, x) = (0, 0)$ as a bifurcation point and $\mu = 0$ as a bifurcation value. This is a saddle-node bifurcation where we have no fixed point on one side of the parameter value and two fixed points on the other side.

Transcritical Bifurcation

In transcritical bifurcation, a fixed point must exist for all values of a parameter and can never be destroyed. However, such a fixed point may change its stability when the control parameter passes through the bifurcation point or critical value. For this reason, the bifurcation is called a transcritical bifurcation or exchange of stability bifurcation.

A one-parameter family of 1D vector fields $(dx/dt) = f(x, \mu)$, $x, \mu \in R^1$, undergoes a transcritical bifurcation at $(x, \mu) = (0, 0)$, if [116]

$$f(0, 0) = 0, \quad \frac{\partial f}{\partial x}(0, 0) = 0, \quad \frac{\partial f}{\partial \mu}(0, 0) = 0,$$

$$\frac{\partial^2 f}{\partial x \partial \mu}(0, 0) \neq 0, \quad \text{and} \quad \frac{\partial^2 f}{\partial x^2}(0, 0) \neq 0. \tag{3.25}$$

Example 3.9

Discuss the occurrence of transcritical bifurcation in the system

$$\frac{dx}{dt} = -\frac{\mu x}{2} + x^2, \quad \frac{dy}{dt} = -2y, \quad x, y, \mu \in R^1.$$

SOLUTION

We have a 2D system with one parameter μ in the first equation. Equilibrium points of the system are $E_1 = (0, 0)$ and $E_2 = (\mu/2, 0)$. The eigenvalues of the Jacobian matrix are $\lambda_1 = -2$, $\lambda_2 = 2x^* - (\mu/2)$. Hence, the equilibrium point E_1 is stable for $\mu > 0$ and unstable for $\mu < 0$. The equilibrium point E_2 is stable for $\mu < 0$ and unstable for $\mu > 0$. Both for $\mu > 0$ and $\mu < 0$, the previously stable (unstable) equilibrium point becomes unstable (stable). The stability is exchanged at a critical value ($\mu = 0$) when the control parameter is varied. This bifurcation is called a transcritical bifurcation.

Pitchfork Bifurcation

The pitchfork bifurcation is common in dynamical systems that have symmetry, that is, invariant under the transformation $T = -T$. Fixed points tend to appear and disappear in symmetrical pairs. In the bifurcation diagram, pitchfork bifurcation looks like a tuning fork. Depending on the outcome of the stability behavior as a bifurcation parameter is varied, pitchfork bifurcation is classified into supercritical and subcritical bifurcations. In a supercritical pitchfork bifurcation, as a bifurcation parameter is varied, the stable equilibrium point (or even periodic orbit) generally becomes unstable at a critical point and gives birth to two new stable equilibrium points (or periodic orbits). In a subcritical pitchfork bifurcation, an unstable equilibrium point (or periodic orbit) becomes stable at a critical value where two more equilibrium points are born which are unstable [46].

A one-parameter family of 1D vector fields $(dx/dt) = f(x, \mu)$, $x, \mu \in R^1$, undergoes a pitchfork bifurcation at $(x, \mu) = (0, 0)$, if [116]

$$f(0, 0) = 0, \quad \frac{\partial f}{\partial x}(0, 0) = 0, \quad \frac{\partial f}{\partial \mu}(0, 0) = 0,$$

$$\frac{\partial^2 f}{\partial x^2}(0, 0) = 0, \quad \frac{\partial^2 f}{\partial x \partial \mu}(0, 0) \neq 0 \quad \text{and} \quad \frac{\partial^3 f}{\partial x^3}(0, 0) \neq 0.$$

(3.26)

Example 3.10

Discuss the occurrence of pitchfork bifurcation in the system

$$\frac{dx}{dt} = \frac{\mu x}{4} - x^3, \quad \frac{dy}{dt} = -4y, \quad x, y, \mu \in R^1.$$

SOLUTION

The same discussion holds as in Example 3.7. Equilibrium points of the system are $E_0 = (0,0)$, $E_1 = (\sqrt{\mu}/2, 0)$, and $E_2 = (-\sqrt{\mu}/2, 0)$. The eigenvalues of the Jacobian matrix are $\lambda_1 = -4$ and $\lambda_2 = (\mu/4) - 3(x^*)^2$. For $\mu < 0$, E_0 is stable and E_1, E_2 are a complex pair. For $\mu > 0$, all the three real equilibrium points exist. The equilibrium point E_0 is unstable (saddle point). The eigenvalues corresponding to E_1, E_2 are -4, $-\mu/2$. Both the equilibrium points are stable nodes. As μ is increased through $\mu = 0$, E_0 loses its stability and two new stable equilibrium points E_1 and E_2 emerge. The bifurcation is a pitchfork bifurcation. The bifurcation is supercritical because the new equilibrium points that emerge are stable.

Remark 3.2

Bifurcation theory of fixed points of maps is very similar to the theory of vector fields. The orbit structure near the fixed point can be reduced to the study of a parametrized family of maps on the 1D center manifold [116].

Suppose a general one-parameter family of C^r ($r \geq 2$) 1D map is given by $x \mapsto f(x, \mu)$, x, $\mu \in R^1$ with $f(0, 0) = 0$,$(\partial f/\partial x)(0, 0) = 1$. The point $(x, \mu) = (0,0)$ is a nonhyperbolic fixed point with eigenvalue 1.

Remark 3.3

For a general one-parameter family of $C^r(r \geq 2)$ 1D maps, the only change in conditions (3.24), (3.25), and (3.26) is $(\partial f/\partial x)(0, 0) = 1$. Other conditions remain the same. In the case of saddle-node bifurcation at $(x, \mu) = (0,0)$, the sign of $[-(\partial^2 f/\partial x^2)(0,0)/(\partial f/\partial \mu)(0,0)]$ decides on which side of $\mu = 0$ the curve of fixed points is located.

Period-Doubling Bifurcation

Consider a 1D map. A one-parameter family $C^r(r \geq 3)$ of 1D maps $x \rightarrow f(x, \mu)$, x, $\mu \in R^1$ undergoes a period-doubling bifurcation, if [116]

$$f(0, 0) = 0, \quad \frac{\partial f}{\partial x}(0, 0) = -1, \quad \frac{\partial f^2}{\partial \mu}(0, 0) = 0,$$

$$\frac{\partial^2 f}{\partial x^2}(0, 0) = 0, \quad \frac{\partial^2 f^2}{\partial x \partial \mu}(0, 0) \neq 0 \quad \text{and} \quad \frac{\partial^3 f^2}{\partial x^3}(0, 0) \neq 0, \tag{3.27}$$

where f^2 is the second iterate defined by $f^2 = (\mu, f(\mu))$. The sufficient conditions for the map to undergo a period-doubling bifurcation are that the map should have a nonhyperbolic fixed point with eigenvalue -1 and that the second iterate of the map should undergo a pitchfork bifurcation at the same nonhyperbolic fixed point. The sign of $[-(\partial^3 f^2/\partial x^3)/(\partial^2 f^2/\partial x \partial \mu)](0, 0)$ tells us on which side of $\mu = 0$ the period-two points lie.

For example, consider the map $x \rightarrow -x - \mu x + x^3 \equiv f(\mu, x)$ [116]. We have the second iterate as

$$f^2(\mu, x) = -(-x - \mu x + x^3)(1 + \mu) + (-x - \mu x + x^3)^3$$

$$= x(1 + \mu)^2 - x^3(1 + \mu)[1 + (1 + \mu)^2] + 3(1 + \mu)^2 x^5 - 3(1 + \mu)x^7 + x^9.$$

We have

$$f(0, 0) = 0, \quad \frac{\partial f}{\partial x}(0, 0) = -1, \quad \frac{\partial f^2}{\partial \mu}(0, 0) = 0,$$

$$\frac{\partial^2 f^2}{\partial x^2}(0, 0) = 0, \quad \frac{\partial^2 f^2}{\partial x \partial \mu}(0, 0) = 2, \quad \frac{\partial^3 f^2}{\partial x^3}(0, 0) = -12.$$

The given map has a nonhyperbolic fixed point at $(x, \mu) = (0, 0)$ with eigenvalue -1. The fixed points are the solutions of $x[x^2 - (2 + \mu)] = 0$. The map has two curves of fixed points $x = 0$ and $x^2 = (2 + \mu)$. Only $x = 0$ passes through the bifurcation point $(x, \mu) = (0, 0)$. $x = 0$ is unstable for $\mu \leq -2$ and $\mu > 0$ and stable for $-2 < \mu < 0$. $x^2 - 2 = \mu$ is unstable for $\mu \geq -2$ and does not exist for $\mu < -2$.

The second iterate is given by $x \rightarrow f^2 (\mu, x) = (1 + \mu)^2 x - 2x^3 + O(4)$. We have $f^2 (0, 0) = 0$, $(\partial/\partial x)f^2 (0, 0) = 1$. The above conditions imply that the second iterate of the given map undergoes a pitchfork bifurcation at $(x, \mu) = (0, 0)$. Except for $(x, \mu) = (0, 0)$, the fixed points of $f^2(\mu, x)$ are not fixed points of $f(\mu, x)$. They must be period two points of $f(\mu, x)$. Hence, $f(\mu, x)$ is said to have undergone a period-doubling bifurcation at $(x, \mu) = (0, 0)$. The map has a nonhyperbolic fixed point at $(x, \mu) = (0, 0)$ with eigenvalue 1. To the $O(4)$ approximation, the map has two curves of fixed points $x = 0$ and $x^2 = (2\mu + \mu^2)/2$. Only $x = 0$ passes through the bifurcation point $(x, \mu) = (0, 0)$.

The saddle-node and the period-doubling bifurcations are called *generic bifurcations*.

Andronov–Hopf Bifurcation (Cycle Birth Bifurcation)

Andronov–Hopf bifurcation is the birth of a limit cycle from an equilibrium point when the equilibrium changes stability via a pair of purely imaginary eigenvalues. The bifurcation can be supercritical or subcritical, resulting in a stable or unstable limit cycle, respectively [44]. When the value of a bifurcation parameter μ is increased, a pair of complex conjugate eigenvalues of the matrix $A(\mu)$ passes from left to right through the imaginary axis of a complex plane at $\mu = 0$, and all other eigenvalues have negative real parts [58]. As a result of supercritical Andronov–Hopf bifurcation, there is a change of stability of the stationary point accompanied by a birth from it of a stable limit cycle. A subharmonic bifurcation cascade of cycles is born as a result of Andronov–Hopf bifurcation in a system "simple" with a quadratic nonlinearity [64]. This bifurcation plays an important role and appears in many dynamical systems like the Lorenz, Rössler, and Chua systems [10], and is a starting point of various cascades of bifurcations leading to chaos. In the case of subcritical Andronov–Hopf bifurcation, an attractor disappears and hence such bifurcation is a crisis. Qualitative analysis of the limit-cycle birth bifurcation for the Rössler system was carried out by Magnitskii [52] and the domain of parameters in which the Andronov–Hopf bifurcation takes place was determined. The transition to chaos is connected to the limit-cycle birth bifurcation which is also the beginning of the Feigenbaum cascade of bifurcations [53].

Example 3.11

Discuss the occurrence of Andronov–Hopf bifurcation in the system

$$\frac{dx}{dt} = (\alpha + i\beta)x - 4x|x|^2, \quad x = x_1 + ix_2, \quad \alpha, \beta \text{ are parameters.}$$

SOLUTION

Let the solution of the system be written as $x_1(t) = u(t) \cos \beta t$, $x_2(t) = u(t) \sin \beta t$, that is, $x = u(t) \exp(i\beta t)$. Then, $u(t)$ satisfies the equation $(du/dt) = u(\alpha - 4u^2)$. For $\alpha < 0$, the system has the unique stable stationary solution which is the focus $u = 0$ (or $x_1 = 0$, $x_2 = 0$). For $\alpha = 0$, the zero solution is a stable focus where $u(t) = (8t + c)^{-1/2} \to 0$ as $t \to \infty$. For $\alpha > 0$, the system has (i) an unstable stationary solution which is the focus $u = 0$ (or $x_1 = 0$, $x_2 = 0$), and (ii) the solution $4u^2 = \alpha$, or $x_1(t) = (\sqrt{\alpha}/2)\cos \beta t$, $x_2(t) = (\sqrt{\alpha}/2)\sin \beta t$. For this solution, $x_1^2 + x_2^2 = (\alpha/4)$ and the trajectory is a stable limit cycle with radius $(\sqrt{\alpha}/2)$. This is a supercritical Andronov–Hopf bifurcation.

Poincaré Map and Poincaré Section

In the study of the three-body problem in celestial mechanics, Poincaré [75] used the idea of replacing a flow of an nth-order continuous time system with an associated $(n-1)$th discrete time system (map). Generally, any discrete time system that is associated with an ODE is referred to as a Poincaré map [116]. The discrete time system is constructed by viewing the phase-space diagram in such a way that the motion is observed periodically. This technique has many advantages like reduction of the dimension of the system by one unit under transition to map. Construction of a Poincaré map requires some knowledge of the geometrical structure of the phase space of the ODE. There are no general analytical methods for computing a Poincaré map associated with arbitrary ODEs. The definition of the Poincaré map is different for autonomous and nonautonomous systems. The definition of a Poincaré map is rarely used in studies as it requires advanced knowledge of the position of a limit cycle. Usually, in constructing a Poincaré map, we choose an $(n-1)$ dimensional surface S, which divides R^n into two regions. If S is chosen properly, then the trajectory under observation repeatedly passes through S. The set of these crossing points is called a Poincaré map. If a system has (i) periodic evolution, or (ii) a quasiperiodic attractor, or (iii) a trajectory in the phase portrait moving on a 2D torus, then the corresponding Poincaré map consists of (i) a repeating set of points, or (ii) points on a closed orbit, or (iii) points moving on an ellipse, respectively [46].

The Poincaré section has one dimension less than the dimension of the phase space. The Poincaré map maps the points of the Poincaré section onto itself. It relates any two consecutive intersection points. Only those intersection points count which come from the same side of the plane. If the Poincaré section is carefully chosen, then no information is lost concerning the qualitative behavior of the dynamics. Consider a 3D state space (X_1, X_2, X_3). For example, choose a section plane, $X_1 - X_2$. Mark the points at which the trajectory crosses this plane successively in the same sense, downward along the negative X_3 axis, say. Thus, we get a sequence of points $P_0, P_1, P_2, \ldots$. This set constitutes the Poincaré section. If the motion is deterministic, then $P_{n+1} = f(P_n)$ is the mapping relating the successive points. This is known as the Poincaré map or the first return map. If the flow was periodic following a limit-cycle attractor, the successive points $P_0, P_1, P_2, \ldots$ of the Poincaré section merge to a single point. If the flow trajectory was on a torus, the Poincaré section is a finite set of points. Finally, if the dynamics is chaotic, the Poincaré section will be a splatter of points covering an area where the successive points of intersection jump erratically all over the area. For dissipative flows, the Poincaré section will generate a discrete map that will contract areas in the plane too [43]. Since the dimension of the attractor for a dissipative 3D chaotic flow must lie between 2 and 3, its Poincaré section will have a dimension between 1 and 2.

Lyapunov Exponents

In the earlier sections, we have briefly commented on the use of Lyapunov exponents. One of the most important concepts for understanding the behavior of a dynamical system is the study of its Lyapunov characteristic exponents (LCE). The LCEs are asymptotic measures of the average rate of growth/decay of small perturbations. They demonstrate the extent of the sensitivity of the system to initial conditions. The value of the maximal LCE is an indicator of the chaotic or regular nature of orbits, while the whole spectrum of LCE is related to entropy (Kolmogorov–Sinai entropy). It measures the rate at which nearby orbits converge or diverge [94].

Consider any two points in a space, $A(x_0)$ and $B(x_0 + \Delta x_0)$, each of which generates an orbit as the solution of the system. The initial separation between the two orbits is Δx_0. The trajectories converge or diverge approximately as $|\Delta x(t)| \approx e^{\lambda t}|\Delta x_0|$. The Lyapunov exponent is the time constant λ in the expression for the distance between two nearby orbits and is defined as

$$\lambda = \lim_{t \to \infty, |\Delta x_0| \to 0} \left[\frac{1}{t} \ln \frac{|\Delta x(x_0, t)|}{|\Delta x_0|} \right]. \tag{3.28}$$

The behavior of the neighboring orbits depends on the value of the Lyapunov exponent. If $\lambda < 0$, then the system is attracted to a fixed point signifying a stable equilibrium. The orbit is a stable periodic orbit. These systems are nonconservative (dissipative) (see Figure 3.8a,b). If $\lambda = 0$, then the system is neutrally stable, signifying periodicity (or quasiperiodicity). The system is in some steady-state mode. If $\lambda > 0$, then the system may be chaotic and unstable. Often, it is sufficient to consider the maximal Lyapunov exponent (MLE) as it determines the overall predictability of the system. The general properties of the Lyapunov exponents associated with dynamical systems and the numerical computation of Lyapunov exponents for 1D map (which is also called Lyapunov number) and flows are given by Lakshmanan and Rajasekhar [46].

The dynamics of an SCA is complex and the future state of the system is not solely decided by the evolution law in a deterministic manner, but is very sensitive to any initial error, which invariably occurs in specifying the initial state. There is loss of predictability of the future state of the system after a finite amount of time has elapsed. This time is inversely proportional to the maximum Lyapunov exponent, which quantifies the sensitivity of the system's asymptotic dynamics with respect to the initial state, that is $t_L = 1/\lambda_1$, where λ_1 is the MLE.

In the following, we describe in brief, a direct method for computing the LCE of order 1. Algorithms are available for computing LCEs of order 2, LCEs of order 3, and the spectrum of LCEs. Modern techniques are available using the Gram–Schmidt orthonormalization method [6], singular-value decomposition method [17,101] and the QR decompositions [9]. A Fortran code for the computation of LCEs is given in Ref. [118]. A code for computation of LCEs

(a)   (b)

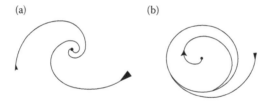

FIGURE 3.8
(a) Attracting fixed point and (b) attracting orbit.

is also available in the algebra platform of *Mathematica*® [84]. A good review of the Lyapunov theory is given by Skokos [94].

Computation of Lyapunov Exponents of Order 1

Consider the system of differential equations

$$\frac{dx}{dt} = f_1(t, x, y, z), \quad \frac{dy}{dt} = f_2(t, x, y, z), \quad \frac{dz}{dt} = f_3(t, x, y, z), \quad (3.29)$$

with given initial conditions $x(0)$, $y(0)$, $z(0)$.

Step 1: Solve the system of equations using the fourth-order Runge–Kutta numerical method. We get the values of x,y,z at time t.

Step 2: Obtain the corresponding variational equations

$$\delta\dot{x} = g_1(\delta t, \delta x, \delta y, \delta z), \quad \delta\dot{y} = g_2(\delta t, \delta x, \delta y, \delta z),$$

$$\delta\dot{z} = g_3(\delta t, \delta x, \delta y, \delta z) \quad (3.30)$$

or

$$\dot{\omega}(t) = A(t)\omega(t),$$

where $A(t)$ is the coefficient matrix and $\omega(t) = (\delta x, \delta y, \delta z)^\mathrm{T}$.

Step 3: *Computation of LCEs of order 1*

Let, $L_1 = \mathbf{R}^3$, $L_2 = \mathbf{R}^2$, and $L_3 = \mathbf{R}$ where $L_3 \subset L_2$, and $L_2 \subset L_1$. Select an initial deviation vector ω_0 from L_1, but not from L_2, such that its norm is 1. Solve Equation 3.30 with step length Δt and using solutions from step 1 and ω_0 to get $\omega_1(\Delta t)$. Similarly, find $\omega_2(\Delta t)$ and $\omega_3(\Delta t)$ by choosing initial deviation vectors from L_2, but not from L_3 and from L_3, respectively, such that the norm of each deviation vector is 1. Compute the Lyapunov exponents

$$\lambda_i = \lim_{t\to\infty}\left[\frac{1}{t}\ln\frac{\|\omega_i(t)\|}{\|\omega_0(t)\|}\right], \quad i = 1, 2, 3, \quad \text{where } \|\omega_0(t)\| = 1.$$

Step 4: At time $t = \Delta t$, normalize $\omega_1(\Delta t)$. Using the solution of Equation 3.29 and the initial deviation vector as $\omega_1(\Delta t)$, solve Equation 3.30 again to obtain $\omega_1(2\Delta t)$. Similarly, compute $\omega_2(2\Delta t)$ and $\omega_3(2\Delta t)$. Repeat the computation k times. Then, the Lyapunov exponents are given by the limits

$$\lambda_j = \lim_{t\to\infty}\left[\frac{1}{t}\ln\frac{\|\omega_j(t)\|}{\|\omega_0(t)\|}\right] \approx \lim_{k\to\infty}\left[\frac{1}{k\Delta t}\sum_{i=1}^{k}\ln\|\omega_j(i\Delta t)\|\right] \quad (3.31)$$

$j = 1, 2, 3$, where $\|\omega_0(t)\| = 1$.

Example 3.12

Compute the Lyapunov exponents of the LV model under the given parameter values and initial conditions.

$$\frac{dR}{dt} = R(a - bF), \quad \frac{dF}{dt} = F(cR - d). \quad R(0) = 10, F(0) = 5.$$

Assume $a = 1.5$, $b = 1.0$, $c = 0.1$, $d = 3.0$, and $\Delta t = 0.1$.

SOLUTION

We solve the system of equations using *Mathematica*. The solutions of $R(t)$ and $F(t)$ are given in Table 3.1.

The corresponding variational equations are

$$\frac{dp}{dt} = (1.5 - F)\,p - Rq, \quad \frac{dq}{dt} = 0.1Fp + q(0.1R - 3).$$

Computation of Lyapunov Exponents of Order 1

The variational equations are also solved using *Mathematica*. Assume the initial vectors as $\omega_1(0) = (1/\sqrt{2},\ 1/\sqrt{2})^T$, and $\omega_2(0) = (1, 0)^T$ for computing λ_1, λ_2, respectively. At every time step, denote the normalized ω by $\hat{\omega}$. The numerical solutions obtained for computing λ_1 are given in Table 3.2 and for computing λ_2 are given in Table 3.3.

Using Equation 3.31, at $t = 1$, we obtain $\lambda_1 = -6.43274$, and $\lambda_2 = -21.8615$. The maximal Lyapunov exponent is $-6.43274 < 0$. The plot of λ_1, λ_2 is given in Figure 3.9.

Remark 3.4

The Lyapunov number λ for 1D map $x_{n+1} = f(\mu, x_n)$ is defined by

$$\lambda = \lim_{n \to \infty} \frac{1}{n} \log \frac{d}{dx} f^n(x_0) = \lim_{n \to \infty} \frac{1}{n} \log \prod_{i=1}^{n} f'(x_i). \quad (3.32)$$

TABLE 3.1
Solutions of $R(t)$ and $F(t)$, Example 3.12

t	$R(t)$	$F(t)$	t	$R(t)$	$F(t)$
0.1	7.40364	4.03538	0.6	4.8548	1.17809
0.2	5.99974	3.19443	0.7	5.08222	0.917104
0.3	5.24883	2.50245	0.8	5.44404	0.716044
0.4	4.88637	1.94974	0.9	5.93543	0.561419
0.5	4.77787	1.51569	1.0	6.55993	0.442645

TABLE 3.2

Solutions for Computing λ_1, Example 3.12

t	0.1	0.2	0.3	0.4	0.5
$\omega_1(t)$	$\begin{pmatrix} -0.046415 \\ 0.591740 \end{pmatrix}$	$\begin{pmatrix} -0.477551 \\ 0.471133 \end{pmatrix}$	$\begin{pmatrix} -0.682493 \\ 0.358537 \end{pmatrix}$	$\begin{pmatrix} -0.738693 \\ 0.260933 \end{pmatrix}$	$\begin{pmatrix} -0.703850 \\ 0.180883 \end{pmatrix}$
t	0.6	0.7	0.8	0.9	1.0
$\omega_1(t)$	$\begin{pmatrix} -0.619985 \\ 0.118156 \end{pmatrix}$	$\begin{pmatrix} -0.515252 \\ 0.0710067 \end{pmatrix}$	$\begin{pmatrix} -0.407502 \\ 0.037317 \end{pmatrix}$	$\begin{pmatrix} -0.307823 \\ 0.014433 \end{pmatrix}$	$\begin{pmatrix} -0.222268 \\ -0.000087 \end{pmatrix}$

TABLE 3.3

Solutions for Computing λ_2, Example 3.12

t	0.1	0.2	0.3	0.4	0.5
$\omega_1(t)$	$\begin{pmatrix} 0.687671 \\ 0.037349 \end{pmatrix}$	$\begin{pmatrix} 0.443422 \\ 0.055939 \end{pmatrix}$	$\begin{pmatrix} 0.261598 \\ 0.061588 \end{pmatrix}$	$\begin{pmatrix} 0.132465 \\ 0.059218 \end{pmatrix}$	$\begin{pmatrix} 0.045619 \\ 0.052409 \end{pmatrix}$
t	0.6	0.7	0.8	0.9	1.0
$\omega_1(t)$	$\begin{pmatrix} -0.008335 \\ 0.043566 \end{pmatrix}$	$\begin{pmatrix} -0.037599 \\ 0.034208 \end{pmatrix}$	$\begin{pmatrix} -0.051169 \\ 0.025862 \end{pmatrix}$	$\begin{pmatrix} -0.053911 \\ 0.018799 \end{pmatrix}$	$\begin{pmatrix} -0.05038 \\ 0.013072 \end{pmatrix}$

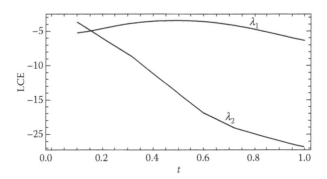

FIGURE 3.9
Lyapunov exponents (LCE of order 1.) Example 3.12.

EXERCISE 3.1

1. Draw the phase plane diagram and time series for the following model, for the parameter values $a_1 = 1.0$, $b_1 = 1.0$, $w = 1.667$, $D = D_1 = 0.334$, $D_2 = 0.5$, $a_2 = 0.4$, $w_3 = 0.05$, $w_1 = 1.667$, $w_2 = 0.05$, $c = 0.01$, $D_3 = 0.5$ [81].

$$\frac{dx}{dt} = a_1x - b_1x^2 - \frac{wxy}{(x+D)},$$

$$\frac{dy}{dt} = -a_2y + \frac{w_1xy}{(x+D_1)} - \frac{w_2yz}{(y+D_2)},$$

$$\frac{dz}{dt} = -cz + \frac{w_3yz}{y+D_3}.$$

2. Draw the projections of the chaotic attractor onto the three different planes of the phase space R^3 (x,y,z) for the following model system using the given initial conditions and the parameter values

$$\frac{dx}{dt} = a_1x - b_1x^2 - \frac{wxy}{(\alpha + \beta y + \gamma x)},$$

$$\frac{dy}{dt} = -a_2y + \frac{w_1xy}{(\alpha + \beta y + \gamma x)} - \frac{w_2yz}{(\alpha_2 + \beta_2 z + \gamma_2 y)},$$

$$\frac{dz}{dt} = cz^2 - \frac{w_3z^2}{y+D_3}. \quad (X_0, Y_0, Z_0) = (1,\ 1,\ 1).$$

$a_1 = 2.2, b_1 = 0.05, w = 0.85, \alpha = 0.45, \beta = 0.2, \gamma = 0.6, a_2 = 1.05, w_1 = 1.65,$
$w_2 = 1.85, w_3 = 0.75, \alpha_2 = 0.85, \beta_2 = 0.95, \gamma_2 = 0.9, c = 0.0273, D_3 = 20.$

3. Draw the 2D scan diagrams using the pairs of the parameters (i) (b, f), (ii) (b, d), where chaos was observed for the following model in the parameter ranges $1.0 \le b \le 7$, $0.0001 \le f \le 1$, $-5 \le \log d \le 0$, and for the parameter values $a = 5, b = 5, m = 0.6, f = 0.2, g = 2.5, s = 2, d = 10^{-4}$, $l = -1.4, n = 5$ with $r_x = s + lx$ at $x = 0.85$ [108]

$$\frac{\partial p}{\partial t} = r_xp(1-p) - \frac{apz}{(1+bp)} + d\frac{\partial^2 p}{\partial x^2},$$

$$\frac{\partial z}{\partial t} = \frac{npz}{(1+bp)} - mz - \frac{fz^2}{(1+g^2z^2)} + d\frac{\partial^2 z}{\partial x^2}.$$

4. Apply *pseudo-prey* method in the following 3D system and find the parameter values for the simulation experiments [108].

$$\frac{dx}{dt} = a_1x - b_1x^2 - \frac{wxy}{(\alpha + \beta y + \gamma x)},$$

$$\frac{dy}{dt} = -a_2y + \frac{w_1xy}{(\alpha + \beta y + \gamma x)} - \frac{w_2yz}{(\alpha + \beta z + \gamma y)},$$

$$\frac{dz}{dt} = cz^2 - \frac{w_3z^2}{y+D_3}.$$

5. Draw the BBS for the following model, for the parameter values
$a_1 = 1.75$, $a_2 = 1.0$, $w = 1$, $D = 10$, $b_1 = 0.05$, $w_1 = 2.0$, $w_3 = 3.75$, $\theta = 0.2$,
$D_1 = 10$, $w_2 = 1.5$, $D_2 = 10$, $c = 0.7$, $D_3 = 20$, $D_4 = 10$.

$$\frac{dx}{dt} = a_1 x - b_1 x^2 - \frac{wxy}{(x + D)},$$

$$\frac{dy}{dt} = -a_2 y + \frac{w_1 xy}{(x + D_1)} - \frac{w_2 yz}{(y + D_2)} - \theta \frac{x}{(x + D_4)} y,$$

$$\frac{dz}{dt} = -cz + \frac{w_3 yz}{(y + D_3)}.$$

6. Draw the bifurcation diagram for the model

$$\frac{dx}{dt} = a_1 x - b_1 x^2 - \frac{wxy}{(x + D)},$$

$$\frac{dy}{dt} = -a_2 y + \frac{w_1 xy}{(x + D_1)} - \frac{w_2 yz}{(y + D_2)} - \theta f(x)y,$$

$$\frac{dz}{dt} = -cz + \frac{w_3 yz}{(y + D_3)},$$

taking θ as control parameter in the ranges (i) $0 \le \theta \le 0.015$, (ii) $0 \le \theta$
≤ 0.3, and (iii) $0 \le \theta \le 0.25$. Assume $f(x) = x$, $x/(x + D_2)$, $x^2/(x^2 + D_2^2)$
and take the set of parameter values as $a_1 = 1.75$, $b_1 = 0.05$, $w = 1$,
$w_1 = 2$, $w_2 = 1.45$, $c = 0.1$, $w_3 = 1$, $a_2 = 1$, $D_3 = 20$, $D = D_1 = D_2 = 10$.

7. Draw the bifurcation diagram for the model

$$\frac{dx}{dt} = x \left[(1 - x) - \frac{y}{x + w_4} \right],$$

$$\frac{dy}{dt} = y \left[-w_5 + \frac{w_6 x}{x + w_4} - \frac{z}{y + (w_8 + w_9 y)z + w_{10}} \right],$$

$$\frac{dz}{dt} = z \left[-w_{11} + \frac{w_{12} y}{y + (w_8 + w_9 y)z + w_{10}} \right],$$

taking w_5 as a control parameter and (i) $w_{11} = 0.03$ and $0.15 < w_5 < 0.5$.
(ii) $w_{11} = 0.06$, $0.25 < w_5 < 0.5$; for the set of parameter values $w_4 = 0.25$,
$w_5 = 0.25$, $w_6 = 0.8$, $w_8 = 0.01$, $w_9 = 0.1$, $w_{10} = 0.28$, $w_{12} = 0.25$.

8. Discuss the saddle node bifurcation of the one-parameter family of 1D vector fields

$$\frac{dx}{dt} = f(x, \mu) = \frac{\mu}{4} - x^2, \quad x, \mu \in R^1.$$

9. Discuss the saddle node bifurcation in the following model:

$$\frac{dx}{dt} = -\mu x + y, \quad \frac{dy}{dt} = -ay + \frac{x^2}{9 + x^2}, \quad \mu, a > 0.$$

10. Discuss the occurrence of transcritical bifurcation in the system

$$\frac{dx}{dt} = f(x, \mu) = -x(x^2 - 4x - \mu), \quad x, \mu \in R^1. \quad -4 < \mu < 0.$$

11. Discuss the existence of pitchfork bifurcation of the dynamical system [121]

   i. $\dfrac{dx}{dt} = f(x, \mu) = \mu x - x^3, \quad x, \mu \in R^1.$

   ii. $\dfrac{dx}{dt} = f(x, \mu) = \mu x + x^3, \quad x, \mu \in R^1.$

12. Examine for saddle node bifurcation of the map $x \mapsto f(x, \mu) = x + (\mu/4) - x^2, \mu \in R^1.$

13. Examine for transcritical bifurcation of the map $x \mapsto f(x, \mu) = x\{1 + (\mu/2)\} - x^2, \mu \in R^1.$

14. Examine for pitchfork bifurcation of the map $x \mapsto f(x, \mu) = x\{1 + (\mu/4)\} - x^3, \mu \in R^1.$

15. Discuss the period-doubling bifurcation in the logistic map [3]

$$x_{n+1} = \mu - x_n^2 = f(x, \mu), \quad x, \mu \in R^1.$$

16. Show that subcritical Hopf bifurcation occurs at the origin as the parameter μ varies for the system

$$\frac{dx}{dt} = \mu x - 2y + xy, \quad \frac{dy}{dt} = \frac{x}{2} + \mu y + y^2.$$

17. Perform the Hopf bifurcation analysis using method of Liu of the positive equilibrium point for the model system given in Problem 2 [108].

18. Draw the Lyapunov exponent bifurcation diagrams for the following model taking ω_8, ω_{12} as control parameters and for the given set of parameter values

$$\frac{dx}{dt} = x(1-x) - \frac{xy}{x+\omega_4},$$

$$\frac{dy}{dt} = -\omega_5 y + \frac{\omega_6 xy}{x+\omega_7} - \frac{yz}{y+\omega_8 z + \omega_9},$$

$$\frac{dz}{dt} = \omega_{10} z^2 - \frac{\omega_{11} z^2}{y+\omega_{12}}.$$

$\omega_4 = 0.48$, $\omega_5 = 1.15$, $\omega_6 = 2.93$, $\omega_7 = 0.54$, $\omega_9 = 0.1$, $\omega_{10} = 0.35$, $\omega_{11} = 0.2$,
(i) $\omega_{12} = 0.25$, $0.01 \le \omega_8 \le 0.6$; and (ii) $\omega_8 = 0.21$, and $0.15 \le \omega_{12} \le 0.3$.

19. Calculate the Lyapunov number for the shift map [32], $x_{n+1} \equiv 2x_n$ (mod 1), $x_n \in [0, 1]$.

20. Compute the Lyapunov exponents of the LV model under the given initial conditions.

$$\frac{dR}{dt} = R(a - bF), \quad \frac{dF}{dt} = F(cR - d). \quad R(0) = 10, \quad F(0) = 10.$$

Assume $a = 4.5$, $b = 0.25$, $c = 0.5$, $d = 4.0$, and $\Delta t = 0.1$ [see Lotka–Volterra Model].

References

1. Afraimovich, V. S. 1989. Attractors. In: Gaponov, A. V., Rabinovich, M. I. and Engelbrechet, J. eds. *Nonlinear Waves*. Springer, Berlin.
2. Afraimovich, V. S., Arnold, V. I., Il'yashenko, Yu S., Shilnikov, L. P. 1989. Bifurcation theory. Dynamical Systems, (Ed. Shilnikov, L. P.) *Encyclopedia of Mathematical Sciences*. 5, Springer, Berlin, Heidelberg.
3. Anishchenko, V. S., Astakhov, V., Neiman, A., Vadivasova, T., Schimansky-Geier, L. 2006. *Nonlinear Dynamics of Chaotic and Stochastic Systems*. Springer–Verlag, Berlin.
4. Baker, G. L., Gollub, J. P. 1996. *Chaotic Dynamics: An Introduction*. Cambridge University Press, USA.
5. Becks, L., Hilker, F. M., Malchow, H., Jurgens, K., Arndt, H. 2005. Experimental demonstration of chaos in a microbial food web. *Nature* 435, 1226–1229.
6. Benettin, G., Galgani, L., Giorgilli, A., Strelcyn, J.-M. 1980. Lyapunov characteristic exponents for smooth dynamical systems and for Hamiltonian systems: A method for computing all of them. Part 2. Numerical application. *Meccanica* 15, 21–30.
7. Bergé, P., Pomeau, Y., Vidal, C. 1984. *Order within Chaos: Towards a Deterministic Approach to Turbulence*. John Wiley & Sons, New York.

8. Charlotte, W. 2009. What are the new implications of chaos for unpredictability? *Br. J. Philos. Sci.* 60(1), 195–220.

9. Chen, Z.-M., Djidjeli, K., Price, W.G. 2006. Computing Lyapunov exponents based on the solution expression of the variational system. *Appl. Math. Comput.* 174, 982–996.

10. Chua, L. O., Komuro, M., Matsumoto, T. 1986. The double scroll family. *IEEE Trans. Circuits Syst.* CAS-33, Pt. 1(2), 1073–1118.

11. Costantino, R. F., Cushing, J. M., Dennis, B., Desharnais, R. A. 1995. Experimentally induced transitions in the dynamic behaviour of insect population. *Nature* 375, 227–230.

12. Costantino, R. F., Desharnais, R. A., Cushing, J. M., Dennis, B. 1997. Chaotic dynamics in an insect population. *Science* 257, 389–391.

13. Curry, J. H. 1978. A generalized Lorenz system. *Commun. Math. Phys.* 60, 193–198.

14. Curry, J. H., Yorke, J. A. 1980. An algorithm for finding closed orbit. *Lect. Notes Math.* 819, 111–133.

15. Cushing, J. M., Costantino, R. F., Dennis, B., Desharnais, R. A., Henson, S. M. 2003. *Chaos in Ecology: Experimental Non-Linear Dynamics*. Elsevier, San Diego.

16. Dennis, B., Desharnais, R. A., Cushing, J. M., Costantino, R. F. 1995. Nonlinear demographic dynamics: Mathematical models, statistical methods, and biological experiments. *Ecol. Monogr.* 65, 261–281.

17. Dieci, L., Lopez, L. 2006. Smooth singular value decomposition on the symplectic group and Lyapunov exponents approximation. *Calcolo* 43, 1–15.

18. Dixon, P. A., Milicich, M. J., Sugihara, G. 1999. Episodic fluctuations in larval supply. *Science* 283, 1528–1530.

19. Eckmann, J. P., Ruelle, D. 1985. Routes to turbulence in dissipative dynamical systems. *Rev. Mod. Phys.* 57, 617–656.

20. Ellner, S., Turchin, P. 1995. Chaos in a noisy world: New methods and evidence from time–series analysis. *Am. Naturalist* 145, 343–375.

21. Feigenbaum, M. J. 1978. Quantitative universality for a class of nonlinear transformations. *J. Stat. Phys.* 19, 25–52.

22. Feigenbaum, M. J. 1979. The universal metric properties of nonlinear transformation. *J. Stat. Phys.* 21(6), 669–706.

23. Ferriere, R., Gatto, M. 1993. Chaotic population dynamics can result from natural selection. *Proc. Roy. Soc. Lond., Ser. B*, 251, 33–38.

24. Feudel, U., Grebogi, C. 1997. Multistability and control of complexity. *Chaos* 7(4), 597–604.

25. Gardini, L., Lupini, R., Messia, M. G. 1989. Hopf-bifercation and transition to chaos in Lotka–Volterra equation. *J. Math. Biol.* 27, 259–272.

26. Grebogi, C. 1994. Linear scaling laws in bifurcations of scalar maps. *Z. Naturforsch*, 49a, 1207–1211.

27. Grebogi, C., Ott, E., Pelikan, S., Yorke, J. A. 1984. Strange attractors that are not chaotic. *Physica D* 13, 261–268.

28. Grebogi, C., Ott, E., Yorke, J. A. 1982. Chaotic attractor in crisis. *Phys. Rev. Lett.* 48, 1507–1510.

29. Grebogi, C., Ott, E., Yorke, J. A. 1983. Crisis, sudden changes in chaotic attractors, and chaotic transients. *Physica* 7D, 181–200.

30. Grebogi, C., Ott, E., Yorke, J. A. 1985. Attractors on an N-torus: Quasiperiodicity versus chaos. *Physica* D15, 354–373.

31. Hanski, I., Turchin, P., Korpimaki, E., Henttonen, H. 1993. Population oscillations of boreal rodents: Regulation by mustelid predators leads to chaos. *Nature* 364, 232–235.

32. Hao, B.-L. 1989. *Elementary Symbolic Dynamics and Chaos in Dissipative Systems.* World Scientific, Singapore.

33. Hastings, A. 2001. Transient dynamics and persistence of ecological systems. *Ecol. Lett.* 4, 215–220.

34. Hastings, A. 2004. Transients: The key to long-term ecological understanding? *Trends Ecol. Evol.* 19, 39–45.

35. Hastings, A., Higgins, K. 1994. Persistence of transients in spatially structured ecological models. *Science* 263, 1133–1136.

36. Hastings, A., Hom, C. L., Ellner, S. P., Turchin, P., Godfray, H. C. J. 1993. Chaos in ecology—Is mother nature a strange attractor? *A. Rev. Ecol. Syst.* 24, 1–33.

37. Higgins, K., Hastings, A., Sarvela, J. N., Botsfort, L. W. 1997. Stochastic dynamics and deterministic skeletons: Population behavior of Dungeness crab. *Science* 276, 1431–1435.

38. Jansen, V. A. A. 1995. Regulation of predator–prey systems through spatial interactions: A possible solution to the paradox of enrichment. *Oikos* 74, 384–390.

39. Jørgensen, S. E. 1979. *Handbook of Environmental Data and Ecological Parameters.* Pergamon Press, New York.

40. Kapitaniak, T. 2000. *Chaos for Engineers: Theory, Applications and Control.* Springer-Verlag, Berlin.

41. Kaufmann, S. A. 1993. *Origins of Order: Self-Organization and Selection in Evolution.* Oxford University Press, Oxford.

42. Kendall, B. E. 2001. Cycles, chaos and noise in predator–prey dynamics. *Chaos, Solitons Fractals* 12, 321–332.

43. Kumar, N. 1996. *Deterministic Chaos: Complex Chance out of Simple Necessity.* University Press, India.

44. Kuznetsov, Yu A. 2006. Andronov–Hopf bifurcation. *Scholarpedia* 10, 1858.

45. Kuznetsov, Yu A., Muratori, S., Rinaldi, S. 1992. Bifurcations and chaos in a periodic predator–prey model. *Intl. J. Bif. Chaos* 2, 117–128.

46. Lakshmanan, M., Rajasekar, S. 2003. *Nonlinear Dynamics: Integrability, Chaos and Pattern.* Springer-Verlag, Germany.

47. Langton, C. G. 1990. Computation at the edge of chaos: Phase-transitions and emergent computation. *Physica* D42, 12–37.

47a. Laplace, P. S. *Oeuvres Complete de Laplace*, Vol 8, 144, Imprimerie Royale, Paris 1843; English translation: Princeton University Press, Princeton 1997.

48. Li, T. Y., Yorke, J. A. 1975. Period three implies chaos. *Am. Math. Mon.* 82(10), 985–992.

49. Liu, W.-M. 1994. Criterion of Hopf-bifurcations without using eigen values. *J. Math. Anal. Appl.* 182, 250–256.

50. Liu, Q. X., Jin, Z., Li, B.-L. 2008. Resonance and frequency-locking phenomena in spatially extended phytoplankton–zooplankton system with additive noise and periodic forces. *J. Stat. Mech.: Theory Exp.* 5, 5011.

51. Lorenz, E. N. 1963. Deterministic non-periodic flow. *J. Atm. Sci.* 20(2), 130–141.

52. Magnitskii, N. A. 1995. Hopf bifurcation in the Rössler system. *Diff. Equ.* 31(3), 538–541.

53. Magnitskii, N. A., Sidorov, S. V. 2006. Bifurcations in nonlinear systems of ordinary differential equations, (Chapter 2). In: *New Methods for Chaotic Dynamics.* (Ed. Chua, L.O), World Scientific, New Jersey.

54. Malchow, H. 1993. Spatio-temporal pattern formation in nonlinear nonequilibrium plankton dynamics. *Proc. Roy. Soc. Lond., Ser. B* 251, 103–109.

55. Malchow, H., Hilker, F. M., Sarkar, R. R., Brauer, K. 2005. Spatiotemporal patterns in an excitable plankton system with lysogenic viral infection. *Math. Comp. Model.* 42, 1035–1048.

56. Malchow, H., Petrovskii, S. V., Medvinsky, A. B. 2002. Numerical study of plankton-fish dynamics in a spatially structured and noisy environment. *Ecol. Model.* 149, 247–255.

57. Malchow, H., Petrovskii, S. V., Venturino, E. 2008. *Spatio-temporal Patterns in Ecology and Epidemiology: Theory, Models and Simulation.* CRC Press, UK.

58. Marsden, J. E., McCracken, M. 1976. *The Hopf Bifurcation and Its Applications.* Springer-Verlag, New York.

59. May, R. M. 1976. Simple mathematical models with very complicated dynamics. *Nature* 261, 459–467.

60. McDonald, S. W, Grebogi, C., Ott, E., Yorke, J. A. 1985. Fractal basin boundaries. *Physica D* 17, 125–153

61. Medvinsky, A. B., Petrovskii, S. V., Tikhonova, I. A., Malchow, H., Li, B.-L. 2002. Spatiotemporal complexity of plankton and fish dynamics. *SIAM Rev.* 44(3), 311–370.

62. Medvinsky, A. B., Tikhonova, I. A., Aliev, R. R., Li, B.-L., Lin, Z.-S., Malchow, H. 2001. Patchy environment as a factor of complex plankton dynamics. *Phys. Rev.* E64, 021915–021922.

63. Morozov, A., Petrovskii, S., Li, B.-L. 2004. Bifurcations and chaos in a predator–prey system with the Allee effect. *Proc. Roy. Soc. Lond., Ser. B* 271, 1407–1414.

64. Navikov, M. D., Pavlov, B. M. 2000. On one nonlinear model with complex dynamics. *Vestnik MSU* 2, 3–7 (in Russian).

65. Newhouse, S., Ruelle, D., Takens, F. 1978. Occurrence of strange axiom A attractors near quasiperiodic flows on Tm, m ≥ 3. *Commun. Math. Phys.* 64(1), 35–40.

66. Nusse, H. E., Yorke, J. A. 1994. *Dynamics: Numerical Explorations. Applied Mathematical Sciences* 101. Springer-Verlag, New York.

67. Ott, E. 1981. Strange attractors and chaotic motions of dynamical systems. *Rev. Mod. Phys.* 53, 655–671.

68. Ott, E. 1993. *Chaos in Dynamical Systems.* Cambridge University Press, Cambridge, UK.

69. Pascual, M. 1993. Diffusion induced chaos in a spatial predator–prey system. *Proc. Roy. Soc. Lond., Ser. B* 251, 1–7.

70. Petrovskii, S. V., Kawasaki, K., Takasu, F., Shigesada, N. 2001. Diffusive waves, dynamical stabilization and spatio-temporal chaos in a community of three competitive species. *Japan J. Indus. Appl. Math.* 18, 459–481.

71. Petrovskii, S. V., Li, B.-L., Malchow, H. 2004. Transition to spatiotemporal chaos can resolve the paradox of enrichment. *Ecol. Compl.* 1(1), 37–47.

72. Petrovskii, S. V., Malchow, H. 1999. A minimal model of pattern formation in a prey–predator system. *Math. Comp. Model.* 29, 49–63.

73. Petrovskii, S. V., Malchow, H. 2001. Wave of chaos: New mechanism of pattern formation in spatio-temporal population dynamics. *Theor. Pop. Biol.* 59, 157–174.

74. Plykin, R. V. 1980. Hyperbolic attractors of diffeomorphisms. *Uspekhi Math. Nauk.* 35, 94–104.

75. Poincaré, H. 1899. *Les Méthodes Nouvelles de la Mécanique Céleste*, 3. Gauthier-Villars, Paris.

75a. Poincaré, H. 1903. *Science and Method*, 1914, Chapter 4, Chance p. 68, English translation by Francis Maitland. University of Toronto Library, Thomas Nelson and Sons, New York. http://wwwarchive.org/detals/sciencemethod00poinuoft

76. Pomeau, Y., Manneville, P. 1980. Intermittent transition to turbulence in dissipative dynamical systems. *Commun. Math. Phys.* 74, 189–197.

77. Rai, V. 2004. Chaos in natural populations: Edge or wedge? *Ecol. Compl.* 1(2), 127–138.

78. Rai, V., Anand, M., Upadhyay, R. K. 2007. Trophic structure and dynamical complexity in simple ecological models. *Ecol. Comp.* 4, 212–222.

79. Rai, V., Schaffer, W. M. 2001. Chaos in ecology. *Chaos, Solitons Fractals*, 12, 197–203.

80. Rai, V., Sreenivasan, R. 1993. Period doubling bifurcations leading to chaos in a model food-chain. *Ecol. Model.* 69, 63–77.

81. Rai, V., Upadhyay, R. K. 2006. Evolving to the edge of chaos: Chance or necessity? *Chaos, Solitons Fractals*, 30, 1074–1087.

82. Rinaldi, S., Muratori, S., Kuznetsov, Y. A. 1993. Multiple attractors, catastrophes and chaos in seasonally perturbed predator–prey communities. *Bull. Math. Biol.* 55, 15–35.

83. Ruelle, D., Takens, F. 1971. On the nature of turbulence. *Commun. Math. Phys.* 20(3), 167–192.

84. Sandri, M. 1996. Numerical calculation of Lyapunov exponents. *Mathematica J.* 6, 78–84.

85. Schaffer, W. M. 1988 Perceiving order in the chaos of nature. In: Boyce, M.S. Ed. *Evolution of Life Histories in Mammals*. Yale University Press, New Haven, CT, pp. 313–350.

86. Scheffer, M. 1991a. Fish and nutrients interplay determines algal biomass: A minimal model. *Oikos* 62, 271–282.

87. Scheffer, M. 1991b. Should we expect strange attractors behind plankton dynamics—And if so, should we bother? *J. Plankton Res.* 13, 1291–1305.

88. Scheffer, W. M., Kot, M. 1985. Do strange attractors govern ecological systems? *Bioscience* 35(6), 342–350.

89. Scheffer, M., Rinaldi, S., Kuznetsov, Yu A., Van Nes, E. H. 1997. Seasonal dynamics of Daphnia and algae explained as a periodically forced predator–prey system. *Oikos* 80, 519–532.

90. Segel, L. A., Jackson, J. L. 1972. Dissipative structure: An explanation and an ecological example. *J. Theor. Biol.* 37, 545–559.

91. Sherratt, J. A., Lewis, M. A., Fowler, A. C. 1995. Ecological chaos in the wake of invasion. *Proc. Natl. Acad. Sci. USA* 92, 2524–2528.

92. Shilnikov, L. P. 1984. Bifurcation theory and turbulence. In: *Nonlinear and Turbulence Processes*, Vol. 2. Harward Academic Publisher, Gordon and Breach.

93. Shulenberger, L., Yalcinyaka, T., Lai, Y. C., Holt, R. D. 1999. Controlling transient chaos to prevent species extinction. *Phys. Lett. A* 260, 156–161.

94. Skokos, Ch. 2010. The Lyapunov characteristic exponents and their computation. *Lect. Notes Phys.* 790, 63–135.

95. Smith, R. H., Daniels, S., Simkiss, K., Bell, E. D., Ellner, S. P., Forest, M. B. 2000. Blowflies as a case study in non-linear population dynamics. In: Perry, J. N., Smith, R. H., Woiwod, I. P., Morse, D. Eds., *Chaos in Real Data*. Kluwer Academic Publishers, Dordrecht, Netherlands, pp. 117–172.

96. Smith, L.A. 2007. *Chaos: A Very Short Introduction*. Oxford University Press, Oxford.

97. Sommerer, J. C., Ott, E. 1993. A physical system with qualitatively uncertain dynamics. *Nature, Lond.* 365, 138–140.

98. Sprott, J.C. 2010. *Elegant Chaos.* World Scientific Publishing Co. Pte. Ltd., Singapore.

99. Steele, J. H., Henderson, E. W. 1981. A simple plankton model. *Am. Natur.* 117, 676–691.

100. Steffen, E., Malchow, H., Medvinsky, A. B. 1997. Effects of seasonal perturbation on a model plankton community. *Envir. Model. Assess.* 2, 43–48.

101. Stewart, D. E. 1997. A new algorithm for the SVD of a long product of matrices and the stability of products. *Electr. Trans. Numer. Anal.* 5, 29–47.

102. Stewart, H. B., Ueda, Y., Grebogi, C., Yorke, J. A. 1995. Double crisis in two-parameter dynamical systems. *Phys. Rev. Lett.* 75, 2478–2481.

103. Strogatz, S. H. 2007. *Nonlinear Dynamics and Chaos.* Westview Press, USA.

104. Tél, T., Gruiz, M. 2006. *Chaotic Dynamics: An Introduction Based on Classical Mechanics.* Cambridge University Press, New York.

105. Turchin, P. 2003. *Complex Population Dynamics: A Theoretical/Empirical Synthesis.* Princeton University Press, Princeton, NJ.

106. Turchin, P., Ellner, S. P. 2000. Living on the edge of chaos: Population dynamics of Fennoscandian voles. *Ecology* 81, 3099–3116.

107. Upadhyay, R. K. 1999. *Dynamical System Studies in Model Ecosystems.* PhD thesis, IIT Delhi, India.

108. Upadhyay, R. K. 2009. Observability of chaos and cycles in ecological systems: Lessons from predator–prey models. *Intl. J. Bifu. Chaos* 19(10), 3169–3234.

109. Upadhyay, R. K., Iyengar, S. R. K., Rai, V. 1998. Chaos: An ecological reality? *Intl. J. Bifu. Chaos* 8, 1325–1333.

110. Upadhyay, R. K., Kumari, N., Rai, V. 2009. Wave of chaos in a diffusive system: Generating realistic patterns of patchiness in plankton-fish dynamics. *Chaos, Solitons Fractals* 40(1), 262–276.

110a. Upadhyay, R. K., Kumari, N., Rai, V. 2009. Exploring dynamical complexity in diffusion driven predator–prey systems: Effect of toxin production by phytoplankton and spatial heterogeneities. *Chaos, Solitons Fractals* 42(1), 584–594.

111. Upadhyay, R. K., Kumari, N., Rai, V. 2008. Wave of chaos and pattern formation in spatial predator–prey systems with Holling type IV predator response. *Math. Model. Nat. Phen.* 3(4), 71–95.

111a. Upadhyay, R. K., Naji, R.K., Nitu Kumari. 2007. Dynamical complexity in some ecological models: Effect of toxin production by phytoplankton. *Nonlinear Anal.: Model. Control* 12(1), 123–138.

112. Upadhyay, R. K., Rai, V. 2009. Complex dynamics and synchronization in two non-identical chaotic ecological systems. *Chaos, Solitons Fractals* 40, 2233–2241.

113. Upadhyay, R. K., Rao, V. S. H. 2009. Short term recurrent chaos and role of toxin producing phytoplankton on chaotic dynamics in aquatic systems. *Chaos, Solitons Fractals* 39, 1550–1564.

114. Vandermeer, J. 1993. Loose coupling of predator–prey cycles: Entrainment, chaos and intermittency in the classical MacArthur consumer-resource equations. *Am. Natur.* 141, 687–716.

115. Vilar, J. M. G., Sole, R. V., Rubi, J. M. 2003. On the origin of plankton patchiness. *Phys. A: Stat. Mech. Appl.* 317, 239–246.

116. Wiggins, S. 1990. *Introduction to Applied Nonlinear Dynamical Systems and Chaos.* Springer-Verlag, New York.

117. Williams, R. 1974. Expanding attractors. *Publ. Math. IHES* 43, 169–203.

118. Wolf, A., Swift, J. B., Swinney, H. L., Vastano, J. A. 1985. Determining Lyapunov exponents from a time series. *Physica D* 16, 285–317.
119. Xiao, J.-H., Li, H.-H., Yang, J.-Z., Hu, G. 2006. Chaotic Turing pattern formation in spatiotemporal systems. *Front. Phys. China* 1, 204–208.
120. Yorke, J. A. http://www-chaos.umd.edu/~yorke

4

Chaotic Dynamics in Model Systems from Natural Science

Introduction

In Chapter 2, we discussed the modeling of systems from natural science. Many models from natural science or ecological systems admit chaotic solutions, but no field evidence has ever been found. Since the pioneering work of Sir Robert May [57–59], deterministic chaos has been studied in models by many authors theoretically [3,34,61,77,95,102,103,107,110], in the laboratory [8,22,40], and in the field [17,30,99,100]. Although chaos was observed by many authors in the models they had considered, it was not observed either in the laboratory studies [10,11] or from the field studies. A number of authors reported the existence of chaos in an appreciable range of a critical parameter of the system [2,34,41,43,45,80,104,105,107,110,111,116,117]. Attempts were also made to examine if the chaotic dynamics is robust with respect to changes in system parameters in a 2D space. If chaos exists in a sufficiently large area in a suitably chosen 2D parameter space, then it may be possible to observe chaos in at least in the laboratory. But, there is no confirmed example of dynamical chaos in an ecological system to date. This lack of evidence is puzzling, as experimental evidence [8] suggests that chaotic dynamics may be possible in real biological populations. Many authors [4,25,93,96] argued that populations evolve away from chaotic dynamics and this could explain the scarcity of evidence in favor of chaos. The Cornell group's experiments carried out so far suggest that this may not be true [9]. The studies of an interdisciplinary team of Cushing et al. [10] suggested that complex dynamics in ecological data could be the result of some simple rules.

For investigating the dynamical behavior of the model systems, the following tools which we have been described in Chapter 3 are used: (i) calculation of the basin boundary structures, (ii) performing 2D parameter scans using two of the parameters in the system as base parameters, (iii) drawing the bifurcation diagrams, (iv) performing time-series analysis and drawing the phase-space diagrams, and (v) generating the spatiotemporal patterns and wave of chaos phenomena for spatial models.

One of the aims of population and community ecology is to understand how a population of a given species influences the dynamics of population of the other species that are members of the same interaction network. Interaction networks in natural ecosystems can be visualized as consisting of simple units known as food chains or food webs that consist of a number of species linked by tropical interactions. Researchers have focused a great deal of their attention on analyzing the dynamical behavior of model food chains. Two continuous time models of interacting species have been extensively studied in literature. These models exhibit only two basic patterns: approach to equilibrium (stable focus) or to a limit cycle. Three species continuous time models are reported to have more complicated patterns. Often, these models form dissipative dynamical systems that can possess three distinct dynamical possibilities like stable focus, limit cycle, and chaos in the phase space [14,42]. Similar results were obtained for multispecies food-web models [47,107]. The models are governed either by difference or differential equations. The difference equation models describe the evolution of biological populations with nonoverlapping generations. The success of these models depends on the identification of general ecological principles [73,98]. For our discussions in this chapter, the following two ecological principles form the basic structure of the model systems:

a. A specialist predator dies out when its favorite food is absent or is in short supply. The predator population decays exponentially in the absence of its prey.

b. The generalist predator switches to an alternative food option as and when it faces difficulty to find its favorite preys. The per capita growth of a generalist predator is limited by dependence on its favorite preys and this limitation is inversely proportional to per capita availability of preys at any instant of time.

Chaos in Single Species Model Systems

Model 1

In section "Model 6" (see Equation 2.34), analysis of the logistic model $x_{n+1} = Ax_n (1 - x_n)$, $0 < x_n < 1$, was given. May [57,58,60] and Thunberg [97] made pioneering studies of the logistic models. The model is a simple example that exhibits all the situations of nonlinear behavior like period-doubling bifurcation, chaos, and so on, depending on the value of the control parameter A. At $A = 4$, stretching out and folding back phenomena occurs, which is essential to chaotic behavior. It maps the interval $(0, 1)$ back onto itself. The fixed points $x^* = 0$ and 0.75 are unstable, while $x^* = 0.345491$ and 0.904508 are

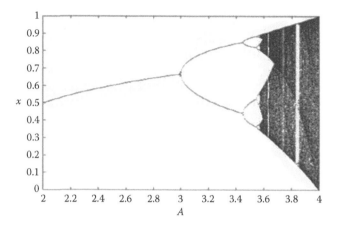

FIGURE 4.1
Bifurcation diagram for the logistic map. (From Thunberg, H. *SIAM Rev.* 43(1), 3–30. Copyright 2001, Society for Industrial and Applied Mathematics (SIAM). Reprinted with permission. All rights reserved.)

unstable period 2 cycles. For $A > 4$, transient chaos occurs and the orbit eventually escapes to infinity for most initial values. The bifurcation diagram plotted by Thunberg [97] is shown in Figure 4.1.

Model 2

Consider the general Ricker model (see section "Model 3," Equation 2.31) (Jorge Duarte et al. [12])

$$P_{n+1} = P_n e^{\lambda[1-(P_n/K)]}\beta(P_n),\qquad(4.1)$$

where the term $\beta(P_n)$ represents a positive density-dependent factor. In Ref. [12], N is used in place of P. Here, λ is the intrinsic growth rate and K is the carrying capacity in the absence of the positive density dependence ($\beta(P_n) = 1$). The dynamics of the model $P_{n+1} = P_n f(P_n)$, where $f(P_n)$ denotes the per-capita growth rate of the population, was studied in detail by Schreiber [90]. Two types of dynamics namely, persistence of population (occurs when $f(0) > 1$ and $f(N) > 0$ for all N) and extinction ($f(N) < 1$ for all N), are possible in the model studied by Schreiber. Jorge Duarte et al. [12] generalized and made a detailed study of the above model under the Allee effect. They have used the computation of Lyapunov exponents and topological entropy to study the model. By introducing the Allee effect, Jorge Duarte et al. showed that two more scenarios are possible, namely bistability (the population persists or becomes extinct depending on the initial conditions) and essential extinction (extinction occurs for all initial conditions). Essential extinction occurs due to transitory chaos.

Model with Predator Saturation

To study the Allee effect due to predator saturation, Jorge Duarte et al. [12] assumed $\beta(P_n)$ in Equation 4.1 as $\beta(P_n) = e^{-m/[1+sP_n]}$, where m is the intensity of predation and the parameter s is proportional to the handling time [32]. The expression for $\beta(P_n)$ represents the probability of escaping predation by a predator with a saturating functional response. Making the substitution $(P_n/K) = x_n$, we obtain

$$x_{n+1} = f(x_n) = x_n \exp\left[\lambda(1 - x_n) - \frac{m}{(1 + sKx_n)}\right]. \tag{4.2}$$

Numerical studies were made by computing the Lyapunov exponent. The Lyapunov exponent for any x_0 is defined as [12]

$$\hat{\lambda}(x_0) = \limsup_{n \to +\infty} \frac{1}{n} \sum_{j=0}^{n-1} \ln\left|f'(x_j)\right|, \quad \text{where } x_j = f^j(x_0).$$

Now, $|f'(x_j)| > 1$ implies that $\hat{\lambda} > 0$, and the system is chaotic. In the present case,

$$f'(x) = e^T[1 + x\{(msK/s_1^2) - \lambda\}], \quad T = \lambda(1 - x) - [m/s_1], \quad s_1 = (1 + sKx).$$

For suitable large values of m and sK, $|f'(x)| > 1$ beyond a certain value of λ. The Lyapunov exponent is positive and the system (4.2) becomes chaotic having sensitive dependence on the initial conditions. The system undergoes the period-doubling cascade route leading to the Feigenbaum scenario [18]. Increase in the growth rate λ causes the emergence of transient chaos and at some critical threshold value the population collapses. Duarte et al. [12] concluded that larger predator handling times destabilize the dynamics of the population. The bifurcation diagram is given in Figure 4.2.

Model with Mating Limitation

Schreiber [90] analyzed model 4.1 with mating limitation also. The model of Schreiber was further analyzed by Jorge Duarte et al. [12] by introducing the Allee effects. The authors assumed $\beta(P_n)$ in Equation 4.1 as $\beta(P_n) = \alpha P_n/(1 + \alpha P_n)$. Here, $\beta(P_n)$ defines the probability of finding a mate and α is the searching capacity of an individual. Setting $(P_n/K) = x_n$, we obtain

$$x_{n+1} = f(x_n) = \frac{\alpha K x_n^2}{1 + \alpha K x_n} e^{\lambda(1-x_n)}. \tag{4.3}$$

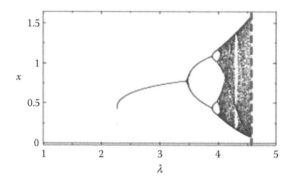

FIGURE 4.2
Bifurcation diagram for model (4.2). $m = 8.0$ and $sK = 16$. (Reprinted from *Nonl. Anal.: Real World Appl.* 13, Duarte, J., Januario, C., Martins, N., Sardayés, J., On chaos, transient chaos and ghosts in single population models with Allee effects, 1647–1661, Copyright 2012, with permission from Elsevier.)

Lyapunov exponents were computed to study the dynamics. In this case,

$$f'(x) = e^T(\alpha Kx/s_1)[1 + (1/s_1) - \lambda x], \quad T = \lambda(1 - x), \quad s_1 = (1 + \alpha Kx).$$

As the individual searching efficiency and the intrinsic growth rate increases, the system goes through the period-doubling cascade route leading to the Feigenbaum scenario. For values of the parameters greater than their critical values, the system crosses into the essential extinction zone. The bifurcation diagram is given in Figure 4.3. The dashed line indicates the parameter value causing the boundary crisis [12].

Rohani et al. [82] have commented about the possible existence of chaos in model and real systems. They opined that if the route to chaos is period

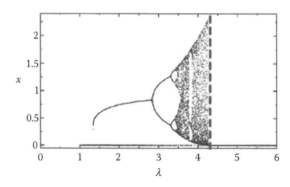

FIGURE 4.3
Bifurcation diagram for model (4.3) with $\alpha K = 2$. (Reprinted from *Nonl. Anal.: Real World Appl.* 13, Duarte, J., Januario, C., Martins, N., Sardayés, J., On chaos, transient chaos and ghosts in single population models with Allee effects, 1647–1661, Copyright 2012, with permission from Elsevier.)

doubling in the model systems, then we may not expect to observe chaotic dynamics in the real populations due to this route's sensitivity to external perturbations. However, if the route to chaos in the model system is a quasiperiodic one, then chaos may persist in the real populations also.

Chaos in Two Species Model Systems

Discrete two population interaction models also exhibit all the scenarios of logistic maps. Further, discrete two population interaction models can give more information about the complex systems than the logistic models.

Jing and Yang [38] investigated the bifurcation and chaos in a discrete-time predator–prey system obtained by the forward Euler method. Bifurcation and center manifold theory were used to derive the existence conditions for flip and Hopf bifurcations. Numerical simulation of the model displays a wide variety of dynamics: chaotic regimes with orbits of different periods, period-doubling and inverse period-doubling bifurcations leading to chaos, interior crisis, boundary crisis, intermittency mechanics, sudden onset or disappearance of chaotic dynamics, attracting and nonattracting chaotic sets. Lyapunov exponents were computed to confirm the dynamical behaviors. The authors considered a generalized Lotka–Volterra system as [38]

$$\dot{x} = x(1 - x - k_2 x^2) - \frac{xy}{1 + ax}, \quad \dot{y} = -y(\delta_0 + \delta_1 y) + \frac{\gamma\, xy}{1 + ax}. \tag{4.4}$$

Particular cases of the model were earlier studied by various authors [5,20,29,36,37,39]. Applying the Euler method, the equations are discretized as

$$x_{n+1} = x_n + \delta x_n \left[(1 - x_n - k_2 x_n^2) - \frac{y_n}{1 + ax_n} \right],$$

$$y_{n+1} = y_n + \delta y_n \left[-(\delta_0 + \delta_1 y_n) + \frac{\gamma\, x_n}{1 + ax_n} \right].$$

The fixed points of the map are $E_0(0, 0)$ and $E_1(K, 0)$, where K is a positive root of $k_2 x^2 + x - 1 = 0$. A unique positive fixed point $E(x^*, y^*)$ exists where x^* is the root of the equation $\delta_1(1 - x - k_2 x^2)(1 + ax)^2 + \delta_0(1 + ax) - \gamma x = 0$, and $y^* = (1 + ax^*)(1 - x^* - k_2 x^{*2})$, where $0 < x < K$ and $\gamma - \delta_0 a > (\delta_0/K)$. Conditions for linear stability were derived. The characteristic equation of the Jacobian matrix is of the form $\lambda^2 + p(x^*, y^*)\lambda + q(x^*, y^*) = 0$. From the Routh–Hurwitz criterion, the necessary and sufficient conditions that the roots are negative or have negative real parts are $p(x^*, y^*) > 0$, and $q(x^*, y^*) > 0$. Since the system has only one unique positive fixed point, the system has no fold bifurcation.

Conditions for the existence of the flip and the Hopf bifurcations were derived. For detailed analysis of the bifurcations, see Jing and Yang [38].

Liu and Xiao [51] have considered the Lotka–Volterra-type predator–prey system as

$$\dot{x} = rx(1-x) - bxy, \quad \dot{y} = (-d + bx)y. \tag{4.5}$$

Applying the forward Euler method, the following discrete time model was obtained:

$$x_{n+1} = x_n + \delta[rx_n(1-x_n) - bx_ny_n], \quad y_{n+1} = y_n + \delta(-d + bx_n)y_n,$$

where δ is the step size. The model system again displays many interesting dynamical behaviors. The fixed points of the map are $E_0(0, 0)$, $E_1(1, 0)$, and $E_2[(d/b), r(b-d)/b^2]$, $b > d$. The authors have shown that the unique positive fixed point can undergo flip bifurcation and Hopf bifurcation. The variational matrix of the system is given by

$$V(x, y) = \begin{bmatrix} 1 + \delta[r(1-2x) - by] & -b\delta x \\ b\delta y & 1 - \delta(d - bx) \end{bmatrix} = \begin{bmatrix} a_{11} & a_{12} \\ a_{21} & a_{22} \end{bmatrix}.$$

The characteristic equation of $V(x, y)$ is $\lambda^2 + p(x, y)\lambda + q(x, y) = 0$, where $p(x, y) = -(a_{11} + a_{22})$, and $q(x, y) = a_{11}a_{22} - a_{12}a_{21}$. It was shown that the fixed point $E_0(0, 0)$ is a saddle point if $0 < \delta < 2/d$, and a source if $\delta > 2/d$. At $\delta = 2/d$, E_0 is a nonhyperbolic fixed point. Thus, the flip (or period-doubling) bifurcation may occur when parameters vary in the neighborhood of $\delta = 2/d$. However, through computations, it was shown that flip bifurcation cannot occur. The fixed point $E_1(1, 0)$ can undergo flip bifurcation when parameters vary in a small neighborhood of positive parameters (b, d, r, δ), where $b \neq d$, $r = 2/\delta$, and $\delta \neq 2/(b-d)$. In this case, the predator becomes extinct and the prey undergoes the period-doubling bifurcation to chaos in the sense of Li–Yorke by choosing the bifurcation parameter as r. Conditions under which the other fixed points may be a source or a sink or a nonhyperbolic point were derived. Also, conditions under which the map undergoes flip bifurcation at the positive equilibrium point $E_2[(d/b), r(b-d)/b^2]$, $b > d$, were derived. For detailed analysis of the bifurcations, see Liu and Xiao [51].

Chaos in Two Species Model Systems with Diffusion

The reaction diffusion systems often exhibit diffusion-driven instability or Turing instability. It occurs when a homogeneous steady state, which is stable for small perturbations in the absence of diffusion, becomes unstable in the presence of diffusion [101]. Turing patterns are stationary structures that appear spontaneously upon breaking the symmetry of the medium, which results only from the coupling between the reaction and the diffusion processes. In the absence of diffusion, these systems tend to a linearly stable uniform steady state. Once Turing patterns arise, they may remain stable until some external perturbation destroys them; but, when perturbation stops, Turing patterns reappear and reorganize themselves. A steady state that is stable in a nonspatial model may become unstable in the corresponding spatial system. Then, after the homogeneity is broken due to linear Turing instability, the nonlinear interactions between the components drive the system into the formation of standing spatial patterns that are irreversible [67]. However, a kind of inverse process (an anti-Turing phenomenon) may take place for some parameter values. That is, a locally unstable equilibrium of the nonspatial system may become dynamically stable in the spatial system. In this case, for certain time and length scale, and equal diffusion coefficients $d_1 = d_2 = 1$, formation of spatial patterns is suppressed and homogeneity is restored [55]. Malchow et al. [55] have also observed the following: (i) Generally, initial conditions affect only the transient stage of pattern formation. (ii) Turing patterns are sensitive to the boundary conditions. (iii) Boundary conditions may affect both the transient stage and large-time asymptotics. In a spatially bounded system, a stationary pattern forms only when the boundary conditions are consistent with the intrinsic properties of the pattern, size, and shape of the domain. Pattern formation results when a parameter changing with time enters into the domain of parameter space where the system becomes unstable due to perturbation.

In real-world systems, the dynamics depend on the interplay between deterministic and stochastic factors since the steady state is continuously affected by spatially heterogeneous stochastic perturbations and fluctuations. Normally, these perturbations decay with time and the system hovers around the steady state, unless the magnitude of these perturbations is very large.

Incorporation of spatial structure is crucial to ecological modeling. Pascual [69] observed diffusion-induced chaos in a spatial prey–predator system, and its robustness was studied by Rai and Jayaraman [76]. The chaotic behavior in these systems was generated when stable limit-cycle oscillators of comparable frequencies couple together to force each other. The origin of these oscillators lies on a spatial gradient of the specific growth rate of the prey. Such gradients occur in natural aquatic environments because of spatial distribution of the nutrients. Normally, four types of systems are used to study the influence of spatial structure on ecological dynamics. They are: (i) reaction–diffusion systems, (ii) coupled map lattices, (iii) cellular automata, and (iv) integro-difference equations. In this chapter, we choose to work with reaction–diffusion systems.

Among ecology's earliest concerns is the development and maintenance of spatiotemporal patterns [53,54]. The dynamics of interacting populations in connection with the spatial phenomena such as the pattern formation and spatio-temporal chaos has recently become a focus of research. Liu et al. [50] studied a spatial phytoplankton–zooplankton system with periodic forcing and additive noise. They considered the Holling type IV predator–prey model with external periodic forcing and colored noise. Zhang et al. [120] demonstrated that two species spatial systems could be useful to explain spatio-temporal behavior of populations whose dynamics is strongly affected by noise and the environmental physical variables. These local movements were modeled by Fickian diffusion terms. Applicability of diffusion terms to describe redistribution of species in space due to random motion of the individuals for any value of population density was shown by Okubo [68]. Since the dispersal rates are assumed to be the same for the prey and predator, the patterns cannot appear due to Turing instability [71,91]. Whether the spatial gradient affecting the growth rate of the prey species can provide the coupling mechanism for the oscillatory predator–prey dynamics was examined by Pascual [69]. Two coupled nonchaotic oscillators can admit chaotic dynamics. It was found that the diffusive movements of the species create a system of coupled oscillators that mutually force each other at incommensurate frequencies. This results in chaotic dynamics. Diffusion and spatial heterogeneity introduce qualitatively new types of behavior in predator–prey interaction. Diffusion on a spatial gradient may drive a cyclic predator–prey system into chaotic behavior. Ecological models of diffusion-driven instability with spatial heterogeneities have been studied for a variety of reasons [65]. The spatio-temporal transition between the regular and chaotic patterns and the phenomenon of wave of chaos was observed by Petrovskii and Malchow [71]. Recently, Upadhyay et al. [108,109] demonstrated that a wave of chaos could be a potential candidate for understanding planktonic patchiness in marine environments. Sherratt et al. [92] have proposed a mechanism for generation of chaos. Waves of invasions by predators may have the effect of producing chaotic states in their wake in oscillatory predator–prey systems driven unstable by diffusive movement of both the species. This mechanism does not require a spatial gradient to couple to diffusion, to create chaos.

Rosenzweig–MacArthur Model with Diffusion

Consider the nondimensionalized form of the Rosenzweig–MacArthur (RM) model with diffusion as (see section "Rosenzweig–MacArthur model" of Chapter 2)

$$\frac{\partial u}{\partial t} = u\left[(1-u) - \frac{v}{u+\alpha}\right] + d_1\frac{\partial^2 u}{\partial x^2}, \tag{4.6}$$

$$\frac{\partial v}{\partial t} = \beta v\left[\frac{u}{u+\alpha} - \delta\right] + d_2\frac{\partial^2 v}{\partial x^2} \tag{4.7}$$

in the domain $D \in \{r = (x, 0), 0 \leq x \leq L\}$, with zero flux boundary conditions at $x = 0$ and $x = L$. All the constants in the equations are positive and d_1, d_2 are the diffusion coefficients assumed to be constants.

Stability analysis: First, consider the case without diffusion ($d_1 = d_2 = 0$). The homogeneous steady-state solution (u^*, v^*) is given by $u^* = \alpha\delta/(1 - \delta)$ and $v^* = (u^* + \alpha)(1 - u^*)$. Perturb the solution of Equations 4.6 through 4.7 about the steady state (u^*, v^*). Setting $u = u^* + U$, $v = v^* + V$ and linearizing, we obtain the system

$$\frac{\partial U}{\partial t} = b_{11}U + b_{12}V + d_1\frac{\partial^2 U}{\partial x^2}, \quad \frac{\partial V}{\partial t} = b_{21}U + b_{22}V + d_2\frac{\partial^2 V}{\partial x^2}, \quad (4.8)$$

where

$$b_{11} = 1 - 2u^* - \frac{\alpha v^*}{(u^* + \alpha)^2} = -\frac{\delta}{1 - \delta}[\delta(1 + \alpha) + \alpha - 1], \quad b_{12} = -\frac{u^*}{u^* + \alpha} = -\delta,$$

$$b_{21} = \frac{\alpha\beta v^*}{(u^* + \alpha)^2} = \beta[1 - \delta(1 + \alpha)], \quad b_{22} = 0.$$

Write the solution of Equation 4.8 in the form $U = se^{\lambda t + ikx}$, $V = we^{\lambda t + ikx}$, where λ and k are the frequency and wave number, respectively. Substitute the expressions for U, V in Equation 4.8. The homogeneous equations in s and w have solution if determinant of the coefficient matrix is zero. We get

$$(\lambda - b_{11} + d_1k^2)(\lambda - b_{22} + d_2k^2) - b_{12}b_{21} = 0, \quad \text{or} \quad \lambda^2 + p\lambda + q = 0, \quad (4.9)$$

where $p = (d_1 + d_2)k^2 - (b_{11} + b_{22})$, $q = d_1d_2k^4 - d_2b_{11}k^2 - b_{12}b_{21}$. Irrespective of the sign of p, the equation has a positive root if $q < 0$. Therefore, diffusion-driven instability occurs when $q(k^2) < 0$. $y = q(k^2)$ is a parabola in the variable $t = k^2$, with vertex at $V[b_{11}/(2d_1), \{\delta b_{21} - (b_{11}^2 d_2)/(4d_1)\}]$, where $\{\delta b_{21} - (b_{11}^2 d_2)/(4d_1)\} < 0$. The minimum occurs at $k_c^2 = b_{11}/(2d_1) > 0$. Now, $q(k^2) = d_1d_2(k^2 - k_1^2)(k^2 - k_2^2) < 0$, for $k_1^2 < k^2 < k_2^2$, where $k_1^2, k_2^2 = [d_2b_{11} \mp \{d_2^2 b_{11}^2 - 4\delta d_1 d_2 b_{21}\}^{1/2}]/(2d_1d_2)$, and $d_2b_{11}^2 - 4\delta d_1 b_{21} > 0$. For example, consider the set of parameter values $\alpha = 0.3$, $\beta = 2$, $\delta = 0.3$, $d_1 = 0.01$, $d_2 = 10$. We obtain $u^* = 9/70$, $v^* = 183/490$, $k_c^2 = 93/14$, $q(k^2) < 0$, for $0.281445 < k^2 < 13.00427$. The graphs of $q(k^2)$ versus k^2 for various values of d_2 are plotted in Figure 4.4 for the above set of parameter values. For all values of k^2 lying in this range, the system is unstable.

A Variant of Rosenzweig–MacArthur Model with Diffusion

Consider the reaction–diffusion model (see Equations 2.67, 2.68) in the 1D case as [13]

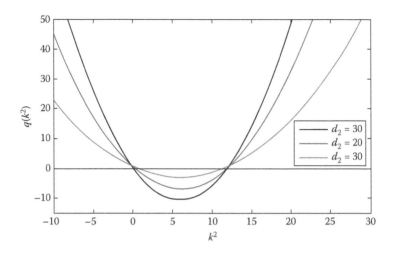

FIGURE 4.4
Graph of the function $q(k^2)$ for k^2 for $d_2 = 10, 20, 30$.

$$\frac{\partial u}{\partial t} = u(1 - u) - \frac{uv}{h_p + u} + \frac{\partial^2 u}{\partial x^2}, \tag{4.10}$$

$$\frac{\partial v}{\partial t} = \frac{buv}{h_p + u} - cv - \frac{v^2 f}{h_z^2 + v^2} + D\frac{\partial^2 v}{\partial x^2}, \tag{4.11}$$

where D is the diffusion coefficient. The initial condition and no-flux boundary conditions are $u(x, 0) > 0$, $v(x, 0) > 0$, for $x \in [0, R]$; and $(\partial u/\partial x) = 0$, $(\partial v/\partial x) = 0$, at $x = 0, R$.

Stability analysis: To study the effect of diffusion on the model system, consider the linearized form of the system about the positive equilibrium point $E^*(u^*, v^*)$. Set $u = u^* + U$, $v = v^* + V$. We obtain the linearized form as

$$\frac{\partial U}{\partial t} = b_{11}U + b_{12}V + \frac{\partial^2 U}{\partial x^2}, \quad \frac{\partial V}{\partial t} = b_{21}U + b_{22}V + D\frac{\partial^2 V}{\partial x^2}, \tag{4.12}$$

where

$$b_{11} = -u^*\left(1 - \frac{v^*}{(h_p + u^*)^2}\right), \quad b_{12} = -\frac{u^*}{h_p + u^*},$$

$$b_{21} = \frac{bh_p v^*}{(h_p + u^*)^2}, \quad b_{22} = \frac{v^* f(v^{*2} - h_z^2)}{(h_z^2 + v^{*2})^2}.$$

Write the solution in the form $U = se^{\lambda t + ikx}$, $V = we^{\lambda t + ikx}$, where λ and k are the frequency and wave number, respectively. Substitute the expressions for U, V in Equation 4.12. The homogeneous equations in s and w have solution if determinant of the coefficient matrix is zero. We get

$$(\lambda - b_{11} + k^2)(\lambda - b_{22} + Dk^2) - b_{12}b_{21} = 0, \quad \text{or} \quad \lambda^2 + p\lambda + q = 0, \quad (4.13)$$

where $p = (D + 1)k^2 - (b_{11} + b_{22}), q = Dk^4 - k^2(Db_{11} + b_{22}) + (b_{11}b_{22} - b_{12}b_{21})$.

By the Routh–Hurwitz criterion, the roots of Equation 4.13 are negative or have negative real parts if $p > 0$, and $q > 0$. A sufficient condition for $p > 0$ is $(b_{11} + b_{22}) < 0$. The sufficient conditions are $v^* < (h_p + u^*)^2$, and $v^{*2} < h_z^2$. For these values, $b_{11} < 0$, $b_{22} < 0$. Also, $b_{12} < 0$ and $b_{21} > 0$. Hence, $q > 0$. Therefore, the positive equilibrium E^* is locally asymptotically stable in the presence of diffusion if the above two conditions are satisfied. The positive equilibrium E^* may or may not be locally asymptotically stable in the absence of diffusion. Suppose that the sufficient conditions are not met. Now, irrespective of the sign of p, diffusive instability can arise if $q < 0$ (one root of Equation 4.13 is positive), that is, if $q(k^2) = Dk^4 - k^2C + B < 0$, where $C = (Db_{11} + b_{22})$, and $B = b_{11}b_{22} - b_{12}b_{21}$. The roots of this equation in k^2 are real and positive when (i) $B > 0$, (ii) $C > 0$, and (iii) $C^2 - 4BD > 0$. Then, $q < 0$ when $k_1^2 < k^2 < k_2^2$, where $k_1^2, k_2^2 = [C \mp \sqrt{C^2 - 4BD}]/(2D)$. Therefore, diffusive instability occurs when these conditions are satisfied. Now, $q(k^2)$ is a quadratic in k^2 and the graph of $y = q(k^2)$ is a parabola opening upwards. The minimum occurs at the vertex of the parabola, that is, for $k^2 = k_m^2$ where $k^2 = k_m^2 = C/(2D)$.

Numerical simulation of model system (4.10) and (4.11): The equilibrium point (u^*, v^*) is the solution of the equations (model without diffusion)

$$u^2 + (h_p - 1)u + v - h_p = 0, \quad [(b - c)u - ch_p](h_z^2 + v^2) - fv(h_p + u) = 0. \quad (4.14)$$

Consider the solution of the system with the given initial and no flux boundary conditions for the following sets of parameter values:

i. $h_p = 0.3$, $b = 2$, $c = 0.8$, $f = 0.0001$, and $h_z = 2.5$. Here, h_p, h_z are the half-saturation constants for phytoplankton and zooplankton densities, respectively. b is the parameter measuring the ratio of product of conversion coefficient with grazing rate to the product of intensity of competition among individuals of phytoplankton with carrying capacity. c is the per capita predator death rate and f is the predation rate of zooplankton by fish population which follows Holling type III functional response. The system of the nonlinear Equations 4.14 is solved by the Newton iteration to obtain $u^* = 0.200000104$, $v^* = 0.400000312$. The sufficient conditions for asymptotic stability are not satisfied. In Equation 4.13, we have $p = (D + 1)k^2 - 0.119994006$, $q = Dk^4 - Ck^2 + B$, $B = 0.383999518$; $C = 0.120000178D - 0.000006172$;

TABLE 4.1

Results of Stability Analysis of Model (4.10) and (4.11)

D	Signs of (p, q)	Roots of Equation 4.13	Range of k^2 and k_m^2	Stability Results
Without diffusion	$(-, +)$	Complex pair with positive real part	—	Unstable
1	$(+, +)$	Complex pair with negative real part	>0.059997	Stable
50	$(+, +)$	Complex pair with negative real part	>0.00235	Stable
100	$(+, +)$	Complex pair with negative real part	>0.001188	Stable
110	$(+, -)$	One positive root	(0.049555, 0.070445), ≈0.06	Unstable
1900	$(+, -)$	One positive root	(0.001708, 0.118292), ≈0.06	Unstable

$C^2 - 4BD > 0$ for $D > 106.7$ and $q < 0$ for $k_1^2 < k^2 < k_2^2$, where $k_1^2, k_2^2 = [C \mp \sqrt{C^2 - 4BD}]/(2D)$. Hence, diffusive instability may set in when $D > 106.7$. Interestingly, k_m^2 is almost same for all $D > 107$. The results of stability analysis are presented in Table 4.1. The plot of $q(k^2)$ versus k^2 is given in Figure 4.5a.

ii. $h_p = 0.3$, $b = 0.2$, $c = 0.1$, $f = 0.0001$, and $h_z = 2$. Using Newton's iteration, we obtain $u^* = 0.299915224$, $v^* = 0.419991515$. We obtain $B = 0.035003776$; $C = 0.050077698D - 0.000009206$; $C^2 - 4BD > 0$, for $D > 55.9$; and $q < 0$ for $k_1^2 < k^2 < k_2^2$, where $k_1^2, k_2^2 = [C \mp \sqrt{C^2 - 4BD}]/(2D)$. Hence, diffusive instability may set in when $D > 55.9$. For example, for $D = 100$, we obtain $k_1^2 = 0.003256$, $k_2^2 = 0.107505$, $k_m^2 = [C/(2D)] = 0.025038$. Again, k_m^2 is almost the same for all $D > 56$. In this case also, we get similar results as given in Table 4.1.

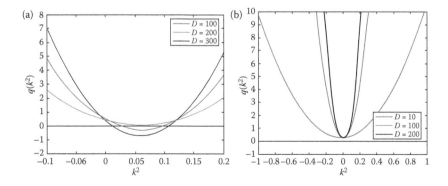

FIGURE 4.5

(a) Plot of $q(k^2)$ versus k^2. Case (i). (b) Plot of $q(k^2)$ versus k^2. Case (iii).

iii. $h_p = 0.8$, $b = 2$, $c = 0.8$, $f = 0.0001$, and $h_z = 2.5$. Using Newton's iteration, we obtain $u^* = 0.2000094429$, $v^* = 0.7999981114$. The sufficient conditions for asymptotic stability $v^* < (h_p + u^*)^2$ and $v^{*2} < h_z^2$ are satisfied. The system is stable in the absence as well as in the presence of diffusion. As D is increased, the value of p increases and the stability of the equilibrium point sets in earlier. The plot of $q(k^2)$ versus k^2 is given in Figure 4.5b.

For the sets of parameter values considered in (i) and (ii), the linearized model system was unstable without diffusion. Introduction of diffusion made the system stable. When diffusion is increased further, the system has again become unstable. However, for the third set of parameter values considered in (iii), the system is asymptotically stable in the absence as well as in the presence of diffusion. This may imply that the choice of parameter values is important to study the effect of diffusion and also whether the system without diffusion is stable or not. Also, it would depend more on the nonlinearity of the system.

In Upadhyay et al. [109], the model was solved using a finite difference technique, semi-implicit in time along with zero flux boundary conditions and nonzero asymmetrical initial condition, $u(x, 0) = 0.2 + 10^{-8}$ $(x - 1200)(x - 2800)$, $v(x,0) = 0.4$. This nonmonotonic form of the initial condition is more realistic from the biological point of view, which determines the initial spatial distribution of the species. The details of the numerical technique are given in Ref. [109]. The model system was solved using MATLAB® 7.0 for the parameter values $h_p = 0.3$, $b = 2$, $c = 0.8$, $f = 0.0001$, and $h_z = 2.5$, up to the time level $t = 1000$, for various values of the diffusivity constant D. Space series generated by Dubey et al. [13] at time $t = 1000$, for (a) $D = 1$, and (b) $D = 1900$, showing the effect of diffusivity constant D on the dynamics of the model system is given in Figures 4.6a,b. Figure 4.6a shows the onset of chaotic

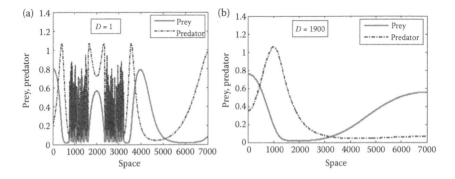

FIGURE 4.6

(a,b) Space series generated at time $t = 1000$, for (a) $D = 1$, (b) $D = 1900$. (With kind permission from Springer Science+Business Media B.V.: *J. Appl. Math. Comp.*, Spatiotemporal pattern formation in a diffusive predator–prey system: An analytical approach, 31, 2009, 413–432, Dubey, B. Kumari, N., Upadhyay, R. K. Copyright 2008, Springer.)

phases in patches at $t = 1000$, $D = 1$. (A patch has a distinct kind of local spatial dynamics than adjacent parts. WOC propagates in both forward as well as in backward directions replacing a patch with local regular dynamics, characterized by a stable limit cycle in the phase plane of the system, with chaotic dynamics. These patches are formed as a result of the action of diffusion on the interacting biological species). As the value of the diffusivity constant D is increased to 10, the irregular jagged pattern occupied the domain $200 \leq x \leq 3800$. As the value of D is increased further, the dynamics displays partial regularity along with the irregular nature. At $D = 1860$, the population dynamics becomes stable after $x = 4000$. For $D \geq 1900$, stable dynamics was observed which supports the analytical findings that the dynamics of a chaotic system may become stable due to diffusion (Figure 4.6b) [13].

2D case: Now, consider the reaction–diffusion model in the 2D space as

$$\frac{\partial u}{\partial t} = u(1 - u) - \frac{uv}{h_p + u} + \left(\frac{\partial^2 u}{\partial x^2} + \frac{\partial^2 u}{\partial y^2} \right), \tag{4.15}$$

$$\frac{\partial v}{\partial t} = \frac{buv}{h_p + u} - cv - \frac{v^2 f}{h_z^2 + v^2} + D\left(\frac{\partial^2 v}{\partial x^2} + \frac{\partial^2 v}{\partial y^2} \right). \tag{4.16}$$

The initial condition and no-flux boundary conditions are $u(x,y,0) > 0$, $v(x, y, 0) > 0$, for $x \in \Omega$, and $(\partial u/\partial n) = 0$, $(\partial v/\partial n) = 0$, on $\partial\Omega$, where n is the outward normal to $\partial\Omega$. The zero flux boundary conditions are used for modeling the dynamics of spatially bounded aquatic ecosystems [89]. It was shown by Dubey et al. [13] that the results on global asymptotic stability in one dimension are also valid in the 2D case. Further, the solutions of the system converge faster to its equilibrium in the case of 2D diffusion in comparison to the 1D case.

Stability analysis: The equilibrium point is the same as in the 1D case (4.10), (4.11). Write the spatiotemporal perturbations around the homogeneous steady state as $u(t, x, y) = u^* + U(t, x, y), v(t, x, y) = v^* + V(t, x, y)$. Substituting into Equations 4.15 and 4.16 and linearizing the system, we obtain

$$\frac{\partial U}{\partial t} = b_{11}U + b_{12}V + \left(\frac{\partial^2 U}{\partial x^2} + \frac{\partial^2 U}{\partial y^2} \right), \quad \frac{\partial V}{\partial t} = b_{21}U + b_{22}V + D\left(\frac{\partial^2 V}{\partial x^2} + \frac{\partial^2 V}{\partial y^2} \right),$$

where $b_{11}, b_{12}, b_{21}, b_{22}$ are same as in the 1D case. Write the solution in the form $U(t, x, y) = se^{\lambda t} \cos(k_x x)\cos(k_y y), V(t, x, y) = we^{\lambda t} \cos(k_x x)\cos(k_y y)$, where λ and k (where $k^2 = k_x^2 + k_y^2$) are the frequency and wave number, respectively. Substitute the expressions for U, V in the above equations. The homogeneous equations in s and w have solution if determinant of the coefficient matrix is zero. We get the characteristic equation as

$$(\lambda - b_{11} + k^2)(\lambda - b_{22} + Dk^2) - b_{12}b_{21} = 0, \quad \text{or} \quad \lambda^2 + p\lambda + q = 0,$$

where the expressions for p and q are the same as in the 1D case and $k^2 = k_x^2 + k_y^2$. The spatially homogeneous steady state is unstable due to heterogeneous perturbation when at least one root of the characteristic equation is positive. The conclusions are same as in the 1D case.

Numerical simulation of the model system (4.15) *and* (4.16): The 2D reaction–diffusion equations were solved using a finite difference technique, semi-implicit in time along with zero flux boundary conditions. The initial values are taken as [65]

$$u_0(x,y) = 0.2 - 2 \times 10^{-7}(x - 0.1y - 225)(x - 0.1y - 675),$$

$$v_0(x,y) = 0.4 - 3 \times 10^{-5}(x - 450) - 1.2 \times 10^{-4}(y - 150).$$

These initial conditions were specifically chosen to be nonsymmetric so that the influence of the corners of the domain can be studied [65]. Since both species exhibit qualitatively similar behavior, except in the early stages of the process when the influence of the initial conditions is dominant, only the prey (phytoplankton) abundance is shown. The details of the numerical technique are given in Upadhyay et al. [109]. A square domain $L \times L$ where $L = 400$ is taken. The set of parameter values are again chosen as in the 1D case: $h_p = 0.3$, $b = 2$, $c = 0.8$, $f = 0.0001$, and $h_z = 2.5$. Using MATLAB 7.0, numerical computations were carried out by Dubey et al. [13]. The results are plotted in Figures 4.7 through 4.9. From the previous discussions, we may expect that an unstable or chaotic system may become stable by increasing the value of the diffusion coefficient D. Prey and predator populations at time level $t = 150$ are presented in Figure 4.7, where a regular spiral pattern was obtained for $D = 1$. As D is increased to 15, the diameter of the spiral obtained at $D = 1$ increased, but the system remained stable.

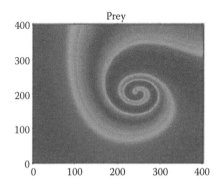
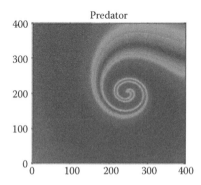

FIGURE 4.7
Prey and predator populations at $t = 150$. $D = 1$. (With kind permission from Springer Science+Business Media B.V.: *J. Appl. Math. Comp.*, Spatiotemporal pattern formation in a diffusive predator–prey system: An analytical approach, 31, 2009, 413–432, Dubey, B. Kumari, N., Upadhyay, R. K. Copyright 2008, Springer.)

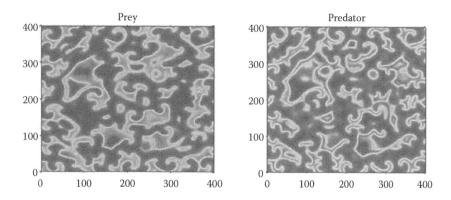

FIGURE 4.8
Prey and predator populations at $t = 1000$. $D = 1$. (With kind permission from Springer Science+Business Media B.V.: *J. Appl. Math. Comp.*, Spatiotemporal pattern formation in a diffusive predator–prey system: An analytical approach, 31, 2009, 413–432, Dubey, B. Kumari, N., Upadhyay, R. K. Copyright 2008, Springer.)

Figure 4.8 represents prey and predator populations at the time level $t = 1000$, where the system is chaotic containing irregular patchy structures for $D = 1$. This initially irregular patchy structure starts evolving to regular structure for $D = 2$, and it partially takes the form of rotating spiral waves for $D = 10$. This regular dynamics persists evolving to more perfect spiral wave pattern for $D = 15$ (see Figure 4.9). A remarkable change in the system dynamics from chaotic to regular is observed, when the diffusivity constant is increased.

Now, consider a typical modification of the initial conditions as discussed by Medvinsky et al. [65]. Assume the initial conditions as $u(x, 0) = u^* + \varepsilon(x - x_1)$ $(x - x_2)$, $v(x, 0) = v^*$, where (u^*, v^*) is the nontrivial state for the coexistence of

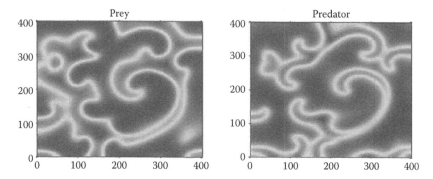

FIGURE 4.9
Prey and predator populations at $t = 1000$. $D = 15$. (With kind permission from Springer Science+Business Media B.V.: *J. Appl. Math. Comp.*, Spatiotemporal pattern formation in a diffusive predator–prey system: An analytical approach, 31, 2009, 413–432, Dubey, B. Kumari, N., Upadhyay, R. K. Copyright 2008, Springer.)

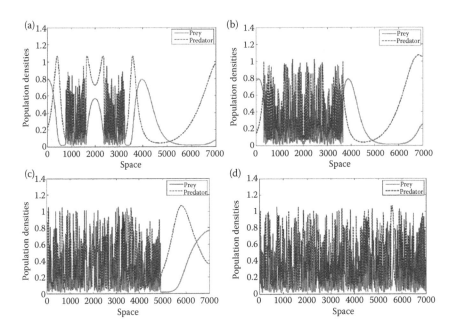

FIGURE 4.10

Growth of chaotic behavior for (a) $t = 1000$, (b) $t = 2000$, (c) $t = 5000$. (d) $t = 10,000$. (Reprinted from *Chaos, Solitons Fractals*, 40(1), Upadhyay, R. K., Kumari, N., Rai, V., Wave of chaos in a diffusive system: Generating realistic patterns of patchiness in plankton-fish dynamics, 262–276, Copyright 2009, with permission from Elsevier.)

prey and predator and ε is a small parameter. The initial conditions are nonmonotonic and are chosen to study whether the dynamics evolves to a complex spatial structure [65].

Numerical simulations using MATLAB 7.0 with the above-mentioned finite difference technique were done by Upadhyay et al. [109], for the same set of parameter values as used earlier with $\varepsilon = 1 \times 10^{-8}$, $x_1 = 1200$, $x_2 = 2800$, and $(u^*, v^*) = (0.2, 0.4)$. The diffusivity coefficient is taken as $D = 1$. The dynamics at $t = 1000$, 2000, 5000, and 10,000 is plotted in Figures 4.10a–d, respectively. A complex spatial dynamics at $t = 1000$ was observed where a smooth regular patch separates chaotic patches. The jagged pattern representing the chaotic behavior of the system grows steadily with time in both the directions, displacing the regular pattern and ultimately occupying the whole region [109]. This phenomenon was termed as wave of chaos by Petrovskii and Malchow [71]. After this, the dynamics of the system does not undergo any further changes.

A Variant of DeAngelis Model with Diffusion and Spatial Heterogeneity

Consider a reaction–diffusion model for a phytoplankton–zooplankton–fish system where at any location (X, Y) and time T, the phytoplankton $P(X, Y, T)$

and zooplankton $Z(X, Y, T)$ populations satisfy the equations [113] (see also Equations 2.71 and 2.72)

$$\frac{\partial P}{\partial T} = R_x P\left(1 - \frac{P}{K}\right) - \frac{vPZ}{\alpha + \beta Z + \gamma P} + D_P \Delta P, \tag{4.17}$$

$$\frac{\partial Z}{\partial T} = \frac{evPZ}{\alpha + \beta Z + \gamma P} - mZ - \frac{FZ^2}{H_Z^2 + Z^2} + D_Z \Delta Z, \tag{4.18}$$

where $\Delta = (\partial^2/\partial X^2)$ in the 1D case, and $\Delta = (\partial^2/\partial X^2) + (\partial^2/\partial Y^2)$ in the 2D case, R_x is the spatially dependent growth rate of phytoplankton due to reproduction, K is the carrying capacity of prey population, v is the rate at which phytoplankton is grazed which follows the Beddington–DeAngelis-type functional response, e is the conversion coefficient from individuals of phytoplankton into individuals of zooplankton, m is the mortality rate of zooplankton, α is the protection provided to phytoplankton by its environment, β is the intensity of interference between individuals of zooplankton, γ determines how fast the per capita feeding rate approaches to its saturation value, F is the predation rate of zooplankton by fish population which follows the Holling type III functional response, H_Z is half-saturation constant for zooplankton species, and D_P and D_Z are the diffusion coefficients of phytoplankton and zooplankton density, respectively.

We introduce the following substitutions and notations to bring the system of equations into a nondimensional form [113],

$$r_x = (R_x/\bar{R}) = s + lx, \; u = P/K, \; v = vZ/K, \; x = X/L, \; y = Y/L, \; t = \bar{R}T,$$

$$\theta = K/(\alpha\bar{R}), \; \theta' = ev\theta, \; m = c\bar{R}, \; \xi = (\beta K)/(\alpha v), \; \eta = (\gamma K)/\alpha, \; g = K/(H_Z v),$$

$$f = (FK)/(H_Z^2 v\bar{R}), \; d_1 = D_P/(L^2\bar{R}), \; d_2 = D_Z/(L^2\bar{R}), \; \bar{R} = R_x(X_0), \; X_0 \in (0, L),$$

where $x = x^*$ depends on the growth rate P. The dimensionless form is given by

$$\frac{\partial u}{\partial t} = r_x u(1 - u) - \frac{\theta uv}{1 + \xi v + \eta u} + d_1 \Delta u, \tag{4.19}$$

$$\frac{\partial v}{\partial t} = -cv + \frac{\theta' uv}{1 + \xi v + \eta u} - \frac{fv^2}{1 + g^2 v^2} + d_2 \Delta v, \tag{4.20}$$

where $\Delta = \partial^2/\partial x^2$ in the 1D case, and $\Delta = (\partial^2/\partial x^2) + (\partial^2/\partial y^2)$ in the 2D case. The initial and boundary conditions are the following: (i) in one dimension: $u(x, 0) > 0, v(x, 0) > 0$, for $x \in [0, L]$; $(\partial u/\partial x) = 0$, and $(\partial v/\partial x) = 0$, at $(0, t)$ and (L, t);

(ii) in two dimensions: $u(x, y, 0) > 0$, $v(x, y, 0) > 0$, for $(x, y) \in \Omega$, and $(\partial u/\partial n) = 0$, $(\partial v/\partial n) = 0$, on $\partial \Omega$. Zero flux boundary conditions imply that no external input is imposed from outside. The prey growth rate, $r_x = s + lx$, is taken as a linear function of x [69]. The parameters θ, θ', ξ, η, c, f, and g denote capture rate, efficiency of biomass conversion from prey to predator, predator interference, how fast the per capita feeding rate approaches its saturation value, the predator's death rate, fish predation, and half-saturation constant, respectively. The stability analysis of the system is given in the following example.

Example 4.1

Discuss the linear stability of the equilibrium points of the model system (4.19), (4.20) without diffusion and spatial heterogeneity with $r_x = s$, a constant.

SOLUTION

The model system is given by

$$\frac{du}{dt} = su(1 - u) - \frac{\theta uv}{1 + \xi v + \eta u}, \quad \frac{dv}{dt} = -cv + \frac{\theta' uv}{1 + \xi v + \eta u} - \frac{fv^2}{1 + g^2 v^2}.$$

The equilibrium points are $E_0(0, 0)$ $E_1(1, 0)$. The positive equilibrium point (u^*, v^*) is the solution of the equations

(i) $s(1 - u) - (\theta v/T) = 0$, (ii) $-c + (\theta' u/T) - [fv/(1 + g^2 v^2)] = 0$.
   $T = (1 + \xi v + \eta u)$.

From (i), we find that when $v = 0$, $u = 1$, and when $u = 0$, $v = s/(\theta - s\xi)$. Since $v > 0$, we require $\theta > s\xi$. Also,

$$v = \frac{s(1 + \eta u)(1 - u)}{\theta - s\xi + s\xi u}, \quad v' = \frac{s[\eta(1 - 2u) - \xi v - 1]}{\theta - s\xi + s\xi u}.$$

Now, $v' > 0$, if $u < 1/2$, and $\eta > (1 + \xi v)/(1 - 2u)$.
From (ii), we find that when $v = 0$, $u = [c/(\theta' - c\eta)] > 0$, for $\theta' > c\eta$. The curve does not cut the positive v-axis. Also,

$$v' = \frac{\theta'(1 + \xi v)(1 + g^2 v^2)^2}{\theta' u\xi(1 + g^2 v^2)^2 + f(1 - g^2 v^2)T^2} > 0, \quad \text{for } g^2 v^2 < 1,$$

(sufficient condition).

The required sufficient conditions are (i) $\theta > s\xi$, (ii) $u < 1/2$, (iii) $\eta > (1 + \xi v)/(1 - 2u)$, (iv) $\theta' > c\eta$, and (v) $g^2 v^2 < 1$.

At $E_0(0, 0)$, the eigenvalues of the variational matrix are $\lambda = s$ and $\lambda = -c$. E_0 is a saddle point. At E_1 (1, 0), the eigenvalues of the variational matrix are $\lambda = -s$, and $\lambda = [\theta'/(1 + \eta)] - c$. If $[\theta'/(1 + \eta)] < c$, $E_1(1, 0)$

is asymptotically stable in the u–v plane. If $\theta' > c(1 + \eta)$, then E_1 is a saddle point with stable manifold locally along the u-direction and unstable manifold locally along the v-direction. At $E^*(u^*, v^*)$, the variational matrix is given by

$$J = \begin{bmatrix} a_{11} & a_{12} \\ a_{21} & a_{22} \end{bmatrix},$$

where

$$a_{11} = -u^*\left[s - \frac{\theta\eta v^*}{T^{*2}}\right], \quad a_{22} = -v^*\left[\frac{\theta'\xi u^*}{T^{*2}} + \frac{f(1 - g^2 v^{*2})}{(1 + g^2 v^{*2})^2}\right],$$

$$a_{12} = -\frac{\theta u^*}{T^{*2}}(1 + \eta u^*), \quad a_{21} = \frac{\theta' v^*}{T^{*2}}(1 + \xi v^*), \quad T^* = 1 + \xi u^* + \eta v^*.$$

The characteristic equation is $\lambda^2 + p\lambda + q = 0$, where $p = -(a_{11} + a_{22})$, $q = a_{11}a_{22} - a_{21}a_{12}$. If $sT^{*2} > \eta\theta v^*$, then $a_{11} < 0$, and if $g^2 v^{*2} < 1$, then $a_{22} < 0$. Also, $a_{12} < 0$, $a_{21} > 0$. Hence, $p > 0$, $q > 0$ under these conditions. By the Routh–Hurwitz criterion, the characteristic equation has negative roots or has roots with negative real parts. Hence, the equilibrium point is asymptotically stable. A stronger sufficient condition is $\theta'\xi > \theta\eta$, $p > 0$.

Example 4.2

Perform the linear stability analysis and study the effect of diffusion on the stability for models (4.19) and (4.20) in the 1D case. Assume the parameter values as $x^* = 0.85$, $s = 0.81$, $l = -0.1$, $\theta = 6.5$, $\theta' = 6.05$, $\xi = 0.02$, $\eta = 2.5$, $c = 0.4$, $g = 0.4$, $f = 0.0001$. Perform numerical simulation for the above set of parameter values with $d_1 = 10^{-4}$, $d_2 = 10^{-3}$, in one dimension. Assume the initial conditions as $u(x,0) = u^* + \varepsilon(x - x_1)(x - x_2)$, $v(x,0) = v^*$, where (u^*, v^*) is the nontrivial state for the coexistence of prey and predator population with $\varepsilon = 10^{-8}$, $x_1 = 1200$, $x_2 = 2800$.

SOLUTION

To study the effect of diffusion on the model system, consider the linearized form of the model system about the positive equilibrium point $E^*(u^*, v^*)$. Setting $u = u^* + U$, $v = v^* + V$, we obtain the linearized form as

$$\frac{\partial U}{\partial t} = b_{11}U + b_{12}V + d_1\frac{\partial^2 U}{\partial x^2}, \quad \frac{\partial V}{\partial t} = b_{21}U + b_{22}V + d_2\frac{\partial^2 V}{\partial x^2}, \quad (4.21)$$

where

$$b_{11} = -(r_x/\theta)(u^*/v^*)[v^*\theta - \eta\, r_x(1 - u^*)^2],$$

$$b_{12} = -(r_x/\theta)(u^*/v^*)(1 - u^*)[\theta - \xi r_x(1 - u^*)],$$

$$b_{21} = \frac{\theta' v^*(1 + \xi v^*)}{T^2},$$

$$b_{22} = -\left[\frac{\theta' u^* v^* \xi}{T^2} + \frac{f v^*(1 - g^2 v^{*2})}{(1 + g^2 v^{*2})^2}\right], \quad T = (1 + \xi v^* + \eta u^*).$$

Write the solution in the form $U = s e^{\lambda t} \cos(n\pi x/L)$, $V = w e^{\lambda t} \cos(n\pi x/L)$, where λ is the frequency and $(L/n\pi)$ is the wavelength, respectively. Substitute the expressions for U, V in Equation 4.21. The homogeneous equations in s and w have solutions if the determinant of the coefficient matrix is zero. We get

$$(\lambda - b_{11} + d_1 l^2)(\lambda - b_{22} + d_2 l^2) - b_{12} b_{21} = 0, \quad \text{or} \quad \lambda^2 + p\lambda + q = 0, \quad (4.22)$$

where

$$p = (d_1 + d_2)l^2 - (b_{11} + b_{22}), \quad l = n\pi/L,$$

$$q = d_1 d_2 l^4 - l^2(d_2 b_{11} + d_1 b_{22}) + (b_{11} b_{22} - b_{12} b_{21}).$$

By the Routh–Hurwitz criterion, the roots of Equation 4.22 are negative or have negative real parts if $p > 0$ and $q > 0$. A sufficient condition for $p > 0$ is $(b_{11} + b_{22}) < 0$. This gives the (strong) conditions (i) $r_x \eta(1 - u^*)^2 < \theta v^*$, and (ii) $g^2 v^{*2} < 1$. For these values, $b_{11} < 0$, $b_{22} < 0$. Also, $b_{21} > 0$. Now, if (iii) $u^* < 1$, and (iv) $\theta > \xi r_x(1 - u^*)$, then $b_{12} < 0$. If $u^* < 1$, and $\theta < \xi r_x(1 - u^*)$, then $b_{12} > 0$. If $u^* > 1$, $b_{12} > 0$. Hence, the sufficient conditions for $q > 0$ are (i), (ii), (iii), and (iv). The positive equilibrium E^* is locally asymptotically stable in the presence of diffusion if the above four conditions are satisfied. If $u^* > 1$, the sufficient conditions for $q > 0$ are (i), (ii), and $b_{11} b_{22} > b_{12} b_{21}$. Further, if the conditions (i) and (ii) only are satisfied, and $b_{11} b_{22} - b_{12} b_{21} < 0$, then the first two terms of q are positive and q can be made positive by taking sufficiently large values of the diffusion coefficients d_1, d_2, and the equilibrium point E^* is locally asymptotically stable. Suppose that the sufficient conditions are not met. Now, irrespective of the sign of p, diffusive instability can arise if $q < 0$ (one root of Equation 4.22 is positive), that is, if $q(l^2) = Dl^4 - Cl^2 + B < 0$, where $D = d_1 d_2$, $C = (d_2 b_{11} + d_1 b_{22})$, and $B = b_{11} b_{22} - b_{12} b_{12}$. The roots of this equation in l^2 are real and positive when $C > 0$, $B > 0$, and $C^2 - 4BD > 0$. Then, $q < 0$ when $l_1^2 < l^2 < l_2^2$, where $l_1^2, l_2^2 = [C \mp \sqrt{C^2 - 4BD}]/(2D)$. Therefore, diffusive instability occurs when these conditions are satisfied. Now, $q(l^2)$ is a quadratic in l^2 and the graph of $y = q(l^2)$ is a parabola opening upwards. The minimum occurs at the vertex of the parabola, that is, for $l^2 = l_m^2$ where $l^2 = l_m^2 = C/(2D)$.

The positive equilibrium point (u^*, v^*) is the solution of the equations $s(1 - u^*) - (\theta v^*/T) = 0, -c + (\theta' u^*/T) - [f v^*/(1 + g^2 v^{*2})] = 0$, (see Equations 4.19 and 4.20). For the given set of parameter values, we solve for u^*, v^* using the Newton iteration method. We obtain $u^* = 0.079429506$,

$v^* = 0.137813271$, $B = b_{11}b_{22} - b_{12}b_{21} = 0.222349094 > 0$, $C = (d_2b_{11} + d_1b_{22}) > 0$, for $(d_2/d_1) > 0.033631079$, $C^2 - 4BD > 0$, for $(d_2/d_1) > 525.11$. For these values of d_1, d_2; $p > 0$. Now, let $d_1 = 0.0001$, $d_2 = 0.055$. Then, $q(l^2) = D(l^2 - 249.556)$ $(l^2 - 161.996) < 0$, for $161.996 < l^2 < 249.556$. The system is unstable for all values of the wavelengths lying in this range. Outside this range for l^2 and $l^2 > 0.073123$; $p > 0$, and $q > 0$. The system is stable. Without diffusion $(r_x = s)$, we find that the eigenvalues are a complex pair with positive real part. The equilibrium point is unstable.

Numerical simulations using MATLAB 7.0 are done for the above set of parameter values with $d_1 = 10^{-4}$, $d_2 = 10^{-3}$. System 4.19 and 4.20 is solved numerically using a semi-implicit (in time) finite difference method. The time and space steps are chosen sufficiently small so that the results are numerically stable. The given nonmonotonic form of the initial conditions is $u(x, 0) = u^* + \varepsilon(x - x_1)(x - x_2)$, $v(x, 0) = v^*$, where $(u^*, v^*) = (0.079429506, 0.137813271)$ is the nontrivial state for the coexistence of prey and predator population. The values of the parameters affecting the system dynamics are taken as $\varepsilon = 10^{-8}$, $x_1 = 1200$, and $x_2 = 2800$. Time series with spatial heterogeneity, when the parameter describing the fish predation is $f = 0.0001$, is plotted in Figure 4.11. The intensity of fish predation has an influence over temporal evolution of the species. Higher values of fish predation may make the evolution of densities stationary. Lower values favor chaotic dynamics. Spatiotemporal patterns of prey density (Figure 4.12a) and predator density (Figure 4.12b) with spatial heterogeneity for $f = 0.0001$ are plotted. From the figures, we observe that the spatial distribution of predators has more erratic density distribution than the prey. Spatiotemporal patterns of prey density (Figure 4.13a) and predator density (Figure 4.13b) without spatial heterogeneity for $f = 0.0001$ are plotted. From Figures 4.12 and 4.13, we observe the regularizing behavior of fish predation.

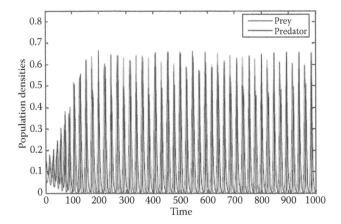

FIGURE 4.11
Time series with spatial heterogeneity for the fish predation parameter $f = 0.0001$.

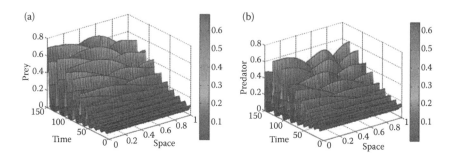

FIGURE 4.12
Spatiotemporal patterns of (a) prey density, and (b) predator density, with with spatial heterogeneity for $f = 0.0001$.

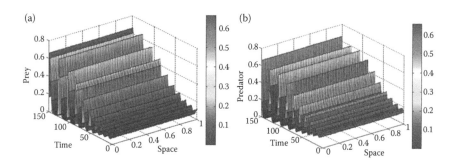

FIGURE 4.13
Spatiotemporal patterns of (a) prey density, and (b) predator density, without spatial heterogeneity for $f = 0.0001$.

Example 4.3

In the 2D case, perform numerical simulation of model (4.19), (4.20) for the same set of parameter values as in Example 4.2, except with $f = 0.02$, and (i) $d_1 = 10^{-5}$, $d_2 = 10^{-4}$, (ii) $d_1 = d_2 = 0.8$. Assume zero-flux boundary conditions and nonzero asymmetrical initial conditions $u(x, y, 0) = 0.002x + 0.1$, $v(x, y, 0) = 0.2(1 - 0.01y) + 0.1$. Take the step lengths of the numerical grid as $\Delta x = \Delta y = 1$ and $\Delta t = 0.1$, and compute up to $t = 4000$ (see also Upadhyay et al. [112]).

SOLUTION

In two dimensions, the system is given by

$$\frac{\partial u}{\partial t} = r_x u(1 - u) - \frac{\theta uv}{1 + \xi v + \eta u} + d_1\left(\frac{\partial^2 u}{\partial x^2} + \frac{\partial^2 u}{\partial y^2}\right), \qquad (4.23)$$

$$\frac{\partial v}{\partial t} = -cv + \frac{\theta' uv}{1 + \xi v + \eta u} - \frac{fv^2}{1 + g^2 v^2} + d_2\left(\frac{\partial^2 v}{\partial x^2} + \frac{\partial^2 v}{\partial y^2}\right), \qquad (4.24)$$

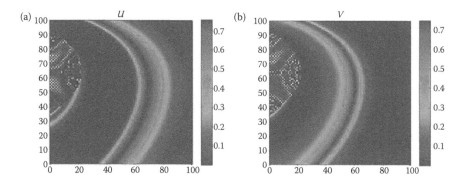

FIGURE 4.14
Turing spatial patterns for (a) prey, and (b) predator densities for $d_1 = 10^{-5}$, $d_2 = 10^{-4}$ at $t = 4000$.

where $u(x, y, 0) > 0$, $v(x, y, 0) > 0$, for $(x, y, 0) \in \Omega$, and $(\partial u / \partial n) = 0$, $(\partial v / \partial n) = 0$, on $\partial \Omega$. The model system is solved using MATLAB 7.0 for the parameter values $x^* = 0.85$, $s = 1.0$, $l = -1.4$, $c = 0.4$, $\theta = 6.5$, $\theta' = 6.05$, $\xi = 0.02$, $\eta = 2.5$, $g = 0.4$, and $f = 0.02$, with (i) $d_1 = 10^{-5}$, $d_2 = 10^{-4}$ and (ii) $d_1 = d_2 = 0.08$. The model equations are solved using a semi-implicit (in time) finite difference method with zero-flux boundary conditions and with nonzero asymmetrical initial conditions. (The initial condition and the biologically realistic parameter values for the simulation experiments were selected on the basis of the results given by Garvie [23] and Medvinsky et al. [65].) The step lengths of the numerical grid are $\Delta x = \Delta y = 1$ and $\Delta t = 0.1$. The sparse banded linear system of algebraic equations was solved using the GMRES algorithm. Turing spatial patterns of prey (Figure 4.14a) and predator (Figure 4.14b) densities for $d_1 = 10^{-5}$, $d_2 = 10^{-4}$ at $t = 4000$ are plotted. The non-Turing patterns of prey (Figure 4.15a) and predator (Figure 4.15b) densities for $d_1 = d_2 = 0.08$ at $t = 4000$ are plotted. In Upadhyay et al. [113], numerical simulations were done for $f = 0.0001$, with (i) $d_1 = 10^{-4}$, $d_2 = 10^{-3}$ and (ii) $d_1 = d_2 = 0.05$, with the remaining parameter values being the same.

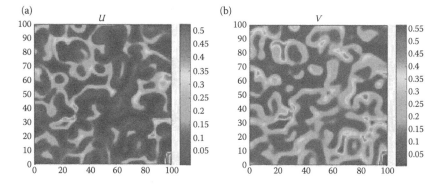

FIGURE 4.15
Non-Turing spatial patterns for (a) prey, and (b) predator densities for $d_1 = d_2 = 0.08$ at $t = 4000$.

Spatiotemporal Dynamics of a Predator–Prey Model Using Holling Type IV Functional Response

Consider the spatiotemporal dynamics of a variant of the model given in Equations 2.73 and 2.74 (see [108,112])

$$\frac{\partial P}{\partial T} = \frac{\xi\, NP}{(H_N + N)} - \phi\, P^2 - \frac{cPZ}{(P^2/i) + P + a} + D_P \Delta P, \tag{4.25}$$

$$\frac{\partial Z}{\partial T} = \frac{bcPZ}{(P^2/i) + P + a} - mZ - \frac{FZ^2}{H_z^2 + Z^2} + D_z \Delta Z, \tag{4.26}$$

where $\Delta = (\partial^2/\partial X^2)$ in the 1D case, and $\Delta = (\partial^2/\partial X^2) + (\partial^2/\partial Y^2)$ in the 2D case, D_P and D_Z are the diffusion coefficients of phytoplankton and zooplankton, respectively, N is the nutrient level of the system which is assumed to be constant, ξ is the maximum per capita growth rate of prey population, H_N is the phytoplankton density at which specific growth rate becomes half of its saturation value. ϕ is the intensity of competition among individuals of phytoplankton population, c is the rate at which phytoplankton is grazed and it follows the Holling type IV functional response, i is a direct measure of the predator's immunity form or tolerance of the prey, a is the half-saturation constant in the absence of any inhibitory effect, b is the conversion coefficient from individuals of phytoplankton into individuals of zooplankton, m is the mortality rate of zooplankton, F is the predation rate of zooplankton population which follows the Holling type III functional response, and H_Z is the zooplankton density at which specific growth rate becomes half its saturation value. Generally, the basic functional response of fish in the presence of a single prey species is taken as type II. Under natural conditions, the type III response of zooplankton density is a more realistic assumption [65,70,88]. The phytoplankton equation is based on the logistic growth formulation, with a Monod type of nutrient limitation. Growth limitations by different nutrients are not treated separately. Instead, it is assumed that there is an overall carrying capacity which is a function of the nutrient level of the system, and the phytoplankton does not deplete the nutrient level [83]. A simple model combining deterministic and stochastic elements is used to explain the phytoplankton and zooplankton oscillations and monotonous relaxation to one of the possible multiple equilibria [94].

In Upadhyay et al. [112], the model was extended to include the spatial heterogeneity and the parameters for nutrient level of the system and fish predation were treated as multiplicative control variables. The authors assumed the following: (i) the fish density and nutrient concentration are environmental factors that can be manipulated [87]. (ii) The grazing rate of zooplankton is dependent on phytoplankton concentration according to type IV functional response [108], while the predation rate of fish on zooplankton is of type III [24].

Substituting $r = \xi N/(H_N + N), a = r/\phi, \alpha = i/a, \beta = bc/r, \gamma = m/r, u = P/a,$ $D = D_Z/D_P, v = cZ/(ar), x = X(r/D_P)^{1/2}, y = Y(r/D_P)^{1/2}, t = rT, f = Fc/(ar^2), \eta = cH_Z/$ (ar) into Equations 4.25, 4.26, we obtain the nondimensional form in the 1D case as

$$\frac{\partial u}{\partial t} = u(1 - u) - \frac{uv}{(u^2/\alpha) + u + 1} + \Delta u, \qquad (4.27)$$

$$\frac{\partial v}{\partial t} = \frac{\beta uv}{(u^2/\alpha) + u + 1} - \gamma v - \frac{fv^2}{v^2 + \eta^2} + D\Delta v, \qquad (4.28)$$

where $u(x, 0) > 0$, $v(x, 0) > 0$, $\Delta = \partial^2/\partial x^2$. We assume no-flux boundary conditions. The equilibrium points are $(0, 0)$, $(1, 0)$. Positive equilibrium point is given by $E^* = (u^*, v^*)$, where u^* is the positive root of the equation

$$T^2(1 - u)^2(\gamma T - \beta u) + f(1 - u)T^2 + \gamma \eta^2 T - \beta \eta^2 u = 0,$$

$$T = (u^2/\alpha) + u + 1, \quad \text{and} \quad v^* = (1 - u)T.$$

Numerical solution of the model system in one space dimension: Consider the domain $0 < x < L_x$, where $L_x = 7000$. For numerical simulations, we choose the values of the parameters as $\alpha = 0.3$, $\beta = 2.33$, $\gamma = 0.25$, $\eta = 2.5$, and $f = 0.02$. With these values, we obtain $u^* = 0.1284$ and $v^* = 1.0314$. The initial conditions are taken in a nonmonotonic form which determines the initial spatial distribution of the species in a real community as in Medvinsky et al. [65]; $u(x, 0) = u^* + \varepsilon(x - x_1)(x - x_2)$, $v(x, 0) = v^*$, where (u^*, v^*) is the nontrivial state for the coexistence of prey and predator and $\varepsilon = 10^{-8}$, $x_1 = 1200$, and $x_2 = 2800$ are the parameters affecting the system dynamics. The dynamics of the prey and predator populations at the time levels $t = 300$ and 1800 for different values of the diffusivity constant, $D = 2$, 1200, and 2400 is plotted as space series in Figure 4.16. We observe the onset of chaotic phase for $D = 2$, which occupies the entire domain for large times. As we increase the value of the diffusivity constant, the dynamics becomes stable. This observation supports the analytical finding that it may be possible that the dynamics of an unstable system can be made stable by considering sufficiently large values of the diffusivity constant. In Upadhyay et al. [112], simulations were done for $f = 0.0001$, and $D = 1, 2000, 3000$ with the remaining parameter values being the same. For these values, we obtain $u^* = 0.126665$, $v^* = 1.030666$.

System in two dimensions: The nondimensionalized form of the equations is

$$\frac{\partial u}{\partial t} = u(1 - u) - \frac{uv}{\left((u^2/\alpha) + u + 1\right)} + \left(\frac{\partial^2 u}{\partial x^2} + \frac{\partial^2 u}{\partial y^2}\right),$$

$$\frac{\partial v}{\partial t} = \frac{\beta uv}{\left((u^2/\alpha) + u + 1\right)} - \gamma v - \frac{fv^2}{v^2 + \eta^2} + D\left(\frac{\partial^2 v}{\partial x^2} + \frac{\partial^2 v}{\partial y^2}\right).$$

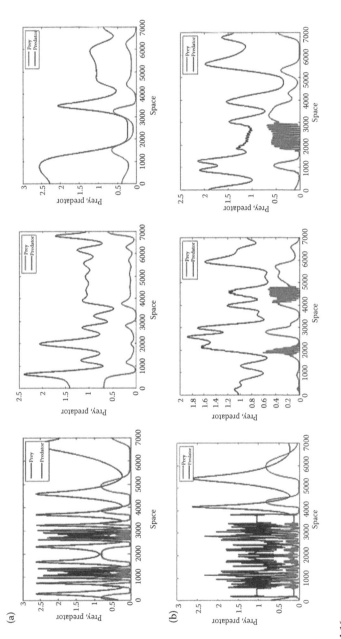

FIGURE 4.16
Space series at times (a) $t = 300$, (b) $t = 1800$, for $D = 2$, 1200, 2400.

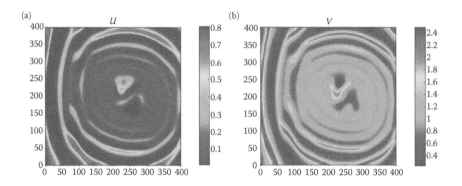

FIGURE 4.17
Spatial patterns of (a) prey and (b) predator densities at time $t = 300, f = 0.02$.

The domain for numerical simulations is taken as $0 < x < L_x$, $0 < y < L_y$, where $L_x = 400$ and $L_y = 400$. The initial distribution of the species is taken as $u_0(x, y) = 0.1284 - \varepsilon_1(x - 0.1y - 225)(x - 0.1y - 675)$, $v_0(x, y) = 1.0314 - \varepsilon_2(x - 450) - \varepsilon_3(x - 150)$. The parameter values are taken as $\alpha = 0.3$, $\beta = 2.33$, $\gamma = 0.25$, $\eta = 2.5$, $f = 0.02$, $D = 2$, $\varepsilon = 2 \times 10^{-7}$, $\varepsilon_2 = 3 \times 10^{-5}$, and $\varepsilon_3 = 1.2 \times 10^{-4}$. Numerical simulation is done using MATLAB 7.0 with a semi-implicit finite difference technique. The uniform mesh step length along the x and the y axis is taken as $h = 1$ and the step length along the t-axis is taken as $\Delta t = 0.33$. Spatial patterns of prey and predator densities at $t = 300, 1200$ are plotted in Figures 4.17a,b and 4.18a,b, respectively. In Upadhyay et al. [112], simulations were done for $f = 0.0001$ and $D = 1$, with the remaining parameter values being the same.

For small times, the prey and predator population densities display regular pattern with spiral centers. The regular spiral patterns grow from their centers, but are destroyed as time increases leading to chaotic patchy patterns. This type of spatiotemporal dynamics was observed in similar predator–prey systems with Holling type II functional response without fish [65] and

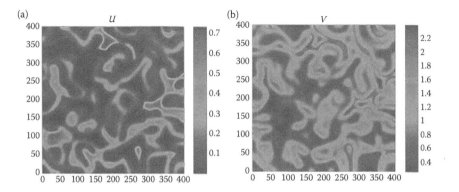

FIGURE 4.18
Spatial patterns of (a) prey and (b) predator densities at time $t = 1200, f = 0.02$.

with fish [109]. It was observed that the dynamics of model system changes and the number of spirals decreases as we increase the values of fish predation f, which shows the stable nature of the dynamics. The system dynamics changes from irregular to regular, when the fish predation and diffusivity constant were increased to sufficiently large values.

Example 4.4

Derive sufficient conditions for diffusion-driven instability for model (4.27) and (4.28) in one dimension.

SOLUTION

Let $E^*(u^*, v^*)$ be the positive equilibrium solution of the system. Write $u = u^* + U$, $v = v^* + V$. Substituting into Equations 4.27 and 4.28, we obtain the linearized form of the equations about $E^*(u^*, v^*)$ as

$$\frac{\partial U}{\partial t} = b_{11}U + b_{12}V + \frac{\partial^2 U}{\partial x^2}, \quad \frac{\partial V}{\partial t} = b_{21}U + b_{22}V + D\frac{\partial^2 V}{\partial x^2}, \quad (4.29)$$

where

$$b_{11} = -u^*\left[1 - \frac{v^*(2u^* + \alpha)}{\alpha T^{*2}}\right], \quad b_{12} = -\frac{u^*}{T^*}, \quad T^* = (u^{*2}/\alpha) + u^* + 1,$$

$$b_{21} = \frac{\beta v^*}{T^{*2}}[1 - (u^{*2}/\alpha)], \quad b_{22} = \left[\frac{\beta u^*}{T^*} - \gamma\right]\left[\frac{v^{*2} - \eta^2}{v^{*2} + \eta^2}\right] = \frac{fv^*}{v^{*2} + \eta^2}\left[\frac{v^{*2} - \eta^2}{v^{*2} + \eta^2}\right].$$

Write the solution in the form $U = se^{\lambda t}\cos(n\pi x/L)$, $V = we^{\lambda t}\cos(n\pi x/L)$, where λ is the frequency and $(L/n\pi)$ is the wavelength, respectively. Substitute the expressions for U, V into Equation 4.29. The homogeneous equations in s and w have solutions if the determinant of the coefficient matrix is zero. We get

$$(\lambda - b_{11} + l^2)(\lambda - b_{22} + Dl^2) - b_{12}b_{21} = 0, \quad \text{or} \quad \lambda^2 + p\lambda + q = 0, \quad (4.30)$$

where

$$p = (D + 1)l^2 - (b_{11} + b_{22}), \quad l = n\pi/L,$$

$$q = Dl^4 - l^2(Db_{11} + b_{22}) + (b_{11}b_{22} - b_{12}b_{21}).$$

By the Routh–Hurwitz criterion, the roots of Equation 4.30 are negative or have negative real parts if $p > 0$, and $q > 0$. A sufficient condition for $p > 0$ is $b_{11} + b_{22} < 0$. Sufficient conditions (strong) are $b_{11} < 0$, $b_{22} < 0$. These conditions are satisfied when $v^* < \eta$, and $v^*(2u^* + \alpha) < \alpha T^{*2}$. Also, $b_{12} < 0$. Now, $b_{21} > 0$, when $u^{*2} < \alpha$. Hence, $q > 0$. Therefore, the positive equilibrium E^* is locally asymptotically stable in the presence of diffusion if the above three conditions are satisfied. Even if any one of the above conditions is not satisfied, by increasing the value of the diffusion coefficient, it may be possible to satisfy $p > 0$ and $q > 0$. However, irrespective of the sign of p, the system is unstable if $q < 0$, as the characteristic equation has a positive root. This implies that there exist parameter values and a range for D, for which the system may be unstable. For example, for the parameter values $\alpha = 0.3, \beta = 2.33, \gamma = 0.25, \eta = 2.5, f = 0.0001, D = 1$, the system is unstable and for the parameter values $\alpha = 0.3, \beta = 2.33, \gamma = 0.25, \eta = 2.5, f = 1.5, D = 10$, the system is stable.

Holling–Tanner Model with Diffusion

Consider the Holling–Tanner model system (see Equations 2.84 and 2.85) with diffusion as [44]

$$\frac{\partial P}{\partial t} = P\left[a_1 - b_1 P - \frac{\omega Z}{P + D_1}\right] + D_P \frac{\partial^2 P}{\partial x^2}, \tag{4.31}$$

$$\frac{\partial Z}{\partial t} = Z\left[a_2 - \frac{\omega_1 Z}{P}\right] + D_Z \frac{\partial^2 Z}{\partial x^2}, \tag{4.32}$$

in the domain $D \in \{r = (x, 0); 0 \le x \le L\}$ with zero flux boundary conditions, $(\partial P/\partial x) = 0, (\partial Z/\partial x) = 0$, at $x = 0$ and $x = L$. All the constants in the equations are positive and D_P, D_Z are diffusion coefficients assumed to be constants.

Stability analysis: First, consider the case without diffusion, that is, $D_P = D_Z = 0$. The nonzero steady-state solution (P^*, Z^*) of Equations 4.31 and 4.32 is given by $Z^* = (P^* a_2)/\omega_1$, and P^* is a positive root of the quadratic equation $b_1 P^2 + P[b_1 D_1 - a_1 + (\omega a_2/\omega_1)] - a_1 D_1 = 0$. Perturb the solution of Equations 4.31 and 4.32 about the steady state (P^*, Z^*) to $(P^* + p, Z^* + z)$, where p and z are small. The linearized system is given by

$$\frac{\partial p}{\partial t} = a_{11} p + a_{12} z + D_P \frac{\partial^2 p}{\partial x^2}, \quad \frac{\partial z}{\partial t} = a_{21} p + a_{22} z + D_Z \frac{\partial^2 z}{\partial x^2}, \tag{4.33}$$

where

$$a_{11} = a_1 - 2 b_1 P^* - \frac{\omega Z^* D_1}{(P^* + D_1)^2} = \frac{\omega Z^*}{(P^* + D_1)} - b_1 P^* - \frac{\omega Z^* D_1}{(P^* + D_1)^2}$$

$$= -P^*\left(b_1 - \frac{\omega Z^*}{(P^* + D_1)^2}\right) = -P^*(b_1 - s_1), \quad s_1 = \frac{\omega Z^*}{(P^* + D_1)^2} = \frac{\omega P^* a_2}{\omega_1 (P^* + D_1)^2}.$$

$$a_{12} = -\frac{\omega P^*}{(P^* + D_1)}, \quad a_{21} = \omega_1 \left(\frac{Z^*}{P^*}\right)^2 = \frac{a_2^2}{\omega_1}, \quad a_{22} = a_2 - 2\omega_1\left(\frac{Z^*}{P^*}\right) = -a_2,$$

since $a_1 - b_1 P^* - [(\omega Z^*)/(P^* + D_1)] = 0$, and $a_2 - [(\omega_1 Z^*)/P^*] = 0$.

Assume the solution of the above equations in the form $p(x, t) = se^{\lambda t}\cos(kx)$, $z(x,t) = we^{\lambda t}\sin(kx)$, where k takes the discrete values, $k = (n\pi/L)$, n is an integer, and λ is the eigenvalue determining the temporal growth. Here, k denotes the wave number. Substitute the expressions for p and z into Equation 4.33. The homogeneous equations have a solution if the determinant of the coefficient matrix is zero. We obtain the characteristic equation as $\lambda^2 + a\lambda + b = 0$, where

$$a = (D_p + D_z)k^2 - (a_{11} + a_{22}) = (D_p + D_z)k^2 + [P^*(b_1 - s_1) + a_2],$$

$$b = D_P D_Z k^4 - (a_{11}D_Z + a_{22}D_P)k^2 + a_{11}a_{22} - a_{12}a_{21}$$

$$= D_P D_Z k^4 + k^2 D_P[P^*(b_1 - s_1)d_c + a_2] + P^*a_2(b_1 - s_1) + \frac{\omega P^* a_2^2}{\omega_1(P^* + D_1)}$$

$$= D_P D_Z k^4 + k^2 D_P[P^*(b_1 - s_1)d_c + a_2] + a_2(P^*b_1 + s_1 D_1),$$

where $d_c = D_Z/D_P$. The steady state (P^*, Z^*) is linearly stable in the presence of diffusion if both the eigenvalues have negative real parts, that is, when $a > 0$ and $b > 0$. To have Turing instability, we require that one or both the eigenvalues have positive real part for some $k \neq 0$. Both the roots are positive or have positive real parts if $a < 0$ and $b > 0$. One of the roots is positive if $a > 0$, and $b < 0$; or when $a < 0$, and $b < 0$. Now, $a > 0$, and $b < 0$, gives

$$(D_p + D_z)k^2 + [P^*(b_1 - s_1) + a_2] > 0, \quad (4.34)$$

$$D_P D_Z k^4 + k^2 D_P[P^*(b_1 - s_1)d_c + a_2] + a_2(P^*b_1 + s_1 D_1) < 0. \quad (4.35)$$

For Equation 4.34 to be satisfied, a sufficient condition is $[P^*b_1 + a_2] > P^*s_1$. For Equation 4.35 to be satisfied, a necessary condition is $[P^*(b_1 - s_1)d_c + a_2] < 0$. This implies that $d_c \neq 1$. We require $d_c > 1$. Write Equation 4.35 as $h(k^2) = k^4 + pk^2 + q$, or as $h(t) = t^2 + pt + q$. For $h(k^2)$ to be negative for some $k \neq 0$, the minimum of $h(k^2)$ must be negative. This would also give bounds on k^2 so that Equation 4.35 is satisfied. The minimum occurs at the vertex of the parabola in the variable $t = k^2$, for $t = -p/2$, and the minimum value is $q - (p^2/4)$. We obtain

$$h_{min} = a_2(P^*b_1 + s_1 D_1) - \frac{1}{4d_c}[P^*(b_1 - s_1)d_c + a_2]^2 < 0, \quad (4.36)$$

at $k^2 = k_m^2 = \dfrac{1}{2D_Z}[P^*(s_1 - b_1)d_c - a_2]$.

At the bifurcation point, that is, at the value of k^2 where the stability changes, $h_{\min} = 0$. We obtain from $h_{\min} = 0$, $q = (p^2/4)$, and the critical wave number is

$$k_c^2 = -\frac{p}{2} = \sqrt{q} = \left[\frac{a_2(P^*b_1 + s_1D_1)}{D_PD_Z}\right]^{1/2}. \tag{4.37}$$

The bounds on k^2 are $k_1^2 < k^2 < k_2^2$, where $k_1^2 = [\{-p - \sqrt{p^2 - 4q}\}/2]$, $k_2^2 = [\{-p + \sqrt{p^2 - 4q}\}/2]$.

Alonso et al. [1] have shown that Turing instabilities can occur due to perturbations of specific spatial wavelengths, and they appear within certain parameter spaces. It is possible to identify these Turing spaces with respect to the parameters and wavelengths and discuss the associated system properties. Linear stability analysis yields intervals for the wavelength k, within which the homogeneous solution is unstable with respect to perturbations with this wavelength. The eigenvalues of the corresponding Jacobian which depend on the wavelength k and the diffusion coefficients D_P, D_Z; have to be positive.

Camara and Alaoui [6] derived the conditions for Hopf and Turing bifurcations in a spatial predator–prey model incorporating the Holling type II and modified Leslie–Gower functional responses.

A Variant of Holling–Tanner Model with Diffusion

Consider a reaction–diffusion model for a phytoplankton–zooplankton–fish system where at any point (x, y) and time t, the phytoplankton population $P(x, y, t)$ is predated by the zooplankton population $H(x, y, t)$, which is in turn predated by fish. The per capita predation rate is described by the Holling type III functional response. The system incorporating the effects of fish predation is given by [114]

$$\frac{\partial P}{\partial t} = P\left[r - B_1P - \frac{B_2PH}{P^2 + D^2}\right] + d_1\Delta P, \tag{4.38}$$

$$\frac{\partial H}{\partial t} = H\left[C_1 - \frac{C_2H}{P} - \frac{FH}{H^2 + D_1^2}\right] + d_2\Delta H, \tag{4.39}$$

$$u(x, y, 0) > 0, \quad v(x, y, 0) > 0,$$

where $\Delta = (\partial^2/\partial X^2)$, or $\Delta = (\partial^2/\partial X^2) + (\partial^2/\partial Y^2)$, and all the parameters are positive. Assume no-flux boundary conditions. r is the prey's intrinsic growth rate in the absence of predation, B_1 is the intensity of competition among individuals of phytoplankton, B_2 is the rate at which phytoplankton is grazed by zooplankton, which follows the Holling type III functional response, C_1 is the predator's intrinsic rate of population growth, and C_2 indicates the number of prey necessary to support and replace each individual predator. The rate equation for the zooplankton population is the logistic growth with carrying capacity proportional to phytoplankton density P/C_2; D, D_1 are the half-satu-ration constants for phytoplankton and zooplankton density, respectively, F is the maximum value of the total loss of zooplankton due to fish predation which also follows the Holling type III functional response; d_1, d_2 are the diffusion coefficients of phytoplankton and zooplankton, respectively. Holling type III functional response is often used to demonstrate cyclic collapses and is suitable for representing the behavior of predator hunting [79,52]. The functional response levels off at some prey density.

One equilibrium point is $(r/B_1, 0)$. The positive equilibrium point, (P^*, H^*), is the solution of the equations

$$r - P\left[B_1 + \frac{B_2 H}{P^2 + D^2}\right] = 0, \quad C_1 - H\left[\frac{C_2}{P} + \frac{F}{H^2 + D_1^2}\right] = 0. \quad (4.40)$$

From the second equation of Equation 4.40, we obtain $P = [C_2 H(H^2 + D_1^2)]/T$, where $T = C_1(H^2 + D_1^2) - FH$. Since, $P > 0$, we require $T > 0$. Write $T = C_1 (H - D_1)^2 + H(2C_1 D_1 - F)$. A sufficient condition for $T > 0$ is $F \le 2C_1 D_1$. From the first equation of Equation 4.40, we get $H = [(r - PB_1)(P^2 + D^2)]/(PB_2)$. Since, $H > 0$, a sufficient condition is $P < (r/B_1)$. For example, consider the parameter values, $r = 1$, $B_1 = 0.2$, $B_2 = 0.91$, $D^2 = 0.3$, $C_1 = 0.22$, $C_2 = 0.2$, $D_1 = 0.1$, $F = 0.02$. Solving Equation 4.40, using the Newton's iteration, we obtain the equilibrium point as $P^* = 1.184922937$, $H^* = 1.205819205$. The two conditions, $F \le 2C_1 D_1$ and $P < (r/B_1)$, are satisfied.

Example 4.5

Discuss the Turing instability and Hopf bifurcation for a variant of the Holling–Tanner reaction–diffusion model for an aquatic system where at any point (X, Y) and time T, the phytoplankton $P(X, Y, T)$ and zooplankton $Z(X, Y, T)$ populations satisfy the equations [113a]

$$\frac{\partial P}{\partial T} = rP\left(1 - \frac{P}{K}\right) - \frac{mPZ}{A + P} + D_P \Delta P, \quad \frac{\partial Z}{\partial T} = sZ\left(1 - \frac{hZ}{P}\right) + D_Z \Delta Z, \quad (4.41)$$

where $\Delta = (\partial^2/\partial X^2) + (\partial^2/\partial Y^2)$, all parameters are positive and have their usual meanings.

SOLUTION

We introduce the following substitutions and notations to bring the system of equations into a nondimensional form

$$u = \frac{P}{K}, \quad v = \frac{mZ}{rK}, \quad t = rT, \quad x = \frac{X}{L}, \quad y = \frac{Y}{L},$$

$$\alpha = \frac{A}{K}, \quad \delta = \frac{s}{r}, \quad \beta = \frac{hr}{m}, \quad D_u = \frac{D_P}{rL^2}, \quad D_v = \frac{D_Z}{rL^2}.$$

We obtain

$$\frac{\partial u}{\partial t} = u(1-u) - \frac{uv}{\alpha + u} + D_u\left(\frac{\partial^2 u}{\partial x^2} + \frac{\partial^2 u}{\partial y^2}\right), \tag{4.42}$$

$$\frac{\partial v}{\partial t} = v\delta\left(1 - \frac{\beta v}{u}\right) + D_v\left(\frac{\partial^2 v}{\partial x^2} + \frac{\partial^2 v}{\partial y^2}\right). \tag{4.43}$$

$$u(x, y, 0) > 0, \quad v(x, y, 0) > 0.$$

Assume no-flux boundary conditions. The system has the equilibrium points $(1, 0)$ and (u^*, v^*). Irrespective of the values of the parameters α, β, positive equilibrium point exists. v^* is the positive solution of the equation $\beta^2 v^2 + [1 + (\alpha - 1)\beta]v - \alpha = 0$, and $u^* = \beta v^*$. To study the Turing instability, consider the linearized form of system about the positive equilibrium point. Write $u = u^* + U$, $v = v^* + V$. The linearized system is given by

$$\frac{\partial U}{\partial t} = a_{11}U + a_{12}V + D_u\left(\frac{\partial^2 U}{\partial x^2} + \frac{\partial^2 U}{\partial y^2}\right),$$

$$\frac{\partial V}{\partial t} = a_{21}U + a_{22}V + D_v\left(\frac{\partial^2 V}{\partial x^2} + \frac{\partial^2 V}{\partial y^2}\right), \tag{4.44}$$

where

$$a_{11} = -u^*\left(1 - \frac{v^*}{(\alpha + u^*)^2}\right), \quad a_{12} = -\frac{u^*}{\alpha + u^*}, \quad a_{21} = \frac{\delta}{\beta}, \quad a_{22} = -\delta.$$

Write the solution of Equation 4.44 in the form $U = se^{\lambda kt + i(k_x x + k_y y)}$, $V = we^{\lambda kt + i(k_x x + k_y y)}$. Substitute the expressions for U, V into Equation 4.44. The homogeneous equations in s and w have solutions if the determinant of the coefficient matrix is zero. We get

$$(\lambda - a_{11} + D_u k^2)(\lambda - a_{22} + D_v k^2) - a_{12}a_{21} = 0, \quad \text{or} \quad \lambda^2 + p\lambda + q = 0, \tag{4.45}$$

where

$$p = (D_u + D_v)k^2 - (a_{11} + a_{22}), \quad k^2 = k_x^2 + k_y^2,$$

$$q = D_u D_v k^4 - k^2(a_{11}D_v + a_{22}D_u) + (a_{11}a_{22} - a_{12}a_{21}) = D_u D_v[t^2 - s_1 t + s_2],$$

$$t = k^2, \text{ and } s_1 = \frac{1}{D_u D_v}\left[u^*\left(\frac{u^*}{\beta(\alpha + u^*)^2} - 1\right)D_v - \delta D_u\right],$$

$$s_2 = \frac{a_{11}a_{22} - a_{21}a_{12}}{D_u D_v} = \frac{\delta u^*}{D_u D_v}\left[1 + \frac{\alpha}{\beta(\alpha + u^*)^2}\right].$$

Irrespective of the sign of p in Equation 4.45, the equation has a positive root if $q < 0$. Therefore, diffusion-driven instability occurs when $q(k^2) < 0$. $y = q(k^2)$ is a parabola in the variable $t = k^2$, with vertex at $V[(s_1/2), \{D_u D_v(4s_2 - s_1^2)/4\}]$, where $4s_2 - s_1^2 < 0$. The minimum occurs at $k_c^2 = (s_1/2) > 0$. The conditions are $s_1 > 0$, and $s_1^2 - 4s_2 > 0$. Now, $q(k^2) = D_u D_v(k^2 - k_1^2)(k^2 - k_2^2) < 0$, for $k_1^2 < k^2 < k_2^2$, where $k_1^2, k_2^2 = 0.5[s_1 \mp \{s_1^2 - 4s_2\}^{1/2}]$. For example, consider the set of parameter values $\alpha = 0.1$, $\beta = 0.25$, $\delta = 0.2$, and $D_u = 0.1$. For $D_v = 3$, we obtain $v^* = 0.1277168$, $u^* = \beta v^* = 0.0319292$, and $k_c^2 = 36.496$. $q(k^2) < 0$, for $0.00699 < k^2 < 72.98509$. For all values of k^2 lying in this range, Turing instability sets in. The graphs of $q(k^2)$ versus k^2 for various values of D_v are plotted in Figure 4.19 for the above set of parameter values [113a]. As D_v increases, the length of the interval (k_{min}^2, k_{max}^2) increases.

We take δ as the bifurcation parameter. Turing bifurcation occurs when $\text{Im}[\lambda(k)] = 0$ and $\text{Re}[\lambda(k)] = 0$ at $k = k_T \neq 0$, where k_T is the critical wave number. This implies that $p = 0$, $q = 0$ at $k = k_T$. Now, $p = 0$ gives $k^2 = (a_{11} + a_{22})/D$, where $D = D_u + D_v$. Substituting $k^2 = (a_{11} - \delta)/D$, in $q = 0$, we get $D_u^2 \delta^2 - \delta C + a_{11}^2 D_v^2 = 0$, where $C = a_{11}(D_u^2 + D_v^2) + D^2 T$, $T = u^*[1 + (\alpha/\beta)(\alpha + u^*)^{-2}]$. The bifurcation parameter is given by $\delta_T = [C \pm \{C^2 - 4a_{11}^2 D_u^2 D_v^2\}^{1/2}]/(2D_u^2)$. At the Turing threshold δ_T, the spatial symmetry of the system is broken and patterns are stationary in time and oscillatory in space with the wavelength $\lambda_T = 2\pi/k_T$, where

$$k_T^2 = [u^* \delta \{\alpha + \beta(\alpha + u^*)^2\}/\{\beta D_u D_v(\alpha + u^*)^2\}]^{1/2}.$$

Hopf bifurcation occurs when $\text{Im}(\lambda(k)) \neq 0$ and $\text{Re}(\lambda(k)) = 0$ at $k = 0$. This implies $p = 0$ at $k = 0$, and $q \neq 0$ at $k = 0$. The second condition is satisfied. Now, $p = 0$ at $k = 0$ gives the critical value of Hopf bifurcation parameter

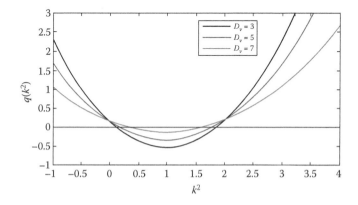

FIGURE 4.19

Graph of $q(k^2)$ versus k^2 for various values of D_v. (From Upadhyay, R. K., Volpert, V., Thakur, N. K. Propagation of Turing patterns in a plankton model. *J. Biol. Dynam.* 6, 524–538, 2012. With permission.)

δ as $\delta_H = u^*[(u^*/\{\beta(\alpha + u^*)^2\}) - 1]$. At the Hopf bifurcation threshold, the temporal symmetry of the system is broken and gives rise to uniform oscillations in space and periodic oscillations in time with the frequency, w_H, and wavelength, λ_H, where

$$w_H = \left[\delta u^*\left\{1 + \frac{\alpha}{\beta(\alpha + u^*)^2}\right\}\right]^{1/2}, \quad \lambda_H = 2\pi\left[\frac{\beta(\alpha + u^*)^2}{\delta u^*\{\alpha + \beta(\alpha + u^*)^2\}}\right]^{1/2}.$$

In Upadhyay et al. [113a], investigations of Turing and Hopf bifurcations and Turing patterns were done which are plotted in Figures 4.20 through 4.22. In Figure 4.20, Turing and Hopf bifurcation diagrams (D_v vs. δ) are plotted for $\alpha = 0.1$, $\beta = 0.25$, and $D_u = 0.1$. Turing and Hopf bifurcation curves separate the parametric space into four domains. These curves intersect at (0.184, 0.2021). Numerical simulations of the system were done in the Turing space for $\alpha = 0.1$, $\beta = 0.25$, $\delta = 0.2$, and $D_u = 0.1$, with domain size 100×100 up to time level $t = 1000$. Typical Turing patterns of prey (first column figures) and predator populations (second column figures) at time $t = 1000$ for $D_v = 3, 7$ are plotted in Figure 4.21. We observe that the stationary stripes–spots mixed patterns in the distribution of phytoplankton and zooplankton density turn to regular stripe patterns over the whole domain by increasing the value of diffusion coefficient of zooplankton. Turing patterns of prey (Figure 4.22a) and predator populations (Figure 4.22b) at $t = 1000$ for $\delta = 0.196$, $\alpha = 0.1$, $\beta = 0.25$, $D_u = 0.1$, $D_v = 3$ are plotted. We observe the shrinking region for stripes–spots mixed patterns as we approach the Turing stability boundary.

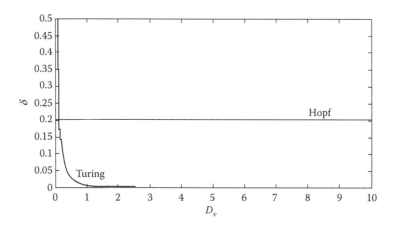

FIGURE 4.20
Turing and Hopf bifurcation curves. (From Upadhyay, R. K., Volpert, V., Thakur, N. K. Propagation of Turing patterns in a plankton model. *J. Biol. Dynam.* 6, 524–538, 2012. With permission.)

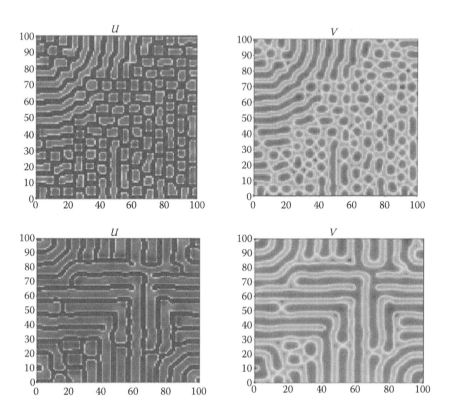

FIGURE 4.21
Turing patterns of prey (first column figures) and predator populations (second column figures) at time $t = 1000$ for $D_v = 3$ (first row), 7 (second row). (From Upadhyay, R. K., Volpert, V., Thakur, N. K. 2012. Propagation of Turing patterns in a plankton model. *J. Biol. Dynam.* 6, 524–538, 2012. With permission.)

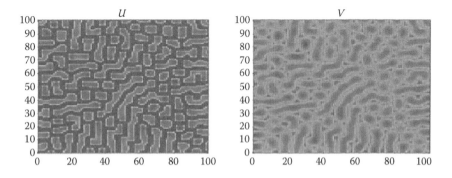

FIGURE 4.22
Turing patterns of (a) prey and (b) predator populations, at time $t = 1000$ for $\delta = 0.196$. (From Upadhyay, R. K., Volpert, V., Thakur, N. K. Propagation of Turing patterns in a plankton model. *J. Biol. Dynam.* 6, 524–538, 2012. With permission.)

Chaos in Multi-Species Model Systems

Ellner [16] and Hastings et al. [33] have given good reviews on chaos in ecology. Gilpin [26] found spiral chaos in a two-prey one-predator system. The system was cast into a set of three coupled nonlinear differential equations representing interactions between populations having overlapping generations. Gilpin found that the model is capable of exhibiting strange attractors in which spiraling trajectories stretch and fold repeatedly to generate the geometry of this attractor. Some authors believe that group selection acts against dynamical chaos [25,96]. Chaotic fluctuating populations are more prone to extinction than equilibrium populations with the consequence that group selection acts to eliminate species which would otherwise evolve in chaos. Their argument is based on the observation that chaotic dynamics involves extinction size population densities. This argument depends on the observation that in simple models, such as the logistic and Ricker maps, minimum population sizes decline as one enters and then moves further into the chaotic region. Schaffer and Kot [86] studied the Gilpin's model further and analyzed its dynamical behavior. These studies along with some later studies [34,63,64] suggested that dynamical chaos could be prevalent in the dynamics of many ecological systems and could be expected to exist in nature. Hastings and Powell [34] considered a model consisting of a resource, a consumer, and a predator and showed that chaos exists in a large range of the parameter that measures the ingestion rate per unit metabolic rate of the consumer species. They also observed that there exists an incubation period for chaos, that is, one is supposed to wait for some time so that the trajectories start evolving on a strange chaotic attractor. McCann and Yodzis [63] disagreed with these results of Hastings and Powell on the ground that the parameter ranges reported were extreme and therefore not realistic. They had obtained biologically realistic conditions for chaos to exist in this model. Specifically, they sought for chaotic solutions in which all the three densities are nonzero. Such a situation was termed as persistent chaos. They had also made the following observations: (i) chaotic dynamics which exist in biologically plausible regions in the parameter space is not very common, and (ii) productive environment is a prerequisite for a system to support a dynamical behavior such as chaos. Ruxton [85] considered the effect of inclusion of a population floor (minimum viable population) on the occurrence of chaos in the model considered by McCann and Yodzis [63]. Ruxton found that the inclusion of this assumption, which renders the Hastings and Powell model (HP model) a bit closer to reality, suppresses chaos. Ruxton further examined the possibility whether suppression of chaos can be counteracted by increasing the primary productivity of the food chain. He found that any reduction in chaos that is caused by immigration or refuge can be compensated by a sufficient increase in the resource renewal rate. It was also concluded that highly enriched systems are the most prospective candidates for chaotic dynamics to exist. Eisenberg and Maszle [15] revisited the HP model and observed that gradual addition of refugia provides a stabilizing influence for which the chaotic dynamics collapsed to stable limit cycles. McCann and Hastings [62] stabilized the food web by eliminating chaotic dynamics and limit-cycles behavior of the HP model by introducing omnivore on top predator and also by increasing the strength of the omnivore. Varriale and Gomes [118] analyzed the HP model in two different approaches. First, they observed the asymptotic states of the system resulting from numerical integration of the equations. Second, they applied the embedding procedure to extract the relevant dynamical exponents from a time series for only one scalar variable. By considering intraspecific density dependence, an important ecological factor in the food-chain model (deterministic and stochastic), Xu and Li [119] showed that the deterministic tritrophic food-chain model stabilized the food-chain system, leading chaotic and cyclic dynamics to a steady state and the famous tea-cup chaotic attractor disappears.

Hastings and Powell Model

Limit-cycle oscillator of the first kind (LCO-1) is obtained when parameters of the Rosenzweig–MacArthur (RM) model are set at limit-cycle values derived from an application of Kolmogorov analysis. The prey–predator interaction in the RM model is formulated in the Volterra scheme, which assumes exponential decay of a predator population in the absence of its sole prey. When we couple two LCO-1s, we get a model studied by Hastings and Powell [34], Rinaldi et al. [81], and Upadhyay et al. [78]. The model is described by the following equations:

$$\frac{dx}{dt} = rx\left(1 - \frac{x}{K}\right) - \frac{wxy}{1 + d_1x} = xF(x,\, y,\, z), \tag{4.46}$$

$$\frac{dy}{dt} = -a_2y + \frac{w_1xy}{1 + d_1x} - \frac{w_2yz}{1 + d_2y} = yG(x,\, y,\, z), \tag{4.47}$$

$$\frac{dz}{dt} = -cz + \frac{w_3yz}{1 + d_2y} = zH(x,\, y,\, z), \tag{4.48}$$

where w_2 measures the maximum value attainable by the per capita functional response of the specialist predator z. Predator z feeds only on y. Parameter c is the decay rate of predator z in the absence of its prey y and w_3 is a measure of its assimilation efficiency. This model was studied by many authors [34,63,64,74,78]. The oscillating predator–prey system (which comprises x and y) induces oscillations between populations y and z. In turn, these oscillations couple to generate chaos.

From Equation 4.46, we get $(dx/dt) \leq rx[1 - (x/K)]$. Hence, $\limsup\limits_{t\to\infty} x(t) \leq K$. Let $u = x + y + z$. Then,

$$\frac{du}{dt} \leq 2rK - rx - a_2y - cz - (w - w_1)\left(\frac{xy}{1 + d_1x}\right)$$

$$- (w_2 - w_3)\left(\frac{yz}{1 + d_2y}\right) \leq 2rK - \delta u,$$

when $w \geq w_1$, $w_2 \geq w_3$, $\delta = \min(r, a_2, c)$. Hence, $\limsup\limits_{t\to\infty} u(t) \leq 2rK/\delta$. Under these conditions, a region of attraction for all solutions is $\Omega = \{(x, y, z): 0 \leq x \leq K, 0 \leq x + y + z \leq 2rK/\delta\}$. The system has three equilibrium points, $E_0(0, 0, 0)$, $E_1(K, 0, 0)$ and $E_2(\hat{x}, \hat{y}, 0)$ where $\hat{x} = a_2/(w_1 - a_2d_1)$ and $\hat{y} = r(1 + d_1\hat{x})(K - \hat{x})/(wK)$, $w_1 > a_2d_1$. The positive equilibrium point $E_2(x^*, y^*, z^*)$ is the solution of the equations $F(x, y, z) = 0$, $G(x, y, z) = 0$, $H(x, y, z) = 0$. Solving the third equation, we get $y^* = c/(w_3 - cd_2)$, $w_3 > cd_2$. The first equation gives $A_1x^2 - A_2x + A_3 = 0$, where $A_1 = rd_1$, $A_2 = r(Kd_1 - 1)$, and $A_3 = K(wy^* - r)$. If $A_3 < 0$, that is when $wy^* - r < 0$, one positive root always exists irrespective of the sign of A_2. Two positive roots exist when $A_2 > 0$, that is, when $Kd_1 - 1 > 0$; $A_3 > 0$, that is, when $wy^* - r > 0$; and $A_2^2 - 4A_1A_3 > 0$. The two equilibrium points are

$$y^* = c/(w_3 - cd_2), x^* = [A_2 \pm \{A_2^2 - 4A_1A_3\}^{1/2}]/(2A_1),$$

$$z^* = (1 + d_2y^*)[w_1x^* - a_2(1 + d_1x^*)]/[w_2(1 + d_1x^*)],$$

where $w_3 > cd_2$, $Kd_1 - 1 > 0$, and $wy^* - r > 0$. For example, consider the following parameter values which are used for numerical simulations later in this section: $r = 1$, $K = 1$, $w = 5$, $d_1 = 2.2$, $a_2 = 0.4$, $w_1 = 5$, $w_2 = 0.1$, $w_3 = 0.1$, $c = 0.01$, $d_2 = 2$. For these parameter values, we get $A_3 = -0.375 < 0$. Irrespective of the sign of A_2, there is one positive root for x^*. The equilibrium point is obtained as $(x^*, y^*, z^*) = (0.767535, 0.125, 12.842501)$.

Local Stability Analysis: The variational matrix J of the system is given by

$$J = \begin{bmatrix} a_{11} & a_{12} & 0 \\ a_{21} & a_{22} & a_{23} \\ 0 & a_{32} & a_{33} \end{bmatrix},$$

where

$$a_{11} = r - \left[\frac{2xr}{K} + \frac{wy}{D^2}\right], \quad a_{12} = -\frac{wx}{D}, \quad a_{21} = \frac{w_1y}{D^2},$$

$$a_{22} = -a_2 + \frac{w_1 x}{D} - \frac{w_2 z}{D_1^2}, \quad a_{23} = -\frac{w_2 y}{D_1}, \quad a_{32} = \frac{w_3 z}{D_1^2}, \quad a_{33} = -c + \frac{w_3 y}{D_1},$$

$$D = 1 + d_1 x, \quad D_1 = 1 + d_2 y.$$

Equilibrium point $E_0(0, 0, 0)$: The eigenvalues of J are r, $-a_2$, $-c$. E_0 is a saddle point with stable manifold locally in the yz-plane and with unstable manifold locally in the x-direction.

Equilibrium point $E_1(K, 0, 0)$: The eigenvalues of J are $-r$, $-c$, $-a_2 + [w_1 K / (1 + d_1 K)]$. When $[w_1 K / (1 + d_1 K)] < a_2$, the equilibrium point is asymptotically stable.

Equilibrium point $E_2(\hat{x}, \hat{y}, 0)$: We find $a_{22} = 0$, and $a_{12} a_{21} = -[(w w_1 \hat{x} \hat{y}) / \hat{D}^3] = -a_2 w \hat{y} / \hat{D}^2$, where we have used the second equation. The eigenvalues of J are $\lambda_1 = -c + [w_3 \hat{y} / \hat{D}_1]$. λ_2, λ_3 are the roots of the equation $\lambda^2 - a_{11} \lambda + [a_2 w \hat{y} / \hat{D}^2] = 0$, where $a_{11} = r - [(2r \hat{x} / K) + (w \hat{y} / \hat{D}^2)]$. For $a_{11} < 0$, the equation has two negative roots or a complex pair with negative real parts. The equilibrium point is asymptotically stable for $[w_3 \hat{y} / \hat{D}_1] < c$, and $a_{11} < 0$.

Equilibrium point $E_3(x^*, y^*, z^*)$: We find $a_{33} = 0$. The characteristic equation of J is $\lambda^3 + s_1 \lambda^2 + s_2 \lambda + s_3 = 0$, where $s_1 = -(a_{11} + a_{22})$, $s_2 = a_{11} a_{22} - a_{12} a_{21} - a_{23} a_{32}$, $s_3 = a_{11} a_{23} a_{32}$, where $a_{11} = r - [(2rx^*/K) + (wy^*/D^{*2})]$, $a_{12} = -wx^*/D^*$, $a_{21} = w_1 y^*/D^{*2}$, $a_{22} = w_2 d_2 y^* z^* / D_1^{*2}$, $a_{23} = -cw_2/w_3$, $a_{32} = w_3 z^* / D_1^{*2}$. By the Routh–Hurwitz criterion, the characteristic equation has negative roots or has one negative root and a complex pair with negative real parts if $s_1 > 0$, $s_2 > 0$, $s_3 > 0$ and $s_1 s_2 - s_3 > 0$. For the parameter values considered above, we obtain one negative root and a complex pair with positive real parts. The equilibrium point is unstable.

Numerical simulations: Numerical simulations are done using the following values of the parameters: $r = 1$, $K = 1$, $w = 5$, $d_1 = 0.5, 2.2, 2.3, 3.0$, $a_2 = 0.4$, $w_1 = 5$, $w_2 = 0.1$, $w_3 = 0.1$, $c = 0.01$, $d_2 = 2$. The chaotic attractor ($d_1 = 3.0$), limit-cycle attractor of period one ($d_1 = 2.2$), limit-cycle attractor of period two ($d_1 = 2.3$), and stable focus attractor ($d_1 = 0.5$) are plotted in Figures 4.23a–c.

Gragnani et al. [27] considered a three species RM model, which is the same as the HP model for $w/d_1 = a_2$, $w_1/d_1 = a_2 e_2$, $w_2/d_2 = a_3$, $w_3/d_2 = a_3 e_3$, $(1/d_1) = b_2$, $(1/d_2) = b_3$.

In Ref. [7], a modification of the HP model was given by introducing a mortality term in the zooplankton population. In these studies, the half-saturation constant of zooplankton was chosen as the key parameter. Mandal et al. [56] had modified the HP model by considering different body sizes of zooplankton, which changes the growth rate and half-saturation constant.

Example 4.6

Discuss the dynamics of the variant of the HP model described by the following equations:

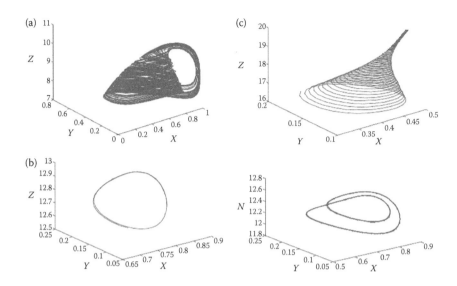

FIGURE 4.23
(a) Chaotic attractor, (b) limit-cycle attractors of period one and two, and (c) stable focus attractor.

$$\frac{dx}{dt} = a_1x - b_1x^2 - \frac{wxy}{x + D} = xF(x, y, z),$$

$$\frac{dy}{dt} = -a_2y + \frac{w_1xy}{x + D_1} - \frac{w_2yz}{y + D_2} - \theta f(x)y = yG(x, y, z),$$

$$\frac{dz}{dt} = -cz + \frac{w_3yz}{y + D_3} = zH(x, y, z).$$

Assume the parameter values as $a_1 = 1.75$, $b_1 = 0.05$, $w = 1$, $D = D_1 = D_2 = 10$, $D_3 = 20$, $c = 0.1$, $a_2 = 1$, $w_1 = 2$, $w_2 = 1.45$, $w_3 = 1$, $\theta = 0.05$, and Holling type II function as $f(x) = x/(x + 10)$.

SOLUTION

From $F(x, y, z) = 0$, we get $b_1x = a_1 - [wy/(x + D)]$. Hence, $x \le a_1/b_1$. The system has the equilibrium points: $E_0(0, 0, 0)$, $E_1 (a_1/b_1, 0, 0)$, $E_2(\hat{x}, \hat{y}, 0)$, where $\hat{x}$ is the positive solution of $(\hat{x} + D_1)[a_2 + \theta f(\hat{x})] - w_1 \hat{x} = 0$, $\hat{y} = (a_1 - b_1 \hat{x})(\hat{x} + D)/w$, $\hat{x} < a_1/b_1$. The positive equilibrium point $E_3(x^*,y^*z^*)$ is the solution of the equations $F(x, y, z) = 0$, $G(x, y, z) = 0$, $H(x, y, z) = 0$. From the third equation, we get $y^* = cD_3/(w_3 - c)$, $w_3 > c$. From the first equation, x^* is the positive solution of the equation $b_1x^2 + (Db_1 - a_1)x + (wy - a_1D) = 0$. From the second equation, we get $z^* = (y^* + D_2)[w_1x^* - (x^* + D_1)(a_2 + \theta f(x^*))]$ $/[w_2(x^* + D_1)]$. Since, $z^* > 0$, we get the condition $w_1x^* > (x^* + D_1)(a_2 + \theta f(x^*))$.

The variational matrix is given by

$$J = \begin{bmatrix} a_{11} & a_{12} & 0 \\ a_{21} & a_{22} & a_{23} \\ 0 & a_{32} & a_{33} \end{bmatrix},$$

where

$$a_{11} = a_1 - 2b_1 x - \frac{wDy}{(x+D)^2} = x\left[-b_1 + \frac{wy}{(x+D)^2}\right],$$

$$a_{12} = -\frac{wx}{x+D}, \quad a_{21} = \frac{w_1 D_1 y}{(x+D_1)^2} - \theta y f'(x), \quad a_{23} = -\frac{w_2 y}{y+D_2}, \quad a_{32} = \frac{w_3 D_3 z}{(y+D_3)^2},$$

$$a_{22} = -a_2 + \frac{w_1 x}{x+D_1} - \frac{w_2 D_2 z}{(y+D_2)^2} - \theta f(x) = \frac{w_2 yz}{(y+D_2)^2}, \quad a_{33} = -c + \frac{w_3 y}{y+D_3}.$$

Equilibrium point $E_0(0, 0, 0)$: The eigenvalues of J are obtained as $a_1, -a_2, -c$ (computing without simplifying the equations). The equilibrium point E_0 is a saddle point with stable manifold locally in the $y-z$ plane and with unstable manifold locally in the x-direction.

Equilibrium point $E_1(a_1/b_1, 0, 0)$: We obtain $a_{11} = -a_1, a_{12} = -[wa_1/(a_1 + b_1 D)]$, $a_{21} = 0, a_{32} = 0, a_{23} = 0, a_{33} = -c, a_{22} = -a_2 + [w_1 a_1/(a_1 + b_1 D_1)] - \theta f(a_1/b_1)$. The eigenvalues of J are $-a_1, -c, a_{22}$. The equilibrium point is stable when $[w_1 a_1/(a_1 + b_1 D_1)] < a_2 + \theta f(a_1/b_1)$.

Equilibrium point $E_2(\hat{x}, \hat{y}, 0)$: With the Holling type II function $f(x) = x/(x + 10)$, we obtain $\hat{x} = 10/(1 - \theta), \hat{y} = (a_1 - b_1 \hat{x})(\hat{x} + D)/w$. The characteristic equation of J is $\lambda^3 + s_1 \lambda^2 + s_2 \lambda + s_3 = 0$, where $s_1 = -(a_{11} + a_{33}), s_2 = a_{11}a_{33} - a_{12}a_{21}, s_3 = a_{12}a_{21}a_{33}, a_{11} = \hat{x}[-b_1 + \{w\hat{y}/(\hat{x} + D)^2\}], a_{12} = -w\hat{x}/(\hat{x} + D), a_{21} = \hat{y}[-\theta f'(\hat{x}) + \{w_1 D_1/(\hat{x} + D_1)^2\}], a_{22} = 0, a_{23} = -w_2 \hat{y}/(\hat{y} + D_2), a_{32} = 0, a_{33} = -c + (w_2 \hat{y})/(\hat{y} + D_3)$. By the Routh–Hurwitz criterion, the characteristic equation has negative roots, or has one negative root and a complex pair with negative real parts if $s_1 > 0, s_2 > 0, s_3 > 0$, and $s_1 s_2 - s_3 > 0$. If these conditions are satisfied, then the equilibrium point is asymptotically stable. For the parameter values given in the problem with $\theta = 0.05$, and Holling type II function, we obtain $\hat{x} = 200/19, \hat{y} = 18135/722$, and the characteristic equation as $\lambda^3 - 0.8085 \lambda^2 + 0.6677\lambda - 0.4216 = 0$. This equation has the positive root 0.7072 and a complex pair with positive real parts. The equilibrium point is unstable.

Equilibrium point $E_3(x^*, y^*, z^*)$: The characteristic equation of J is $\lambda^3 + s_1 \lambda^2 + s_2 \lambda + s_3 = 0$, where $s_1 = -(a_{11} + a_{22}), s_2 = -(a_{12}a_{21} + a_{23}a_{32} - a_{11}a_{22})$, $s_3 = a_{11}a_{23}a_{32}, a_{11} = x^*[-b_1 + \{wy^*/(x^* + D)^2\}], a_{12} = -wx^*/(x^* + D), a_{21} = y^* [-\theta f'(x^*) + \{w_1 D_1/(x^* + D_1)^2\}], a_{22} = w_2 y^* z^*/(y^* + D_2)^2, a_{23} = -w_2 y^*/(y^* + D_2), a_{32} = w_3 D_3 z^*/(y^* + D_3)^2, a_{33} = 0$. By the Routh–Hurwitz criterion, the equilibrium point is asymptotically stable if the characteristic equation has negative roots, or has one negative root and a complex pair with negative real parts, that is, if $s_1 > 0, s_2 > 0, s_3 > 0$ and $s_1 s_2 - s_3 > 0$. For the parameter values given in the problem with $\theta = 0.05$, and Holling type II function, we obtain $x^* = 33.9896616$, $y^* = 20/9$, and $z^* = 4.271152707$, and the characteristic equation as $\lambda^3 + 1.568232\lambda^2 - 0.090007\lambda + 0.07572 = 0$. This equation has the negative root -1.6507 and a complex pair with positive real parts. The equilibrium point is unstable.

Simple Prey–Specialist Predator–Generalist Predator Interaction

Consider a model system, in which the prey population of size x serves as the only food for the specialist predator population of size y. This population, in turn, serves as a favorite food for the generalist vertebrate predator population of size z. This interaction is represented by the following system of a simple prey–specialist predator–generalist predator interaction [107,110]

$$\frac{dx}{dt} = a_1 x - b_1 x^2 - \frac{wxy}{(x + D)},$$ (4.49)

$$\frac{dy}{dt} = -a_2 y + \frac{w_1 xy}{(x + D_1)} - \frac{w_2 y^2 z}{(y^2 + D_2^2)},$$ (4.50)

$$\frac{dz}{dt} = cz^2 - \frac{w_3 z^2}{(y + D_3)}.$$ (4.51)

A typical ecological situation presented by the food-chain model (4.49) through (4.51) is given in Figure 4.24.

The equations for the rate of change of population size for prey and specialist predator are written following the Volterra scheme (predator population dies out exponentially in the absence of its prey). The interaction between this predator y and the generalist predator z is modeled by the Leslie–Gower scheme, where the loss in a predator population is proportional to the reciprocal of per capita availability of its most favorite food. a_1 is the intrinsic growth rate of the prey population x, a_2 is the intrinsic death rate of the predator y in the absence of the only food x, c measures the rate of self-reproduction of generalist predator z, and the square term signifies the fact that mating frequency is directly proportional to the number of male

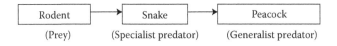

Rodent   →   Snake   →   Peacock

(Prey)   (Specialist predator)   (Generalist predator)

FIGURE 4.24
Typical ecological situation presented by the food-chain model (4.49) through (4.51).

as well as female individuals. w, w_1, w_2, and w_3 are the maximum values which per capita growth rate can attain. b_1 measures the strength of intra-specific competition among the individuals of the prey species x. D and D_1 quantify the extent to which the environment provides protection to the prey x and can be thought of as a refuge or a measure of the effectiveness of the prey in evading a predator's attack. D_2 is the value of y at which per capita removal rate of y becomes $w_2/2$, D_3 represents the residual loss in the z population due to severe scarcity of its favorite food y. The third term $w/(x + D)$ on the right-hand side of Equation 4.49 is obtained by consider-ing the probable effect of the density of the prey's population on predators attack rate. If this term is multiplied by x, it gives the attack rate on the prey per predator. Denote $f(x) = wx/(x + D)$. It is reasonable to assume that the attack rate would be a function of the parasite's ability to attack the hosts. When $x \to \infty$, $f(x) \to w$, which is the maximum that it can reach. The third term $w_2 y^2 z/(y^2 + D_2^2)$ on the right-hand side of Equation 4.50 represents the per capita functional response of the vertebrate predator z and was first introduced by Takahashi in the classical book by May [59]. The term $w_2 yz/(y + D_2)$ represents the per capita functional response of the invertebrate predator z. The generalist predator z is a sexually reproducing species. The interaction terms appearing in the rate equations restore to some extent the symmetry which characterizes the Lotka–Volterra model. To incorporate the above fact, y^2 is introduced in the last term of Equation 4.50 as their functional response is of Holling type III. Aziz-Alaoui [2] and Upadhyay and Rai [110] considered the above model, where the third term on the right-hand side of Equation 4.50 was taken as $w_2 yz/(y + D_2)$.

Origin of the model system: The first two equations of both the phases (linear and nonlinear) are the classical RM predator–prey type used to interpret the dynamical behavior of certain predator–prey communities. The third term of the second equation in both the phases is due to the middle predator y being an invertebrate. For discussing the stability, bifurcation, and chaotic behavior, many authors [34,42] consider the third equation as

$$\frac{dz}{dt} = \left(-d_1 + \frac{d_2 y}{d_3 + y} \right) z,$$

that is, a system in which x is the number of logistic-type prey, y is the number of Holling type II specialist predator, and z is the number of Holling type II generalist predator. An interesting formulation for discussing the predator dynamics was given by Leslie [46] and reported in the book by Pielou [72]. This equation was taken as

$$\frac{dz}{dt} = cz \left(1 - \frac{z}{my} \right),$$

where c and m are parameters. In this formulation, the growth of the predator population is taken as $dz/dt = cz(1 - z/K)$, of logistic type, where the measure of the environmental carrying capacity K is assumed to be proportional
to the prey abundance, that is, $K = my$. Then, the logistic equation becomes

$$\frac{dz}{dt} = cz\left(1 - \frac{z}{m_1 + my}\right),$$

where the additional constant m_1 normalizes the residual reduction in
the predator population z because of severe scarcity of the favorite food.
Simplifying, we obtain

$$\frac{dz}{dt} = cz - \frac{c}{m}\left(\frac{z^2}{(m_1/m) + y}\right) = cz - \frac{w_3 z^2}{y + D_3},$$

where $D_3 = m_1/m$ and $w_3 = c/m$. Let us now assume that a generalist predator z predates on predator y. Even though generalist predators have their
favorite preys, they switch over to other preys when these are in short supply.
Taking this into account and using the Holling type III functional response,
the growth rate equation for the predator z can be written as

$$\frac{dz}{dt} = cz - \frac{w_3 z^2}{y}.$$

Most generalist predators are sexually reproducing. In sexually reproducing populations, a behavioral phenomenon known as sexual selection [66]
is common. Since sexual selection depends on the success of certain individuals over others of the same sex and involves behavioral traits such as
choosiness and species recognition, it is natural to expect that the growth of
a sexually reproducing population will be proportional to the number densities of the two sexes. It is known from population genetics that evolution
attempts to maintain the ratio of the number densities of the two sexes to
unity. To accommodate these facts, the above equation is modified to

$$\frac{dz}{dt} = cz^2 - \frac{w_3 z^2}{y + D_3},$$

where w_3 measures the limitation on growth of the generalist predator by its
dependence on its favorite prey y.

The system has the equilibrium points $E_0(0,0,0)$, $E_1(a_1/b_1, 0, 0)$, and $E_2(\hat{x}, \hat{y}, 0)$,
where $\hat{x} = [a_2 D_1/(w_1 - a_2)]$, $w_1 > a_2$, $\hat{y} = [(a_1 - b_1\hat{x})(\hat{x} + D)]/w$, $\hat{x} < a_1/b_1$. The
positive equilibrium point (x^*, y^*, z^*) is the solution of the equations

$$a_1 - b_1 x - \frac{wy}{(x+D)} = 0, \quad -a_2 + \frac{w_1 x}{(x+D_1)} - \frac{w_2 yz}{(y^2 + D_2^2)} = 0, \quad z\left[c - \frac{w_3}{(y+D_3)}\right] = 0.$$

The third equation gives $y^* = (w_3 - cD_3)/c$. Since $y^* > 0$, we require $w_3 > cD_3$. The first equation gives $a_1 - b_1 x^* = wy^*/(x^* + D)$. Since, $y^* > 0$, we have $x^* < a_1/b_1$. Substituting for y^* in the first equation, we get $b_1 x^{*2} + (b_1 D - a_1)$ $x^* + s_1 - a_1 D = 0$, where $s_1 = w(w_3 - cD_3)/c$. Irrespective of the sign of the coefficient of x^*, there exists a positive root if $s_1 - a_1 D < 0$, that is, for $w(w_3 - cD_3) < a_1 Dc$. Solving the second equation for z, we get

$$z^* = \frac{(y^{*2} + D_2^2)}{w_2 y^*}\left(-a_2 + \frac{w_1 x^*}{(x^* + D_1)}\right).$$

Since $z^* > 0$, we require $[w_1 x^*/(x^* + D_1)] > a_2$ or $x^* > a_2 D_1/(w_1 - a_2)$. Under the above conditions, a positive equilibrium point (x^*, y^*, z^*) exists. If $b_1 D - a_1 < 0$, $s_1 - a_1 D > 0$, and $[(b_1 D - a_1)^2 - 4b_1(s_1 - a_1 D)] > 0$, then we get two positive solutions for x, that is we get two positive equilibrium points.

Local stability of equilibrium points: The Jacobian matrix of the system is given by

$$J = \begin{bmatrix} a_{11} & a_{12} & 0 \\ a_{21} & a_{22} & a_{23} \\ 0 & a_{32} & a_{33} \end{bmatrix},$$

where

$$a_{11} = a_1 - 2b_1 x - \frac{wDy}{(x+D)^2} = x\left[-b_1 + \frac{wy}{(x+D)^2}\right],$$

$$a_{21} = \frac{w_1 D_1 y}{(x+D_1)^2}, \quad a_{23} = -\frac{w_2 y^2}{y^2 + D_2^2}, \quad a_{32} = \frac{w_3 z^2}{(y+D_3)^2}, \quad a_{12} = -\frac{wx}{x+D},$$

$$a_{22} = -a_2 + \frac{w_1 x}{x+D_1} - \frac{2w_2 D_2^2 yz}{(y^2+D_2^2)^2} = \frac{w_2 yz(y^2 - D_2^2)}{(y^2+D_2^2)^2}, \quad a_{33} = 2cz - \frac{2w_3 z}{y+D_3}.$$

Equilibrium point $E_0(0, 0, 0)$: The eigenvalues of J are obtained as $a_1, -a_2, 0$. (Computing without simplifying the equations.) E_0 is a nonhyperbolic fixed point. Furthermore, as one eigenvalue is positive and another one is negative, E_0 is always unstable.

Equilibrium point $E_1(a_1/b_1, 0, 0)$: We obtain $a_{11} = -a_1$, $a_{21} = 0$, $a_{23} = 0$, $a_{32} = 0$, $a_{33} = 0$, $a_{22} = -a_2 + [w_1 a_1/(a_1 + b_1 D_1)]$, $a_{12} = -[wa_1/(a_1 + b_1 D_1)]$. The eigenvalues of J are $0, -a_1, a_{22}$, and $a_{22} < 0$, when $[w_1 a_1/(a_1 + b_1 D_1)] < a_2$. In this case, E_1 is a nonhyperbolic fixed point.

Equilibrium point $E_2(\hat{x},\ \hat{y},\ 0)$: The eigenvalues of J are 0, and the roots of the quadratic $\lambda^2 - a_{11}\lambda - a_{12}a_{21} = 0$. Note that $a_{12} < 0$, $a_{21} > 0$. If $a_{11} < 0$, the equation has two negative roots or a complex pair with negative real parts. The equilibrium point E_2 is asymptotically stable. If $a_{11} > 0$, the equation has two positive roots or a complex pair with positive real parts. Then, the equilibrium point E_2 is unstable.

Equilibrium point $E_3(x^*, y^*, z^*)$: The characteristic equation of J is $\lambda^3 + s_1\lambda^2 + s_2\lambda + s_3 = 0$, where $s_1 = -(a_{11} + a_{22})$, $s_2 = (a_{11}a_{22} - a_{12}a_{21} - a_{23}a_{32})$, $s_3 = a_{11}a_{23}a_{32}, a_{11} = x^*[-b_1 + \{wy^*/(x^* + D)^2\}], a_{12} = -wx^*/(x^* + D), a_{21} = w_1D_1y^*/(x^* + D_1)^2$, $a_{22} = w_2y^*z^*(y^{*2} - D_2^2)/(y^{*2} + D_2^2)^2$, $a_{23} = -w_2y^{*2}/(y^{*2} + D_2^2), a_{32} = w_3z^{*2}/(y^* + D_3)^2$, $a_{33} = 0$. Note that $a_{12} < 0$, $a_{21} > 0$, $a_{23} < 0$, and $a_{32} > 0$. By the Routh–Hurwitz criterion, the equilibrium point is asymptotically stable if the characteristic equation has negative roots, or has one negative root and a complex pair with negative real parts, that is, if $s_1 > 0$, $s_2 > 0$, $s_3 > 0$, and $s_1s_2 - s_3 > 0$. We have the following results for the parameter values used for simulation.

a. $w = 1$, $a_2 = 1$, $w_1 = 2$, $D = D_1 = D_2 = 10$, $D_3 = 20$, $a_1 = 2.0$, $b_1 = 0.05$, $w_2 = 1.45$, $w_3 = 1$ and $c = 0.0257$. We obtain $x^* = 30.70954752$, $y^* = 4860/257$, $z^* = 8.489772746$, and the characteristic equation as $\lambda^3 + 0.898683\lambda^2 - 0.113277\lambda + 0.06392645 = 0$. This equation has one negative root, -1.0620, and a complex pair with positive real parts. The equilibrium point is unstable.

b. $w = 1$, $a_2 = 1$, $w_1 = 2$, $D = D_1 = D_2 = 10$, $D_3 = 20$, $a_1 = 1.93$, $b_1 = 0.06$, $w_2 = 0.405$, $w_3 = 1$, and $c = 0.03$. We obtain $x^* = 25.99254935$, $y^* = 40/3$, $z^* = 22.85645275$, and the characteristic equation as $\lambda^3 + 1.167617\lambda^2 + 0.109780\lambda + 0.157459 = 0$. The conditions $s_1 > 0$, $s_2 > 0$, $s_3 > 0$ are satisfied, but $s_1s_2 - s_3 < 0$. This equation has one negative root, -1.1869, and a complex pair with positive real parts. The equilibrium point is unstable.

Simulation experiments: In Upadhyay et al. [109b], the equations were integrated numerically using the fourth-order Runge–Kutta method with a time step size of 0.001, with the initial conditions taken as $x(0) = 6.0$, $y(0) = 1.0$, and $z(0) = 1.0$. The values of the following parameters were kept constant: $a_1 = 2.0$, $b_1 = 0.05$, $w_2 = 1.45$, $w_3 = 1$, and $c = 0.0257$. The values of the other parameters were taken as $w = 1$, $D = D_1 = D_2 = 10$, $a_2 = 1$, $w_1 = 2$, and $D_3 = 20$. The dynamical outcomes and the ranges in which they occur were determined. It was found that the system supports stable focus and stable limit-cycle attractors in reasonably large parameter regimes, whereas chaos was exhibited in narrow parameter regimes (see Table 1 in [109b]). They concluded the following: (i) The parameters a_1, c, w_2, and w_3 are responsible for the chaotic behavior of the system. (ii) The biology of the bottom predator (which is sexually reproducing and generalist) may be a crucial factor in the governance of the dynamical consequences of the given food chain.

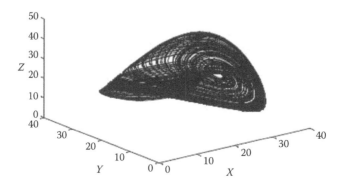

FIGURE 4.25
Chaotic attractor for model system (4.49), (4.50), (4.51).

Simulation is done using the parameter values $a_1 = 1.93$, $a_2 = 1$, $b_1 = 0.06$, $w = 1$, $w_2 = 0.405$, $w_1 = 2$, $D = D_1 = D_2 = 10$, $D_3 = 20$, $w_3 = 1$, and $c = 0.03$. Chaotic solutions are observed for these parameter values as plotted in Figure 4.25. Upadhyay and Rai [110] had plotted a figure for the parameter values $a_1 = 2$, $a_2 = 1$, $b_1 = 0.05$, $w = 1$, $w_2 = 1.45$, $w_1 = 2$, $D = D_1 = D_2 = 10$, $D_3 = 20$, $w_3 = 1$, and $c = 0.0257$.

Single Predator, Two Competing Prey Model

The extinction and coexistence of interacting species are of great importance and were studied extensively [48,49]. The effect of predation on two competing prey species was studied in Ref. [35]. It was shown that the competing prey species could coexist even with identical resource requirements if each prey species has invasion potential for the complementary predator–prey subcommunity. It was also observed that the outcomes depend critically on the prey species capability of invading the complementary subcommunity formed by the predator species and the other prey. The competitive exclusion principle [31] states that two competing species can coexist only if they exploit their environment differently. Huang et al. [36] investigated the qualitative behavior of a predator–prey system and obtained the criteria for the existence of a limit cycle for the system. In Ref. [9], Cramer et al. numerically showed that coexistence is possible in case of equal predation. Fujii [21] studied the stability analysis for two prey–one predator systems and showed that there exists a stable limit cycle in some ranges of parameters. Vance [115] and Grasman et al. [28] have also studied the one predator–two prey model communities. Feng [19] derived the criteria for the existence of a steady state and its stability for one predator–two prey model in terms of the natural growth rates of these three species.

Upadhyay and Iyengar [106] have also studied the dynamics of a single predator and two competing preys. Consider a Gause-type model for a predator species X and two competing prey species N_1, N_2. The predation effect of X on the two competing prey species N_1 and N_2 is governed by the system of ordinary differential equations [106]

$$\frac{dN_1}{dt} = N_1 \left[r_1 \left(1 - \frac{N_1}{K_1} \right) - \frac{b_{13} X}{1 + a_1 N_1} - \alpha_{12} N_2 \right], \tag{4.52}$$

$$\frac{dN_2}{dt} = N_2 \left[r_2 \left(1 - \frac{N_2}{K_2} \right) - \frac{b_{23} X}{1 + a_2 N_2} - \alpha_{21} N_1 \right], \tag{4.53}$$

$$\frac{dX}{dt} = X \left[\frac{\beta_{31} N_1}{1 + a_1 N_1} + \frac{\beta_{32} N_2}{1 + a_2 N_2} - \delta - \gamma X \right], \tag{4.54}$$

where $(r_1, K_1, \alpha_{12}, b_{13}, \beta_{31})$, $(r_2, K_2, \alpha_{21}, b_{23}, \beta_{32})$ are the (intrinsic growth rate, carrying capacity, interspecific competition coefficient, feeding rates per predator per unit prey consumed, assimilation, or conversion efficiency of the predator) of the prey species N_1 and N_2, respectively; and (δ, γ) are the death rate and interspecific interference coefficient of the predator species X in the absence of the preys. The values of all the parameters are positive.

The system has seven equilibrium points:

(i) $E_0(0, 0, 0)$, (ii) $E_1(K_1, 0, 0)$, (iii) $E_2(0, K_2, 0)$, (iv) $E_3(\hat{N}_1, \hat{N}_2, 0)$, where $\hat{N}_1 = K_1 r_2$ $(r_1 - K_2 \alpha_{12})/Q$, $\hat{N}_2 = K_2 r_1 (r_2 - K_1 \alpha_{21})/Q$, $Q = r_1 r_2 - K_1 K_2 \alpha_{12} \alpha_{21}$.

(v) *Existence of* $E_4(\bar{N}_1, 0, \bar{X})$: $\bar{N}_1$, $\bar{X}$ are the positive solutions of

$$X = r_1 \left(1 - \frac{N_1}{K_1} \right) \left(\frac{1 + a_1 N_1}{b_{13}} \right), \quad \frac{\beta_{31} N_1}{1 + a_1 N_1} = \delta + \gamma X.$$

Eliminating X from the two equations, we get $f(\bar{N}_1) = A \bar{N}_1^3 + B \bar{N}_1^2 + C \bar{N}_1 + E = 0$, where $A = \gamma r_1 a_1^2$, $B = \gamma r_1 a_1 (2 - K_1 a_1)$, $C = \gamma r_1 (1 - 2a_1 K_1) + K_1 b_{13} (\beta_{31} - \delta a_1)$, $E = -K_1(\gamma r_1 + \delta b_{13})$. As all the parameters are positive, we have $A > 0$, $E < 0$. For $a_1 > 2/K_1$, we have $B < 0$. Therefore, irrespective of the sign of C, there is one positive root of the equation. For example, for the values of the parameters used in the numerical simulation $\gamma = 0.005$, $r_1 = 4.8$, $K_1 = 95$, $b_{13} = 0.27$, $\beta_{31} = 1.5$, $a_1 = 1.4$, $\delta = 1.0$, we obtain the equation $0.04704 \bar{N}_1^3 - 4.4016 \bar{N}_1^2 - 3.795 \bar{N}_1 - 27.93 = 0$. Using the Newton–Raphson method, we obtain the positive root as $\bar{N}_1 \approx 94.492$. From the second equation, we obtain $\bar{X} \approx 12.678$. It may be noted that the values of δ lie in [0.25, 5.85]. Alternately, note that $0 < \bar{N}_1 < K_1$. We have $f(0) = E < 0$, $f(K_1) = K_1 b_{13}[K_1(\beta_{31} - \delta a_1) - \delta] > 0$, for $(K_1 \beta_{31}) > \delta(K_1 a_1 + 1)$. If this condition is satisfied, then $f(0)f(K_1) < 0$, and there exists a positive root in $(0, K_1)$. For the parameter values considered above, this condition is satisfied.

(vi) *Existence of $E_5(0, \bar{N}_2, \bar{X})$: $\bar{N}_2$, $\bar{X}$* are the positive solutions of

$$X = r_2\left(1 - \frac{N_2}{K_2}\right)\left(\frac{1 + a_2N_2}{b_{23}}\right), \qquad \frac{\beta_{32}N_2}{1 + a_2N_2} = \delta + \gamma X.$$

Eliminating X from the two equations, we get $f(\bar{N}_2) = A\bar{N}_2^3 + B\bar{N}_2^2 + C\bar{N}_2 + E = 0$, where $A = \gamma r_2 a_2^2$, $B = \gamma r_2 a_2(2 - K_2a_2)$, $C = \gamma r_2(1 - 2a_2K_2) + K_2b_{23}(\beta_{32} - \delta a_2)$, $E = -K_2(\gamma r_2 + \delta b_{23})$. As all the parameters are positive, we have $A > 0$, $E < 0$. For $a_2 > 2/K_2$, we have $B < 0$. Therefore, irrespective of the sign of C, there is one positive root of the equation. For example, for the values of the parameters used in the numerical simulation $\gamma = 0.005$, $r_2 = 4$, $K_2 = 99$, $b_{23} = 0.1$, $\beta_{32} = 1.5$, $a_2 = 0.25$, $\delta = 1.0$, we obtain the equation $0.00125\,\bar{N}_2^3 - 0.11375\,\bar{N}_2^2 + 11.405\,\bar{N}_2 - 11.88 = 0$. Using the Newton–Raphson method, we obtain the positive root as $\bar{N}_2 \approx 1.0525$. From the second equation, we obtain $\bar{X} \approx 49.9753$. Alternately, note that $0 < \bar{N}_2 < K_2$. We have $f(0) = E < 0$, $f(K_2) = K_2b_{23}[K_2(\beta_{32} - \delta a_2) - \delta] > 0$, for $(K_2\beta_{32}) > \delta\,(K_2a_2 + 1)$. If this condition is satisfied, then $f(0)f(K_2) < 0$, and there exists a positive root in $(0, K_2)$. For the parameter values considered above, this condition is satisfied.

(vii) *Existence of $E_6(N_2^*, N_2^*, X^*)$: N_1^*, N_2^*, X^** are the positive solutions of

$$N_2 = \frac{1}{\alpha_{12}}\left[r_1\left(1 - \frac{N_1}{K_1}\right) - \frac{b_{13}X}{1 + a_1N_1}\right], \qquad N_1 = \frac{1}{\alpha_{21}}\left[r_2\left(1 - \frac{N_2}{K_2}\right) - \frac{b_{23}X}{1 + a_2N_2}\right],$$

$$X = \frac{1}{\gamma}\left[\frac{\beta_{31}N_1}{1 + a_1N_1} + \frac{\beta_{32}N_2}{1 + a_2N_2} - \delta\right].$$

Eliminate X from the first and third equations. Denote this equation by $f_1(N_1, N_2) = 0$. Again, eliminate X from the second and third equations. Denote this equation by $f_2(N_1,N_2) = 0$. We note that $0 < N_1 < K_1$, and $0 < N_2 < K_2$. Now, we study the following cases:

a. Set $N_2 = 0$ and consider the equation $f_1(N_1, 0) = 0$. We find $f_1(0, 0) < 0$. Also, $f_1(K_1, 0) > 0$ if $(\beta_{31}K_1)/(1 + a_1K_1) > \delta$. Hence, there is a root of $f_1(N_1, 0) = 0$ in the interval $(0, K_1)$. We find that $(df_1/dN_1)(N_1, 0) > 0$ always. Therefore, the root $N_1 \in (0, K_1)$ is unique.

b. Set $N_1 = 0$ and consider the equation $f_1(0, N_2) = 0$. We find $f_1(0, 0) < 0$. Also, $f_1(0, K_2) > 0$ if $(\beta_{32}K_2)/(1 + a_2K_2) > \delta$ and $(K_2\alpha_{12} - r_1) + (b_{13}/\gamma)$ $[\{(\beta_{32}K_2)/(1 + a_2K_2)\} - \delta] > 0$. Hence, there is a root of $f_1(0, N_2) = 0$ in the interval $(0, K_2)$. We find that $(df_1/dN_2)(0, N_2) > 0$ always. Therefore, this root $N_2 \in (0, K_2)$ is unique.

c. Similarly, we consider the cases corresponding to the equation $f_2(N_1, N_2) = 0$. Using the above analysis, we find that the following condition is also to be satisfied.

$$(K_1\alpha_{21} - r_2) + (b_{23}/\gamma)[\{(\beta_{31}K_1)/(1 + a_1K_1)\} - \delta] > 0.$$

We find that $(df_2/dN_1)(N_1,0) > 0$ and $(df_2/dN_2)(0,N_2) > 0$ always. The root $N_2 \in (0,K_2)$ is unique.

d. Using the equations $f_1(N_1,N_2) = 0$ and $f_2(N_1,N_2) = 0$, we find that $(dN_2/dN_1) < 0$ for all N_1, N_2 if the following conditions are satisfied:

$$\frac{r_2}{K_2} + \frac{b_{23}}{\gamma(1 + a_2N_2)^2}\left[\beta_{32}\left(\frac{1 - a_2N_2}{1 + a_2N_2}\right) - \left(\frac{\beta_{31}a_2N_1}{1 + a_1N_1}\right) + \delta a_2\right] > 0,$$

$$\frac{r_1}{K_1} + \frac{b_{13}}{\gamma(1 + a_1N_1)^2}\left[\beta_{31}\left(\frac{1 - a_1N_1}{1 + a_1N_1}\right) - \left(\frac{\beta_{32}a_1N_2}{1 + a_2N_2}\right) + \delta a_1\right] > 0.$$

Therefore, there exists a point (N_1^*, N_2^*) at which the two isoclines $f_1(N_1, N_2) = 0$ and $f_2(N_1, N_2) = 0$ intersect. Knowing the values of N_1^*, N_2^*, the value of X^* is computed from the third equation. Since $X^* > 0$, we require the condition

$$\delta < \left[\frac{\beta_{31}N_1^*}{1 + a_1N_1^*} + \frac{\beta_{32}N_2^*}{1 + a_2N_2^*}\right].$$

Note that most of the conditions derived above are sufficient conditions. Computationally, when the existence of (N_1^*, N_2^*) is known, we can find the values of these unknowns by solving the equations $f_1(N_1, N_2) = 0$ and $f_2(N_1, N_2) = 0$ using Newton's iteration method, for a given set of parameter values. Consider the following parameter values used for simulations: $\gamma = 0.001$, $r_1 = 2$, $\alpha_{21} = 0.01$, $a_1 = 2.5$, $a_2 = 1.8$, $\delta = 1.4$, $K_1 = 100$, $K_2 = 145$, $b_{13} = 1.6$, $b_{23} = 1.5$, $\beta_{31} = 7$, $\beta_{32} = 4$, $\alpha_{12} = 0.001$, $r_2 = 4.5$. With these values of the parameters, we obtain $N_1^* \approx 99.5493, 0.5013$; $N_2^* = 144.8629$; $X^* = 2.8025$, 1.4411, 0.8995.

Local stability of the fixed point $E_6(N_1^*, N_2^*, X^*)$: The community matrix J of the system is given by

$$J = \begin{bmatrix} a_{11} & a_{12} & a_{13} \\ a_{21} & a_{22} & a_{23} \\ a_{31} & a_{32} & a_{33} \end{bmatrix},$$

where

$$a_{11} = N_1\left[-\frac{r_1}{K_1} + \frac{b_{13}a_1X}{(1 + a_1N_1)^2}\right], \quad a_{22} = N_2\left[-\frac{r_2}{K_2} + \frac{b_{23}a_2X}{(1 + a_2N_2)^2}\right],$$

$$a_{33} = -X\gamma, \quad a_{12} = -N_1\alpha_{12}, \quad a_{13} = -N_1b_{13}/(1 + a_1N_1),$$

$$a_{21} = -N_2\alpha_{21}, \quad a_{23} = -N_2 b_{23}/(1 + a_2 N_2), \quad a_{31} = X\beta_{31}/(1 + a_1 N_1)^2,$$

$$a_{32} = X\beta_{32}/(1 + a_2 N_2)^2.$$

The eigenvalues of this matrix are the roots of the equation $\lambda^3 + A_1\lambda^2 + A_2\lambda + A_3 = 0$, where $A_1 = [a_{11} + a_{22} + a_{33}]$, $A_2 = a_{22}a_{33} - a_{23}a_{32} + a_{11}a_{22} + a_{11}a_{33} - a_{21}a_{12} - a_{31}a_{13}$, $A_3 = -[a_{11}(a_{22}a_{33} - a_{23}a_{32}) - a_{21}(a_{12}a_{33} - a_{13}a_{32}) + a_{31}(a_{12}a_{23} - a_{13}a_{22})]$. For E_6 to be locally asymptotically stable, we require the eigenvalues to be negative or complex with negative real parts. From the Routh–Hurwitz criterion, we obtain the required conditions as $A_1 > 0$, $A_2 > 0$, $A_3 > 0$, $A_1 A_2 > A_3$. The set $\Omega = \{(N_1, N_2, X): 0 \leq N_1 \leq K_1, 0 \leq N_2 \leq K_2, 0 \leq X \leq (\beta_{31}P_1(K_1) + \beta_{32}P_2(K_2) - \delta)/\gamma\}$ is a region of attraction for all solutions initiating in the interior of the positive orthant.

In Ref. [106], numerical simulations were done for the parameter values $\gamma = 0.001$, $r_1 = 2$, $\alpha_{21} = 0.01$, $a_1 = 2.5$, $a_2 = 1.8$, $\delta = 1.4$, $K_1 = 100$, $K_2 = 145$, $b_{13} = 1.6$, $b_{23} = 1.5$, $\beta_{31} = 7$, $\beta_{32} = 4$, $\alpha_{12} = 0.001$, $r_2 = 4.5$. The plane projection of a chaotic attractor obtained for these parameter values is plotted in Figure 4.26.

2D parameter scans: Four groups of the parameters—(i) (r_1, α_{12}), $2.0 \leq r_1 \leq 8.0$, $0.006 \leq \alpha_{12} \leq 2.0$; (ii) (r_2, b_{23}), $2.0 \leq r_2 \leq 8.0$, $0.001 \leq \alpha_{12} \leq 0.11$; (iii) $(\alpha_{12}, \alpha_{21})$, $0.001 \leq \alpha_{12} \leq 1.0$; $0.05 \leq \alpha_{21} \leq 1.25$; (iv) (β_{32}, b_{23}), $1.5 \leq \beta_{32} \leq 5.0$; $0.001 \leq b_{23} \leq 0.05$)—were formed for the simulation studies. The parameter values common for all the simulation experiments were taken as $K_1 = 95$, $a_1 = 1.4$, $K_2 = 99$, $a_2 = 0.25$, $\delta = 5.3$, $\gamma = 0.005$. The regions of parameters for the extinction of the prey species were presented in tabular form (see Table 4.2 of Ref. [106]). From these results, the following conclusions were drawn:

FIGURE 4.26
X–Y Projection ($X = N_1$, $Y = N_2$) of strange chaotic attractor. (From Upadhyay, R. K., Iyengar, S. R. K. 2006. Extinction and coexistence of competing prey species in ecological systems. *J. Comp. Methods Sci. Eng.* 6, 131–150. Copyright 2006, IOS Press. Reprinted with permission.)

a. In groups (i), (ii), and (iv), prey N_1 becomes extinct.

b. In group (iii), extinction of either prey N_1 or prey N_2 is possible in some subintervals.

The regions of parameters for the coexistence of the species either on a stable limit cycle or on a stable focus were presented in tabular form (see Table 4.3 of Ref. [106]). From these results, the following conclusions were drawn:

a. In group (i), the species coexist on a limit cycle.

b. In group (ii), the species coexist on a stable focus and a limit cycle.

c. In groups (iii) and (iv), the species coexist either on a stable focus or on a limit cycle.

EXERCISE 4.1

1. (*A variant of the RM model with diffusion*) Consider a three species phytoplankton–zooplankton–fish model which satisfies the reaction–diffusion equations [109]

$$\frac{\partial P}{\partial T} = P\left[R_x\left(1 - \frac{P}{K}\right) - \frac{\upsilon Z}{H_P + P} \right] + D\frac{\partial^2 P}{\partial X^2},$$

$$\frac{\partial Z}{\partial T} = Z\left[\frac{e\upsilon P}{H_P + P} - \delta - \frac{FZ}{H_Z^2 + Z^2} \right] + D\frac{\partial^2 Z}{\partial X^2},$$

with positive initial conditions and zero flux boundary conditions $(\partial P/\partial X) = (\partial Z/\partial X) = 0$ at $X = 0$ and $X = L$ for all T. Nondimensionalize the equations by substituting $p = (P/K)$, $z = \upsilon(Z/K)$, $x = (X/L)$, $t = \bar{R}T$, where $\bar{R} = R_x(X_0)$ for some $X_0 \in (0, L)$, (see Pascual [69]). Linearize the system about the equilibrium point, assume that the solution is in the form $U = se^{\lambda t + ikx}$, $V = we^{\lambda t + ikx}$, and discuss the local stability of the system.

2. For the model in Problem 1, compute the positive equilibrium point using the parameter values $n = 6.5$, $a = 6.5$, $b = 5$, $m = 0.6$, $g = 2.5$, $s = 2$, $l = -1.4$, $d = 0.0001$, and $f = 0.0001$. Assuming the initial conditions as (i) $p = 0.2$ and $z = 0.4$ and (ii) $p = 0.21$ and $z = 0.41$, plot the time series and space–time plots.

3. Discuss the Turing instability of model system (4.38) and (4.39) in one and two dimensions. Assume the parameter values as $r = 1$, $B_1 = 0.2$, $B_2 = 0.91$, $D^2 = 0.3$, $C_1 = 0.22$, $C_2 = 0.2$, $D_1 = 0.1$, $F = 0.02$.

4. Perform numerical simulations of the problem 3 (system (4.38) and (4.39)) in one dimension and two dimensions. Assume the parameter values as $r = 1$, $B_1 = 0.2$, $B_2 = 0.91$, $D^2 = 0.3$, $C_1 = 0.22$, $C_2 = 0.2$, $D_1 = 0.1$, $F = 0.1$. The initial distribution is given as the constant-gradient distribution [55]: $P(x, 0) = P^*$, $H(x, 0) = H^* + \varepsilon x + \delta$, where $\varepsilon = -1.5 \times 10^{-6}$, $\delta = 10^{-6}$ are parameters and $(P^*, H^*) \approx (1.8789, 1.3984)$ is the nontrivial equilibrium point. Study also the case $\varepsilon = -1.5 \times 10^{-3}$, $\delta = 10^{-6}$. In one dimension, draw the space versus density plot. In two dimensions, assume $D_1 = 0.01$, $d_1 = 0.05$, $d_2 = 1$. Plot the Turing patterns of population P for $F = 0.0001$, 0.01, 0.035, 0.065, 0.0768.

5. Consider a situation where a toxin-producing phytoplankton population (prey) of size x is predated by individuals of specialist predator zooplankton population y. This zooplankton population, in turn, serves as a favorite food for the generalist predator, molluscs, population of size z. This interaction is represented by the following system of a simple prey–specialist predator–generalist predator interaction (*A variant of Upadhyay–Rai model* [103])

$$\frac{dx}{dt} = a_1 x - b_1 x^2 - \frac{wxy}{x + D} = xF(x,y,z),$$

$$\frac{dy}{dt} = -a_2 y + \frac{w_1 xy}{x + D_1} - \frac{w_2 yz}{y + D_2} - \theta f(x)y = yG(x,y,z)$$

$$\frac{dz}{dt} = cz^2 - \frac{w_3 z^2}{y + D_3} = zH(x,y,z),$$

where all parameters are positive. Discuss the local stability of the model system. Assume the parameter values as $w = 1$, $a_2 = 1$, $w_1 = 2$, $D = D_1 = D_2 = 10$, $D_3 = 20$, $a_1 = 1.93$, $b_1 = 0.06$, $w_2 = 0.405$, $w_3 = 1$, $c = 0.03$, $\theta = 0.05$, and Holling type II function $f(x) = x/(x + 10)$.

6. (*Variant of HP model with Holling type IV functional response*) [105]. Consider the following system as a model simulating a tritrophic food chain consisting of three species and characterized by a Holling type-IV functional response, where $x(t)$ is the population density of the lowest trophic level species (prey), $y(t)$ is the population density of the middle trophic level species (intermediate predator), and $z(t)$ is the population density of highest trophic level species (top predator). The intermediate predator y feeds on the prey x according to the Holling type-IV functional response and the top predator z preys on y according to the same functional response.

$$\frac{dx}{dt} = x\left[a_1 - b_1 x - \frac{wy}{(x^2/i) + x + a}\right],$$

$$\frac{dy}{dt} = y\left[-a_2 + \frac{w_1 x}{(x^2/i) + x + a} - \frac{w_2 z}{(y^2/i) + y + a}\right],$$

$$\frac{dz}{dt} = z\left[-c + \frac{w_3 y}{(y^2/i) + y + a}\right],$$

where all the parameters are positive. Discuss the local stability of the model system. Test the local stability for the parameter values: $a = 1$, $a_1 = 1$, $a_2 = 0.2$, $b_1 = 1$, $w = 1.95$, $w_1 = 1.38$, $w_2 = 2.85$, $w_3 = 1.6$, $i = 0.3$, and $c = 0.25$.

7. (*Variant of Upadhyay–Rai model with a Holling type IV functional response*) [110a]. Consider a situation where a prey population x is predated by individuals of population y. The population y, in turn, serves as a favorite food for individuals of population z. This interaction is represented by the following system of a simple prey–middle predator–top predator interaction

$$\frac{dx}{dt} = a_1 x - b_1 x^2 - \frac{wxy}{x^2/i_1 + x + a},$$

$$\frac{dy}{dt} = -a_2 y + \frac{w_1 xy}{x^2/i_1 + x + a} - \frac{w_2 yz}{y^2/i_2 + y + a},$$

$$\frac{dz}{dt} = cz^2 - \frac{w_3 z^2}{y + d}.$$

Discuss the local stability of the model system. Test the local stability for the parameter values: $w = 1$, $w_1 = 1.8$, $a = 8.2$, $i_1 = 45$, $i_2 = 30$, $a_2 = 0.9$, $w_3 = 1.57$, $d = 35$, $a_1 = 2.25$, and for (i) $b_1 = 0.075$, $w_2 = 0.045$, $c = 0.027$; (ii) $b_1 = 0.088$, $w_2 = 0.15$, $c = 0.026$. Plot the attractors in both cases.

8. For the model in Problem 6, draw the chaotic attractor and plot the corresponding time series. Taking c as a control parameter draw the bifurcation diagram for the parameter values $a_1 = 1$, $b_1 = 1$, $w = 1.95$, $i = 0.3$, $a = 1$, $a_2 = 0.2$, $w_1 = 1.38$, $w_2 = 2.85$, $c = 0.25$, $w_3 = 1.6$.

References

1. Alonso, D., Bartumeus, F., Catalan, J. 2002. Mutual interference between predators can give rise to Turing spatial patterns. *Ecology* 83(1), 28–34.
2. Aziz-Alaoui, M. A. 2002. Study of a Leslie–Gower type tri-trophic population model. *Chaos, Solitons and Fractals* 14(8), 1275–1293.
3. Beddington, J. R., Free C. A., Lawton, J. H. 1976. Dynamical complexity in predator–prey model framed in simple difference equations. *Nature* 225, 58–60.
4. Berryman, A. A., Millstein, J. A. 1989. Are ecological systems chaotic—And if not, why not? *Trends Ecol. Evol.* 4, 26–28.
5. Calio, G., Jing, Z. J., Sy, P. 1997. Qualitative analysis of a mathematical model of a predator–prey system. In: *Proceedings of Functional and Global analysis*, University of the Philippines, Diliman, Quezon City.
6. Camara, B., Aziz-Alaoui, M. A. 2009. Turing and Hopf patterns formation in a predator–prey model with Leslie–Gower-type functional response. *Dynam. Continuous, Discrete Impulsive Systems* 16, 479–488.
7. Chattopadhyay, J., Sarkar, R. R. 2003. Chaos to order: Preliminary experiments with population dynamics models of three trophic levels. *Ecol. Model.* 163, 45–50.
8. Costantino, R. F., Desharnais, R. A., Cushing, J. M., Dennis, B. 1997. Chaotic dynamics in an insect population. *Science* 257, 389–391.
9. Cramer, N. F., May, R. M. 1971. Interspecific competition, predation and species diversity: A comment. *J. Theor. Biol.* 34, 289–293.
10. Cushing, J. M., Costantino, R. F., Dennis, B., Desharnais, R. A., Henson, S. M. 2003. *Chaos in Ecology: Experimental Non-linear Dynamics*. Elsevier, San Diego.
11. Cushing, J. M., Henson, S. M., Desharnais, R. A., Dennis, B., Costantino, R. F., King, A. A. 2001. A chaotic attractor in ecology: Theory and experimental data. *Chaos, Solitons Fractals* 12(2), 219–234.
12. Duarte, J., Januario, C., Martins, N., Sardayés, J. 2012. On chaos, transient chaos and ghosts in single population models with Allee effects. *Nonl. Anal.: Real World Appl.* 13, 1647–1661.
13. Dubey, B., Kumari, N., Upadhyay, R. K. 2009. Spatiotemporal pattern formation in a diffusive predator–prey system: An analytical approach. *J. Appl. Math. Comp.* 31, 413–432.
14. Edwards, A. M., Bees, M. A. 2001. Generic dynamics of a simple plankton population model with a non-integer exponent of closure. *Chaos, Solitons Fractals* 12(2), 289–300.
15. Eisenberg, J. N., Maszle, D. R. 1995. The structural stability of a three-species food chain model. *J. Theor. Biol.* 176, 501–510.
16. Ellner, S. 1991. Detecting low-dimensional chaos in population dynamics data: A critical review. In: *Chaos and Insect Ecology*, Logan, J. A. and Hain, F. P. eds. Virginia Experimental Station Information Series, 91–93, Blacksburg, VA.
17. Ellner, S., Turchin, P. 1995. Chaos in a noisy world: New methods and evidence from time-series analysis. *Am. Naturalist* 145, 343–375.
18. Feigenbaum, M. J. 1980. Universal behavior in nonlinear systems. *Los Alamos Sci.* 1, 4.
19. Feng, W. 1993. Coexistence, stability and limiting behavior in a one-predator–two-prey model. *J. Math. Anal. Appl.* 179, 592–609.

20. Freedman, H. I. 1976. Graphical stability, enrichment, and pest control by a natural enemy. *Math. Biosci.* 31, 207–225.
21. Fujii, K. 1977. Complexity–stability relationship of two prey one predator species system model: Local and global stability. *J. Theor. Biol.* 69, 613–623.
22. Fussmann, G. F., Ellner S. P., Shertzen, K. W., Hairston, Jr. N. G. 2000. Crossing the Hopf bifurcation in a live predator–prey system. *Science* 290, 1358–1360.
23. Garvie, M. R. 2007. Finite difference schemes for reaction–diffusion equations modeling predator–prey interactions in MATLAB. *Bull. Math. Biol.* 69(3), 931–956.
24. Gentleman, W., Leising, A., Frost, B., Strom, S., Murray, J. 2003. Functional responses for zooplankton feeding on multiple resources: A review of assumptions and biological dynamics. *Deep-Sea Res Pt II* 50, 2847–2875.
25. Gilpin, M. E. 1975. *Group Selection in Predator–prey Communities*. Princeton University Press, Princeton, NJ.
26. Gilpin, M. E. 1979. Spiral chaos in a predator–prey model. *Am. Naturalist* 113, 307–308.
27. Gragnani, A., De Feo, O., Rinaldi, S. 1998. Food chains in the chemostat: Relationships between mean yield and complex dynamics. *Bull. Math. Biol.* 60, 703–719.
28. Grasman, J, van Den Bosch, F., van Herwaarden, O. A. 2001. Mathematical conservation ecology: A one predator–two prey system as case study. *Bull. Math. Biol.* 63, 259–269.
29. Hainzl, J. 1988. Stability and Hopf bifurcation in a predator–prey system with several parameters. *SIAM J Appl. Math* 48, 170–180.
30. Hanski, I., Turchin, P., Korpimaki, E., Henttonen, H. 1993. Population oscillations of boreal rodents: Regulation by mustelid predators leads to chaos. *Nature* 364, 232–235.
31. Harding, G. 1960. The competitive exclusion principle. *Science* 131, 1292–1298.
32. Hassell, M. P., Lawton, J. H., Beddington, J. R. 1976. The components of arthropod predation. I. The prey death rate. *J. Anim. Ecol.* 45, 135–164.
33. Hastings, A., Hom, C. L., Ellner, S., Turchin, P., Godfray, H. C. J. 1993. Chaos in Ecology: Is mother nature a strange attractor? *Ann Rev. Ecol. Syst.* 24, 1–33.
34. Hastings, A., Powell, T. 1991. Chaos in a three-species food-chain. *Ecology* 72, 896–903.
35. Hsu, S. B. 1981. Predator-mediated coexistence and extinction, *Math. Biosci.* 54, 231–248.
36. Huang, L., Tang, M., Yu, J. 1995. Qualitative analysis of Kolmogorov-type models of predator–prey systems. *Math. Biosci.* 130, 85–97.
37. Jing, Z. J. 1989. Local and global bifurcations and applications in a predator–prey system with several parameters. *Syst. Sci. Math Sci.* 2, 337–352.
38. Jing, Z., Yang, J. 2006. Bifurcation and chaos in discrete-time predator–prey system. *Chaos, Solitons Fractals* 27, 259–277.
39. Kazarinoff, N. D., van den Driessche, P. 1978. A model predator–prey system with functional response. *Math. Biosci.* 39, 125–134.
40. King, A. A., Costantino, R. F., Cushing, J. M., Henson, S. M., Desharnais, R. A., Dennis, B. 2004. Anatomy of a chaotic attractor: Subtle model predicted patterns revealed in population data. *Proc. Natl. Acad. Sci. USA* 101, 408–413.
41. King, A. A., Schaffer, W. M., Gordon, C., Treat, J., Kot, M. 1996. Weakly dissipative predator–prey systems. *Bull. Math. Biol.* 58, 835–859.

42. Klebanoff, A., Hastings, A. 1994. Chaos in three species food chains. *J. Math. Biol.* 32, 427–451.

43. Kooi, B. W., Boer, M. P. 2003. Chaotic behavior of a predator–prey system. *Dyn. Cont. Discrete Impulsive Syst. B: Appl. Algorithms* 10, 259–272.

44. Kumari, N. 2009. *Modeling the Dynamical Complexities in Diffusion Driven Ecological Systems*. PhD thesis, Indian School of Mines, Dhanbad, India.

45. Kuznetsov, Y. A., Muratori, S., Rinaldi, S. 1992. Bifurcations and chaos in a periodic predator–prey model. *Int. J. Bifur. Chaos* 2, 117–128.

46. Leslie, P. H. 1948. Some further notes on the use of matrices in population mathematics. *Biometrika* 35, 213–245.

47. Letellier, C., Aziz-Alaoui, M. A. 2002. Analysis of the dynamics of a realistic ecological model. *Chaos, Solitons and Fractals* 13, 95–107.

48. Levins, R. 1970. Extinction: Some mathematical problems in biology. *Lectures on Mathematics in the Life Sciences*, 2, 77–107, Providence, RI.

49. Levin, R., Culver, D. 1971. Regional co-existence of species and competition between rare species. *Proc. Natl. Acad. Sci., USA* 68, 1246–1248.

50. Liu, Q.-X., Li, B.-L., Jin, Z. 2007. Resonance and frequency-locking phenomena in spatially extended phytoplankton–zooplankton system with additive noise and periodic forces. Preprint: arXiv: 0705.3724v2.

51. Liu, X., Xiao, D. 2007. Complex dynamic behaviour of a discrete-time predator–prey system. *Chaos, Solitons Fractals* 32, 80–94.

52. Ludwig, D., Jones, D., Holling, C. 1978. Qualitative analysis of an insect outbreak system: The spruce budworm and forest. *J. Anim. Ecol.* 47, 315–332.

53. Malchow, H. 1993. Spatio-temporal pattern formation in non-linear non-equilibrium plankton dynamics. *Proc. Roy. Soc. Lond. Series B* 251, 103–109.

54. Malchow, H. 1996. Nonlinear plankton dynamics and pattern formation in an ecohydrodynamic model system. *J. Marine Systems* 7(2/4), 193–202.

55. Malchow, H., Petrovskii, S. V., Venturino, E. 2008. *Spatiotemporal Patterns in Ecology and Epidemiology: Theory, Models and Simulation*. CRC Press, Boca Raton.

56. Mandal, S., Ray, S., Roy, S., Jørgensen, S. E. 2006. Order to chaos and vice versa in an aquatic ecosystem. *Ecol. Model.* 197, 498–504.

57. May, R. M. 1975. Biological populations obeying difference equations: Simple points, stable cycles, and chaos. *J. Theor. Biol.* 49, 511–524.

58. May, R. M. 1976. Simple mathematical models with very complicated dynamics. *Nature* 261, 459–467.

59. May, R. M. 1987. Chaos in natural populations. *Proc. Roy. Soc. London Series* A27, 419–428.

60. May, R. M. 2001. *Stability and Complexity in Model Ecosystems*. Princeton University Press, Princeton, NJ.

61. May, R. M., Oster, G. F. 1976. Bifurcation and dynamic complexity in simple ecological models. *Am. Naturalist* 110, 578–599.

62. McCann, K., Hastings, A. 1997. Re-evaluating the omnivory—Stability relationship in food webs. *Proc. Roy. Soc. Lond. Ser. B* 264, 1249–1254.

63. McCann, K., Yodzis, P. 1994. Biological conditions for chaos in a three-species food-chain. *Ecology* 75(2), 561–564.

64. McCann, K., Yodzis, P. 1995. Bifurcation structure of a three-species food chain model. *Theor. Popul. Biol.* 48, 93–125.

65. Medvinsky, A. B., Petrovskii, S. V., Tikhonova, I. A., Malchow, H., Li, B.-L. 2002. Spatio temporal complexity of plankton and fish dynamics. *SIAM Rev.* 44, 311–370.

66. Moller, A. P., Legendre, S. 2001. Allee effect, sexual selection and demographic stochasticity. *Oikos* 92, 27–34.

67. Nicolis, G., Prigogine, I. 1977. *Self-Organization in Non-Equilibrium Systems*. John Wiley & Sons, New York.

68. Okubo, A. 1980. *Diffusion and Ecological Problems: Mathematical Models*. Springer, Berlin.

69. Pascual, M. 1993. Diffusion induced chaos in a spatial predator–prey system. *Proc. Roy. Soc. Lond. Ser. B* 251, 1–7.

70. Petrovskii, S. V., Malchow, H. 1999. A minimal model of pattern formation in a prey–predator system. *Math. Comp. Model.* 29, 49–63.

71. Petrovskii, S. V., Malchow, H. 2001. Wave of chaos: New mechanism of pattern formation in spatio-temporal population dynamics. *Theor. Popul. Biol.* 59, 157–174.

72. Pielou, E. C. 1977. *Mathematical Ecology: An Introduction*. Wiley International, New York.

73. Rai, V. 2004. Chaos in natural populations: Edge or wedge? *Ecol. Compl.* 1(2), 127–138.

74. Rai, V., Anand, M. 2004. Is dynamic complexity of ecological systems quantifiable? *Intl. J. Ecol. Environ. Sci.* 30, 123–130.

75. Rai, V., Anand, M., Upadhyay, R. K. 2007. Trophic structure and dynamical complexity in simple ecological models. *Ecol. Comp.* 4, 212–222.

76. Rai, V., Jayaraman, G. 2003. Is diffusion-induced chaos robust? *Curr. Sci.* 84, 925–929.

77. Rai, V., Sreenivasan, R. 1993. Period doubling bifurcations leading to chaos in a model food-chain. *Ecol. Model.* 69, 63–77.

78. Rai, V., Upadhyay, R. K. 2004. Chaotic population dynamics & biology of the top-predator. *Chaos, Solitons Fractals* 21, 1195–1204.

79. Real, L. A. 1977. The kinetic of functional response. *Am. Nat.* 111, 289–300.

80. Rinaldi, S., Muratori, S. 1993. Conditioned chaos in seasonally perturbed predator–prey models. *Ecol. Model.* 69, 79–97.

81. Rinaldi, S., Muratori, S., Kuznetsov, Y. 1993. Multiple attractors, catastrophes and chaos in seasonally perturbed predator prey communities. *Bull. Math. Biol.* 55, 15–35.

82. Rohani, P., Miramontes, O., Hassell, M. P. 1994. Quasiperiodicity and chaos in population models. *Proc. Roy. Soc. Lond. B.* 258, 17–22.

83. Rosenzweig, M. L. 1971. Paradox of enrichment: Destabilization of exploitation ecosystems in ecological time. *Science* 171, 385–387.

84. Rosenzweig, M. L., MacArthur, R. H. 1963. Graphical representation and stability conditions of predator–prey interactions. *Am. Naturalist* 97, 209–223.

85. Ruxton, G. D. 1996. Chaos in a three-species food chain with a lower bound on the bottom population. *Ecology* 77(1), 317–319.

86. Schaffer, W. M., Kot, M. 1985. Do strange attractors govern ecological systems? *Bioscience* 35(6), 342–350.

87. Scheffer, M. 1991a. Fish and nutrients interplay determines algal biomass: A minimal model. *Oikos* 62, 271–282.

88. Scheffer, M. 1991b. Should we expect strange attractors behind plankton dynamics—And if so, should we bother? *J. Plankton Res.* 13, 1291–1305.

89. Scheffer, M. 1998. *Ecology of Shallow Lakes*. Population and Community Biology Series 22. Chapman & Hall, London.

90. Schreiber, S. J. 2003. Allee effects, extinctions and chaotic transients in simple population models. *Theor. Popul. Biol.* 64, 201–209.

91. Segel, L. A., Jackson, J. L. 1972. Dissipative structures: An explanation and an ecological example. *J. Theor. Biol.* 37, 545–559.

92. Sherratt, J. A., Lewis, M. A., Fewler, A. C. 1995. Ecological chaos in the wake of invasion. *Proc. Natl. Acad. Sci. USA* 92, 2524–2528.

93. Shertzer, K. W., Ellner, S. P. 2002. Energy storage and the evolution of population dynamics. *J. Theor. Biol.* 215, 183–200.

94. Steele, J. H. and Henderson, E. W. 1981. A simple plankton model. *Am. Naturalist* 117, 676–691.

95. Stone, L., He, D. 2007. Chaotic oscillations and cycles in multi-trophic ecological systems. *J. Theor. Biol.* 248, 382–390.

96. Thomas, W. R., Pomerantz, M. J., Gilpin, M. E. 1980. Chaos, asymmetric growth and group selection for dynamical stability. *Ecology* 61, 1312–1320.

97. Thunberg, H. 2001. Periodicity versus chaos in one-dimensional dynamics. *SIAM Rev.* 43(1), 3–30.

98. Turchin, P., Ellner, S. P. 2000. Living on the edge of chaos: Population dynamics of Fennoscandian voles. *Ecology* 81, 3099–3116.

99. Turchin, P., Ellner, S. P. 2000. Modeling time-series data. *Chaos in Real Data.* In: Perry, J. N., Smith, R. H., Woiwod, I. P., and Morse, D. R. eds. Kluwer Academic Publishers, Dordrecht, Netherlands, pp. 33–48.

100. Turchin, P., Hanski, I. 1997. An empirically-based model for the latitudinal gradient in vole population dynamics. *Am. Naturalist* 149, 842–874.

101. Turing, A. M. 1952. On the chemical basis of morphogenesis. *Philos. Trans. Roy. Soc. Lond. Ser. B* 237, 37–72.

102. Upadhyay, R. K. 2000. Chaotic behaviour of population dynamic systems in ecology. *Math. Comp. Model.* 32, 1005–1015.

103. Upadhyay, R. K. 2008. Chaotic dynamics in a three species aquatic population model with Holling type II functional response. *Nonlinear Anal.: Model. Control* 13(1), 103–115.

104. Upadhyay, R. K., Bairagi, N., Kundu, K., Chattopadhyay, J. 2008. Chaos in Eco-epidemiological problem of Salton Sea and its possible control. *Appl. Math. Comput.* 196, 392–401.

105. Upadhyay, R. K., Banerjee, M., Parshad, R. D., Raw, S. N. 2011. Deterministic chaos versus stochastic oscillation in a prey–predator–top predator model. *Math. Model. Anal.* 16(3), 343–364.

106. Upadhyay, R. K., Iyengar, S. R. K. 2006. Extinction and coexistence of competing prey species in ecological systems. *J. Comp. Methods Sci. Eng.* 6, 131–150.

107. Upadhyay, R. K., Iyengar, S. R. K., Rai, V. 1998. Chaos: An ecological reality? *Intern. J. Bifur. Chaos* 8, 1325–1333.

108. Upadhyay, R. K., Kumari, N., Rai, V. 2008. Wave of chaos and pattern formation in spatial predator prey systems with Holling type IV predator response. *Math. Model. Nat. Phenomena* 3(4), 71–95.

109. Upadhyay, R. K., Kumari, N., Rai, V. 2009. Wave of chaos in a diffusive system: Generating realistic patterns of patchiness in plankton-fish dynamics. *Chaos, Solitons Fractals* 40(1), 262–276.

109a. Upadhyay, R. K., Kumari, N., Rai, V. 2009. Exploring dynamical complexity in diffusion driven predator–prey systems: Effect of toxin production by phytoplankton and spatial heterogeneities. *Chaos, Solitons Fractals* 42(1), 584–594.

109b. Upadhyay, R. K., Mukhopadhyay, A., Iyengar, S. R. K. 2007. Influence of environmental noise on the dynamics of a realistic ecological model. *Fluct. Noise Lett.* 7(1), L61–L77.

110. Upadhyay, R. K., Rai, V. 1997. Why chaos is rarely observed in natural populations? *Chaos, Solitons Fractals* 8(12), 1933–1939.

110a. Upadhyay, R. K., Rai, V., Raw, S. N. 2011. Challenges of living in the harsh environments: A mathematical modeling study. *Appl. Math. Comp.* 217, 10105–10117.

111. Upadhyay, R. K., Raw, S. N., Rai, V. 2010. Dynamical complexities in a tri-trophic hybrid food chain model with Holling type II and Crawley–Martin functional responses. *Nonlinear Anal.: Model. Control* 15(3), 361–375.

112. Upadhyay, R. K., Thakur, N. K., Dubey, V. 2010. Nonlinear non-equilibrium pattern formation in a spatial aquatic system: Effect of fish predation. *J. Biol. Sys.* 18(1), 129–159.

113. Upadhyay, R. K., Thakur, N. K., Rai, V. 2011. Diffusion-driven instabilities and spatiotemporal patterns in an aquatic predator–prey system with Beddington–DeAngelis type functional response. *Int. J. Bif. Chaos* 21(3), 663–684.

113a. Upadhyay, R. K., Volpert, V., Thakur, N. K. 2012. Propagation of Turing patterns in a plankton model. *J. Biol. Dynam.* 6(2), 524–538.

114. Upadhyay, R. K., Wang, W., Thakur, N. K. 2010. Spatiotemporal dynamics in a plankton system. *Math. Model. Nat. Phenom.* 5(5), 101–121.

115. Vance, R. R. 1978. Predation and resource partitioning in one-predator–two-prey model. *Am. Nat.* 112, 797–813.

116. Vandermeer, J. 1993. Loose coupling of predator–prey cycles: Entrainment, chaos and intermittency in the classical MacArthur consumer-resource equations. *Am. Naturalist* 141, 687–716.

117. Vandermeer, J., Stone, L., Blasius, B. 2001. Categories of chaos and fractal basin boundaries in forced predator–prey models. *Chaos, Solitons Fractals* 12(2), 265–276.

118. Varriale, M. C., Gomes, A. A. 1998. A study of a three species food chain. *Ecol. Model.* 110, 119–133.

119. Xu, C., Li, Z. 2002. Influence of intraspecific density dependence on a three-species food chain with and without external stochastic disturbances. *Ecol. Model.* 155, 71–83.

120. Zhang, L., Wang, W., Xue, Y., Jin, Z. 2008. Complex dynamics of Holling-type IV predator–prey model. Preprint arXiv:0801.4365v1.

5

Modeling of Some Engineering Systems

Introduction

Advances in various disciplines have established that nonlinear mathematics can be put to novel applications in different branches of engineering. Nonlinear phenomena play a crucial role in the design and control of engineering systems and structures. In this chapter, we shall deal with the applications in some mechanical and electrical systems. A few examples are pendulum-like systems, mass–spring systems, jerk systems, nonlinear electronic circuit systems, and so on.

Analysis of a mechanical system is the first step for the understanding of pendulum vibrations and resonance. The pendulum can be used as a dynamic absorber mounted on high buildings, bridges, or chimneys. The geometrical nonlinearities introduced by the pendulum may lead to a rapid increase in the oscillations of both the structure and the pendulum, which ultimately may lead to chaotic dynamics. Various types of nonlinear electronic circuits (consisting of either real nonlinear physical devices such as nonlinear diodes, Chua's diode, capacitors, inductors, resistors, or devices) constructed with ingenious piecewise linear circuit elements have been utilized to study complex dynamics and to understand the highly nonlinear behaviors of some electronic systems [33]. Chaos and bifurcation control promise to have a major impact on time and energy critical engineering applications [8].

Models in Mechanical Systems

Most basic mechanical systems are combinations of three elements: inertia, spring, and damping elements [3]. (i) Mass and moment of inertia constitute the inertia elements. (ii) A linear spring element is deformed by an external force and the deformation is proportional to the force. A spring element is classified as a translational or a rotational (torque) element. For a linear spring, the spring constant k is defined by (change in force)/(change in displacement) for a translational spring, and (change in torque)/(change in angular displacement) for a rotational spring. (iii) A damping element dissipates (damps) the heat energy. If $\dot{x}_1$, $\dot{x}_2$ are the velocities at the ends of the element, then the damping force is proportional to $\dot{x}_1 - \dot{x}_2$, that is, $F = \alpha(\dot{x}_1 - \dot{x}_2)$, where α is the constant of proportionality. For a rotational element, the damping force is given by $F = \beta(\dot{\theta}_1 - \dot{\theta}_2)$, where $\dot{\theta}_1, \dot{\theta}_2$ are the angular velocities at the ends of the element and β is the constant of proportionality. A damping element always provides resistance in the system.

Newton's second law states that (total force) = mass × acceleration, that is, $\Sigma F = ma$. For rotational motion of a rigid body about a fixed axis, the law states that (total torque) = (moment of inertia) × (angular acceleration), that is, $\Sigma T = Ja$. Newton's law provides us with the required second-order differential equation whose solution governs the path taken by the system. For example, for free vibrations without damping of a mass (m)–spring system, the equation of motion is governed by the equation $m\ddot{x} + kx = 0$, or $\ddot{x} + \omega^2 x = 0$, where $\omega = \sqrt{k/m}$, is the natural frequency. If the initial conditions are $x(0) = x_0$ and $\dot{x}(0) = v_0$, then the periodic motion (simple harmonic motion) of the spring is given by $x(t) = (v_0/\omega)\sin(\omega t) + x_0\cos(\omega t)$. For free vibrations with damping (with β as the constant of proportionality), the equation of motion is governed by the equation $m\ddot{x} + \beta\dot{x} + kx = 0$ or $\ddot{x} + p\dot{x} + qx = 0$, where $p = \beta/m$, $q = k/m$. Damping oscillations are obtained when $p^2 - 4q < 0$. The damping factor is proportional to $e^{-p/2}$.

To formulate the equation of motion for a conservative system (total energy, that is, (kinetic energy) + (potential energy) remains constant or does not involve friction or damping), we may also adopt the energy method which consists of the following two steps: (i) find the equation governing the total energy of the system, called *energy equation*, and (ii) differentiate the energy equation to obtain the equation of motion. For example, consider again the mass(m)–spring system without damping. For this system,

Potential energy = $PE = (kx^2/2)$,   Kinetic energy = $KE = (m\dot{x}^2/2)$,
Total energy = $(kx^2 + m\dot{x}^2)/2$.

Since the system is conservative, we get

$$\frac{1}{2}\frac{d}{dt}[kx^2 + m\dot{x}^2] = 0, \quad \text{or} \quad \dot{x}[kx + m\ddot{x}] = 0, \quad \text{or} \quad \ddot{x} + \omega^2 x = 0,$$

where $\omega = \sqrt{k/m}$, is the natural frequency.

Example 5.1

The length of a pendulum is 1 m, which is pivoted at one end point. At the other end, a mass of 1 kg is attached. One end of a spring of stiffness

0.5 N-m/rad is pivoted and the other end is attached in a straight line to the pendulum at a distance of 0.75 m from the pivot in the same vertical plane. In this vertical plane, the pendulum is displaced by an angle θ from the equilibrium position. Using the energy method, determine the frequency of free oscillations for small amplitude.

SOLUTION

The pendulum is given in Figure 5.1.

Potential energy consists of two parts: (L = length of pendulum)

$$PE \text{ due to elastic deformation} = \frac{1}{2}kl^2 \approx \frac{1}{2}k(0.75\,\theta)^2 = \frac{9}{64}\theta^2.$$

$$PE \text{ due to gravitation} = mgL(1 - \cos\theta) = g(1 - \cos\theta).$$

$$KE = \frac{1}{2}m(\dot{x})^2 = \frac{1}{2}m(L\dot{\theta})^2. \text{ Total energy} = \frac{9}{64}\theta^2 + g(1 - \cos\theta) + \frac{1}{2}(\dot{\theta})^2.$$

By energy method, we get

$$\frac{1}{64}\frac{d}{dt}\left[9\theta^2 + 64g(1 - \cos\theta) + 32(\dot{\theta})^2\right] = 0, \quad \text{or}$$

$$2\dot{\theta}(9\theta + 32g\sin\theta + 32\ddot{\theta}) = 0, \quad \text{or} \quad \ddot{\theta} + (1/32)(32g + 9)\theta = 0.$$

Hence, $\omega = \sqrt{(1/32)(32g + 9)}$.

In the remaining part of this section, we consider the dynamics of the nonlinear driven pendulum and bronze ribbon experiments which exhibit chaotic dynamics in an experimental setup.

Consider the motion of a damped and driven pendulum. A visualization of this system is an adult pushing a child on a swing. The differential equation governing this system is [24]

$$\frac{d^2\vartheta}{dt^2} + R\frac{d\vartheta}{dt} + M\sin\vartheta = A\sin\Omega t, \tag{5.1}$$

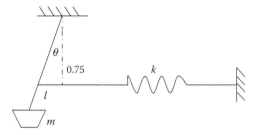

FIGURE 5.1
Pendulum in a vertical plane.

where R, M, and A are the coefficients of friction, restoring moment ($M = mgl$, where m, l are the mass and length of the pendulum, respectively, and g is the gravitational acceleration), and amplitude of the harmonically driving torque, respectively, and Ω is the frequency. The equation of motion of such dynamical systems can have typical types of solution depending on the initial conditions or on the other parameters in the system. For example, in the case of an oscillator, if the damping is strong compared to the driving force, the motion will stop after sometime, that is, the solution is constant. If the driving force is of the order of damping, then the forces produce a balance and the solution starts oscillating. The motion becomes periodic after sometime. For certain values of the driving force, the time period may become infinite (the motion never repeats itself even though it is still oscillatory). In such cases, the different initial conditions give very different motions, and the motion is chaotic.

In the following, we briefly discuss some important experimental contributions of Hübinger et al. [24] and Dressier et al. [16] to understand the effect of nonlinearity on mechanical systems. Both experiments started with a simple mechanical system and applied periodic forcing. These nonchaotic elements combine to produce chaos. The experimental systems are a driven pendulum and a driven bronze ribbon. The researchers built a mechanical pendulum that could be forced externally by torque at its pivot and a metal ribbon whose oscillations can be forced by placing it in an alternating magnetic field. Both the systems exhibit chaotic orbits in a two-dimensional Poincaré map [1].

Hübinger et al. [24] investigated experimentally the performance of the Ott, Grebogi, and Yorke feedback control method [48] to control chaotic motion. A driven pendulum and a driven bronze ribbon were taken as the experimental setups. In both experiments, unstable periodic orbits are characterized by large Lyapunov exponents. They have extracted the control vectors for the feedback control from the experimental data. In the first experiment, the damping is due to mechanical friction and forcing is applied through an electrical motor at the pivot point of the pendulum. The motor applies sinusoidal force so that the torque alternates in a clockwise/anticlockwise direction (see Figure 5.2). The electric motor at the pivot, controlled by a parallel computer via the digital–analog converter, provides the periodic forcing. The motion of the mechanical pendulum is governed approximately by the differential equation (see Hübinger et al. [24])

$$I\frac{d^2\vartheta}{dt^2} + R\frac{d\vartheta}{dt} + M\sin\vartheta = A\sin\Omega t + B, \tag{5.2}$$

where ϑ, I, R, and M denote the angle of deflection, coefficient of inertia, coefficient of friction, and restoring moment, respectively. A is the amplitude of the harmonically driving torque, Ω is its frequency, and B is a constant torque. A and B are proportional to the voltages U_A and U_B, respectively, which are

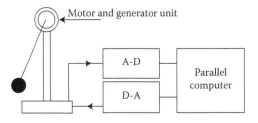

FIGURE 5.2
Setup of a pendulum experiment. (Reprinted with permission from Hübinger, B. et al. Controlling chaos experimentally in systems exhibiting large effective Lyapunov exponents. *Phys. Rev. E* 50(2), 932–948. Copyright 1994, American Physical Society.)

applied to the electric motor of the pendulum as $U(t) = U_A \sin \Omega t + U_B$, to generate the driving torque. The maximal amplitude which can be applied to the electric motor is $U_A = 6$ V, whereas the driving frequency Ω can be varied from 0.5 to 1.6 Hz [24].

Using the notations and change of variables as $\Omega_0 = \sqrt{M/I}, \omega = \Omega/\Omega_0$, $\psi = \Omega t = \omega \Omega_0 t = \omega \tau$, and $\tau = \Omega_0 t$, we obtain the nondimensional form of Equation 5.2 as

$$\frac{d^2\vartheta}{d\tau^2} + \gamma \frac{d\vartheta}{d\tau} + \sin \vartheta = a \sin \omega \tau + b, \quad \text{where } \gamma = \frac{R}{\sqrt{IM}}, a = \frac{A}{M}, b = \frac{B}{M}. \quad (5.3)$$

Here, Ω_0 is the eigenfrequency of the unforced pendulum for small amplitude. Writing this equation as a first-order system, we obtain

$$\frac{d\vartheta}{d\tau} = \theta, \quad \frac{d\theta}{d\tau} = -\gamma\theta - \sin \vartheta + a \sin \psi + b, \quad \frac{d\psi}{d\tau} = \omega. \quad (5.4)$$

The study of the pendulum motion was carried out in the state space of the variables, ϑ, $(d\vartheta/dt) = \theta$ and ψ. The control parameter was set as $U_B = 0$, and the parameter values were chosen as $U_A = 6$ V, $\Omega = 0.83 \times 2\pi$ Hz, such that the pendulum moves into a chaotic regime. The attractor of the chaotic motion of the pendulum in these coordinates is shown in Figure 5.3 for a Poincaré section corresponding to $\psi = 0$. (For details, see Hubinger et al. [24].)

The second experimental setup of Hubinger et al. [24] was inspired by the magnetoelastic buckled beam experiment of Moon [43]. The setup consists of a horizontally cantilevered elastic unbent bronze beam. Two small permanent magnets were attached to the free end of the ribbon (see Figure 5.4). Two bigger permanent magnets produce a magnetic field in which the bronze beam is located. The resulting magnetic force destabilizes the straight unbent ribbon. Two stable equilibrium positions of the beam appear. The system is driven by two coils placed around the free end of the beam and supplied with an AC voltage. The beam then starts to vibrate. A wire strain gauge was used to measure the curvature near the base of the ribbon. The gauge gives a voltage

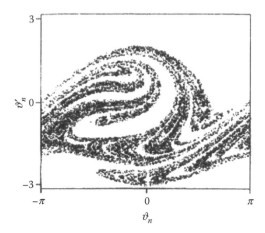

FIGURE 5.3

Chaotic attractor of the driven pendulum. (Reprinted with permission from Hübinger, B. et al. Controlling chaos experimentally in systems exhibiting large effective Lyapunov exponents. *Phys. Rev. E* 50(2), 932–948. Copyright 1994, American Physical Society.)

signal, which has a relation to the deflection of the tip. The voltage was transferred to the PC using an analog–digital converter. The output voltage was sampled. When the period of excitation was chosen as $T = 1$ s, and the value of the control parameter was taken as zero, the ribbon vibrates chaotically. A deterministic structure of chaotic attractor is presented in Figure 5.5 [24]. For

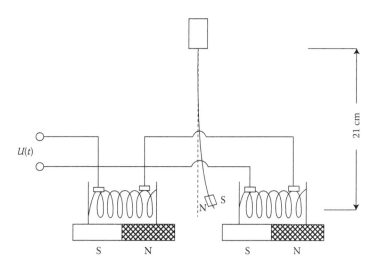

FIGURE 5.4

Experimental set up of the chaotic bronze ribbon. (Reprinted with permission from Hübinger, B. et al. Controlling chaos experimentally in systems exhibiting large effective Lyapunov exponents. *Phys. Rev. E* 50(2), 932–948. Copyright 1994, American Physical Society.)

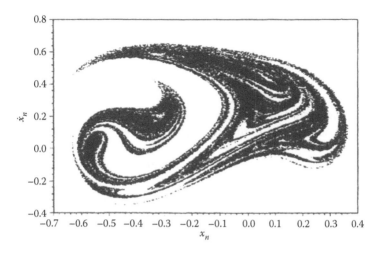

FIGURE 5.5
Chaotic attractor from the bronze ribbon experiment. (Reprinted with permission from Hübinger, B. et al. Controlling chaos experimentally in systems exhibiting large effective Lyapunov exponents. *Phys. Rev. E* 50(2), 932–948. Copyright 1994, American Physical Society.)

the detailed analysis of the determination of control vectors from measurements and other results, see Hubinger et al. [24].

Dressier et al. [16] had also conducted the bronze ribbon experiment to track the unstable periodic orbits and stabilize them. The authors had used a variant of the Ott, Grebogi, and Yorke feedback control method [48] to control chaotic motion. The method consists of the following steps. (i) At each tracking step, feedback control vectors were extracted from the experimental data. (ii) Unstable periodic orbits were redetermined using an adaptive orbit correction. (iii) The unstable periodic orbit was tracked into a particular parameter regime where the chaotic attractor had lost its stability and a new stable periodic orbit was born. The experimental setup of the chaotic bronze ribbon was the same as given in Figure 5.4.

The experimental nonlinear pendulum problem was also discussed by Pereira-Pinto et al. [51]. Aline Souza de Paula et al. [50] also analyzed the periodic response, chaos, and transient chaos in an experimental nonlinear pendulum. Both authors used the Runge–Kutta fourth-order method for numerical computations. The authors have shown that there is a good agreement between the experimental and numerical results. Experimental investigations of the nonlinear pendulum were earlier performed by Blackburn and Baker [4], and Deserio [15]. Time-series analysis of the dynamics of the nonlinear pendulum was carried out by Franca and Savi [18]. The authors had determined the values of the parameters using their experimental setup. The pendulum consists of a disc with a lumped mass that is connected to a rotary motion sensor. To drive this setup and provide torsional stiffness to the system, it was attached to a string–spring device. This device was connected to an electric motor. To provide adjustable damping, the setup was also connected to a magnetic device. The analysis of the nonlinear pendulum motions produced rich dynamic responses such as chaos, transient chaos, and so on. It was assumed that system dissipation can be written as a combination of linear viscous dissipation and dry friction to derive the equation of motion. Let ϕ denote the angular position of the experimental pendulum. The equation of motion was derived as

$$I\ddot{\phi} + \varsigma\dot{\phi} + (kd^2/2)\phi + (2\mu/\pi)\arctan(q\dot{\phi}) + (mgD/2)\sin\phi$$
$$= (kd/2)\left[\sqrt{a^2 + b^2 - 2ab\cos(\omega t)} - (a - b)\right], \tag{5.5}$$

where q is a large number and the meanings of the parameters are as follows: ω, the forcing frequency related to motor rotation; a, the position of the guide of the string with respect to the motor; b, the length of the excitation crank of the motor; D, the diameter of the metallic disc; d, the diameter of the driving pulley; m, the lumped mass; ς, the linear viscous damping coefficient; μ, the dry friction coefficient; g, the gravity acceleration; I, the inertia of the disk-lumped mass; and k, the string stiffness. The equation of motion (5.5) is written as a system of two first-order equations and the Runge–Kutta method applied for its solution. The state space and the strange chaotic attractor are presented in Figure 5.6. The set of parameter values used by the authors [50] is $\omega = 5.61$ rad/s, $a = 0.16$ m, $b = 0.06$ m, $d = 0.048$ m, $m = 0.0147$ kg, $g = 9.81$ m/s^2, $\mu = 1.272 \times 10^{-4}$ N m, $I = 1.738 \times 10^{-4}$ kg m^2, $D = 0.095$ m, $k = 2.47$ N/m, $\varsigma = 2.368 \times 10^{-5}$ kg m^2/s, and $q = 10^6$.

By changing the forcing frequency to $\omega = 5.61$ rad/s from $\omega = 5.1$ rad/s, at which it shows the periodic response, chaos appears. Figures 5.6a,b present a state space of chaotic response for both numerical and experimental results, showing a close agreement. Lyapunov exponents were also computed. Employing the algorithm due to Wolf et al. [77], the spectrum was obtained as [50], $\lambda = (+0.4483, -0.5732)$. Figures 5.6c,d represent the experimental and numerical strange attractors.

Nonlinear Oscillators

One way to produce chaos in a system governed by ODEs is to apply periodic forcing to a nonlinear oscillator. Li and Moon [36] derived criteria for chaotic dynamics of a three-well potential oscillator with homoclinic and heteroclinic orbits. A simple damped pendulum with sinusoidal forcing is given by $\ddot{x} + b\dot{x} + \sin x = \sin\Omega t$, where b is a friction parameter or damping term and Ω is the frequency of the forcing. Writing as a system, we get $\dot{x} = y, \dot{y} = \ddot{x} = \sin\Omega t - by - \sin x$. For the initial conditions $(x, y, t) = (0, 1, 0)$,

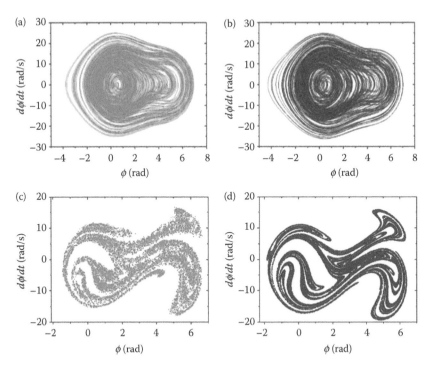

FIGURE 5.6
Chaotic (strange) attractor for experimental and numerical results. (Reprinted from *J. Sound Vib.* 294, Paula, A. S. de, Savi, M. A., Pereira-Pinto, F. H. I., Chaos and transient chaos in an experimental nonlinear pendulum, 585–595, Copyright 2006, with permission from Elsevier.)

and $b = 0.05$, $\Omega = 0.8$, the damped forced pendulum exhibits a strange attractor [62]. The attractor has a noninteger dimension >2, but <3. Dissipative chaotic flows generally have such a chaotic attractor [62]. The appearance of a strange attractor in the dynamics is a requirement of chaotic phenomenon and was called the "fingerprints of chaos" [52]. van der Pol observed chaos in a periodically forced neo-bulb relaxation oscillator for the model $\ddot{x} + b(x^2 - 1)\dot{x} + x = A\sin\Omega t$, for $(b, A, \Omega) = (1, 1, 0.45)$ and 2-tori for $(1, 0.9, 0.5)$ [62]. This model was earlier studied by Cartwright and Littlewood [7] and was shown by Levinson [35] to have chaotic solution. The critical point $(0, 0)$ of the van der Pol equation [71,72], $\ddot{x} + b(x^2 - 1)\dot{x} + x = 0$, is an unstable node if $b \geq 2$ and an unstable focus for $0 < b < 2$. Thus, the critical point $(0, 0)$ is always unstable. Chaotic behavior has also been found in the generalized van der Pol's equation $\ddot{x} + b(x^2 - 1)\dot{x} + cx + dx^3 = A\cos\Omega t$, for $b = 0.2$, $c = 0$, $d = 1$, $A = 17$, $\Omega = 4$ [30]. The van der Pol equation has been used to model many real-world problems such as heartbeats [73], earthquakes [6], vocal fold vibrations [32], well drill chatter [76], bipolar disorders [61], sunspot cycles [49], and many other oscillatory phenomena.

The more widely studied system in which damping is linear, but restoring force contains a cubic nonlinearity, was introduced by George Duffing [17] and is given by

$$\ddot{x} + a\dot{x} + b_1 x + b_2 x^3 = A \sin \Omega t. \tag{5.6}$$

The system models various phenomena [75] such as magnetoelastic buckled beam [44], a nonlinear electronic circuit [69], and so forth. Bifurcation and chaos in a three-well Duffing system with one external force were investigated in detail by Huang and Jing [23]. Depending on the signs of b_1 and b_2, the system is classified as follows [62]:

i. If $b_1 < 0$, $b_2 < 0$, then the solutions are unbounded and thus cannot lead to chaos.

ii. If $b_1 > 0$, $b_2 > 0$, the system models a spring that gets stiffer as it is stretched or compressed (*stiffening-spring* system).

iii. If $b_1 > 0$, $b_2 < 0$, the system models a spring that gets weaker when it is stretched or compressed (*softening-spring* system). It could also model a pendulum driven to large amplitude, but not so large that it goes "over to top."

iv. If $b_1 < 0$, $b_2 > 0$, the system models a ball rolling in a trough with two dips separated by a bump. It is called *Duffing's two-well oscillator*.

Cases (ii), (iii), and (iv) can exhibit chaos for appropriately chosen values of the parameters, although the two-well case is the most common since its solutions are always bounded by the strong cubic restoring force. Writing Equation 5.6 as a system, we get $\dot{x} = y$, $\dot{y} = A \sin \Omega t - ay - b_1 x - b_2 x^3$. With the initial conditions $(x, y, t) = (-1, -0.6, 0)$, and $a = 1$, $A = 1$, $b_1 = -1$, $b_2 = 1$, $\Omega = 0.8$, the damped two-well oscillator exhibits a strange attractor [62]. A variant of this equation without a linear term in x was studied by Banatto et al. [5]. Savi and Pacheco [54] discussed the nonlinear dynamics of a Duffing oscillator with two-degrees of freedom. The prospect of chaotic behavior is of concern and since the equations of motion are associated with a five-dimensional system, the analysis was performed by considering two Duffing oscillators, both with single-degree of freedom, coupled by a spring–dashpot system. It is possible to construct chaotic forced oscillators that are combinations of different oscillators like the Rayleigh–Duffing oscillator [20]. The Duffing–van der Pol oscillator and other combinations were studied by Ueda [70]. The Rayleigh and van der Pol equations can be combined to give a limit cycle that is a perfect circle in state space [60]. The equation of motion of the damped and driven Morse oscillator [29] can be represented in terms of dimensionless variables as $\ddot{x} + a\dot{x} + be^{-x}(1 - e^{-x}) = A\cos\Omega t$. This model has been widely used as a model for the laser isotope separation, infrared multiphoton excitation, dissociation of van der Waal's complexes, and so on. A typical study of this Morse oscillator for $a = 0.8$, $b = 8$, $A = 3.5$, $\Omega = 2$ exhibits chaotic phenomena through Feigenbaum's period-doubling route when the angular frequency Ω is varied [33]. Recently, Sun and Sprott [64] investigated a sinusoidally driven system with a signum nonlinearity, analytically as well as by numerical simulation. The authors considered the equation of motion as $\ddot{x} + a\dot{x} - x + b\,\text{sgn}(x) = A\sin\Omega t$, where $\text{sgn}(x)$ is the signum function. A strange attractor plotted in Figure 5.7 is observed by choosing the parameter values as $a = 1.095$, $b = 1.1$, $A = 1.1$, $\Omega = 1$. A symmetric "onion-like" strange attractor was observed by Sun and Sprott [64] for the parameter values $a = 1.05$, $b = 1$, $A = 1.1$, $\Omega = 1$. They had also obtained the largest Lyapunov exponent as 0.1241, and the Kaplan–Yorke dimension as $D_{KY} = 2.134$. A number of chaotic oscillators were described in the literature [9]. Tamaševicius et al. [65] suggested the following regarding the use of oscillators: (i) The oscillator should not be higher than a third-order system and preferably autonomous. (ii) It should contain a simple, single nonlinear unit. (iii) Smooth, monotonous, and nonlinear functions are preferred to piecewise linear, nonmonotonous ones. (iv) The oscillator should operate at kilohertz frequencies to simplify the measuring procedures.

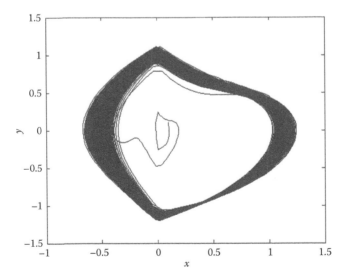

FIGURE 5.7
Chaotic attractor for sinusoidally driven system with a signum nonlinearity.

Example 5.2

Obtain the frequency response relations for the Duffing oscillator $\ddot{x} + a\dot{x} + b_1 x + b_2 x^3 = A\sin\Omega t, 0 < b_2 \leq 1$.

SOLUTION

Assume the solution in the form $x(t) = \beta\sin(\Omega t + \delta)$, where δ is a phase constant to be fixed. Substituting $x(t)$ and setting the coefficients of sin Ωt and cos Ωt to zero separately (neglecting the higher harmonic terms and higher-order terms in δ), we obtain

$$[\beta(b_1 - \Omega^2) + (3/4)b_2\,\beta^3]\cos\delta - a\beta\,\Omega\sin\delta = A, \tag{5.7}$$

$$[\beta(b_1 - \Omega^2) + (3/4)b_2\,\beta^3]\sin\delta + a\beta\,\Omega\cos\delta = 0. \tag{5.8}$$

Multiplying Equation 5.7 by cos δ, Equation 5.8 by sin δ and adding, we get

$$[\beta(b_1 - \Omega^2) + (3/4)b_2\,\beta^3] = A\cos\delta.$$

Multiplying Equation 5.7 by sin δ, Equation 5.8 by cos δ and subtracting, we get $a\beta\Omega = -A\sin\delta$. Eliminating δ from these equations, we get

$$[\beta(b_1 - \Omega^2) + (3/4)b_2\beta^3]^2 + (a\beta\Omega)^2 = A^2, \text{ or}$$

$$\beta^2[\{(b_1 - \Omega^2) + (3/4)b_2\beta^2\}^2 + a^2\Omega^2] = A^2. \tag{5.9}$$

For a given set of parameter values, (a, b_1, b_2), the plot of amplitude β in terms of Ω is a sixth degree curve (frequency–response curve) or a cubic in $\beta^2 = t > 0$. The resonance is nonlinear. In terms of t, Equation 5.9 can be written as

$$9b_2^2 t^3 + 24b_2(b_1 - \Omega^2)t^2 + 16[(b_1 - \Omega^2)^2 + a^2\Omega^2]t - 16A^2 = 0.$$

For all values of $b_1 \geq \Omega^2$, there is one change in sign in the coefficients, that is, there is one positive root for t, that is, only one value for the amplitude β. For all $b_1 < \Omega^2$, the equation has three positive roots or one positive root. Multiple valued solutions may be obtained in this case. The multiple valued nature of the response curve has significance because it leads to jump phenomena [21]. For $b_2 = 0$, $a = 0$, the exact solution of the equation is given by

$$x(t) = \beta\sin(\sqrt{b_1}t + \theta) + \frac{A}{b_1 - \Omega^2}\sin\Omega t,$$

where β and θ are the integration constants. This solution contributes to the secondary response.

Example 5.3

Obtain the frequency–response relations for the generalized van der Pol's equation $\ddot{x} + b(x^2 - 1)\dot{x} + cx + dx^3 = A\cos\Omega t$.

SOLUTION

Assume the solution in the form $x(t) = \beta\cos(\Omega t + \delta)$, where δ is a phase constant to be fixed. Substituting $x(t)$ and setting the coefficients of $\sin\Omega t$ and $\cos\Omega t$ to zero separately (neglecting the higher harmonic terms and higher-order terms in δ), we obtain

$$[\beta(c - \Omega^2) + (3/4)d\beta^3]\cos\delta - (b/2)\beta(\beta^2 - 2)\sin\delta = A, \qquad (5.10)$$

$$[\beta(c - \Omega^2) + (3/4)d\beta^3]\sin\delta + (b/2)\beta(\beta^2 - 2)\cos\delta = 0. \qquad (5.11)$$

Multiplying Equation 5.10 by $\sin\delta$, Equation 5.11 by $\cos\delta$ and subtracting, we get $(b/2)\beta(\beta^2 - 2) = -A\sin\delta$. Multiplying Equation 5.10 by $\cos\delta$, Equation 5.11 by $\sin\delta$ and adding, we get $\beta(c - \Omega^2) + (3/4)d\beta^3 = A\cos\delta$. Eliminating δ from these equations, we get

$$\beta^2\{[(c - \Omega^2) + (3/4)d\beta^2]^2 + (b^2/4)(\beta^2 - 2)^2\} = A^2. \qquad (5.12)$$

For a given set of parameter values, (b, c, d), the plot of amplitude β in terms of Ω is a sixth degree curve (frequency–response curve) or a cubic in $\beta^2 = t > 0$. The resonance is nonlinear. In terms of t, Equation 5.12 can be written as

$$(9d^2 + 4b^2)t^3 + 8[3d(c - \Omega^2) - 2b^2]t^2 + 16[(c - \Omega^2)^2 + b^2]t - 16\,A^2 = 0.$$

For all values of $3d(c - \Omega^2) - 2b^2 \geq 0$, there is one change in sign in the coefficients, that is, there is one positive root for t, that is, only one value for the amplitude β. For all $3d(c - \Omega^2) - 2b^2 < 0$, the equation has three positive roots or has one positive root. Multiple valued solutions may be obtained in this case.

Chaos in Mass–Spring System

Vasegh and Khellat [74] consider the mass–spring friction system given in Figure 5.8. The frictional force including viscous friction is assumed to be proportional to the square of the velocity, $a\dot{x}^2$. The spring force is a nonlinear function of x modeled as $bx|x|$, which is similar to the classical nonlinear spring defined by bx^3. The external force is taken as $A\cos\Omega t$. The equation of motion is

$$m\ddot{x}(t) + a_1\dot{x}^2(t) + b_1x(t)|x(t)| = A_1\cos\omega t, \quad \text{or}$$

$$\ddot{x}(t) + a\dot{x}^2(t) + bx(t)|x(t)| = A\cos\omega t, \qquad (5.13)$$

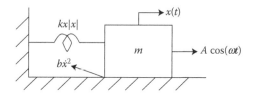

FIGURE 5.8
A nonlinear mass–spring model. (From Vasegh, N., Khellat, F. 2009. *PhysCon-2009*, Catania, Italy, pp. 1–4. Reprinted with permission from Professor Dr. F. Khellat.)

where a, b, and A are nonnegative parameters. For $b = 0$, we can obtain the exact solution. Let $A = 0$. Then, the origin is the only equilibrium point of system (5.13). Linearization about the origin shows that it is a nonhyperbolic fixed point. To study system (5.13) when $A = 0$, consider the following two cases:

Case 1: Let $b \neq 0$, $a = 0$: In this case, choose the Lyapunov function as $V(x, \dot{x}) = 2bx^2|x| + 3\dot{x}^2$. Along the trajectory of the system, $\dot{V}(x, \dot{x}) = 6bx\dot{x}|x| + 6\dot{x}\ddot{x} = 0$. Since the origin does not lie on the close curve $2bx^2|x| + 3\dot{x}^2 = c^2$, it is a periodic solution of the system for the initial conditions $x(0) = x_0$, $\dot{x}(0) = \dot{x}_0$ and $c^2 = 2bx_0^2|x_0| + 3\dot{x}_0^2$.

Case 2: Let $b \neq 0$, $a \neq 0$: In this case, origin is a center. All solutions starting in some neighborhood of the origin are periodic [53].

Periodic solutions in the above cases are not asymptotically stable solutions. This is because by changing the initial conditions, other solutions are observed. Also, this shows that model (5.13) has quasiperiodic solutions. It is not unexpected as the quasiperiodic solutions occur in the forced linear systems where the free system has a periodic solution with different (not rational multiple of each other) period from the corresponding forced one. A quasiperiodic solution is obtained for $a = 1$, $b = 1$, $A = 1$, and $\omega = 1.9\pi$. For the parameter values $a = 0.1$, $b = 8$, $A = 1$, and $\omega = 0.1\pi$, chaos was observed in the phase plane $(x, \dot{x})$. (see Figure 5.9). [74].

Chaotic Oscillations in Duffing–Holmes Oscillator

A Duffing–Holmes oscillator is a chaotic system describing the dynamics of the forced vibrations of a buckled elastic beam. A Duffing–Holmes oscillator is given by the second-order nonautonomous differential equation $\ddot{x} + a\dot{x} - x + x^3 = A \sin\Omega t$, where a, A, and Ω are the damping coefficient, amplitude, and frequency of the external driving force, respectively [1,22,38]. Writing it as a system, we get $\dot{x} = y$, $\dot{y} = A\sin\Omega t - ay + x - x^3$. Suitable choices of the control parameters a, A, and Ω produce periodic and chaotic oscillations. Numerical results obtained by Longsuo [38], for $a = 0.25$, $A = 0.27$,

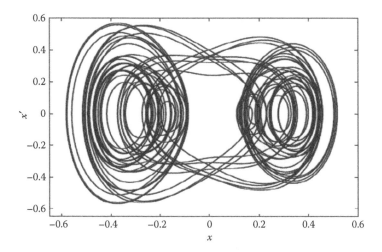

FIGURE 5.9
Phase portrait of the model. (From Vasegh, N., Khellat, F. 2009. *PhysCon-2009*, Catania, Italy, pp. 1–4. Reprinted with permission from Professor Dr. F. Khellat.)

and $\Omega = 1.0$, with initial values taken as (1, 0), are presented as a phase portrait in Figure 5.10 (Lissajous figures).

Stabilization of unstable periodic orbits in the Duffing–Holmes chaotic oscillator, namely the orbits in the side wells of the two-well potential, was demonstrated experimentally by Tamaševicius et al. [66] using the resonant negative feedback method. Based on the forced oscillator given by the Duffing–Holmes differential equation $\ddot{x} + a\dot{x} - x + x^3 = A\sin\Omega t$, Silva and Young [56,57] suggested a nonautonomous circuit providing chaotic oscillations with a noise-like spectrum. Low-frequency version of the oscillator was discussed in detail by Kandangath [25]. Tamasevieiute et al. [67] suggested

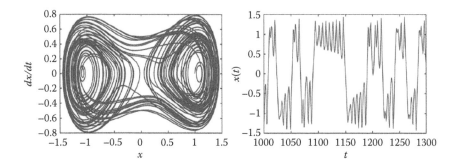

FIGURE 5.10
Phase portrait and time history of Duffing–Holmes oscillator. (From Longsuo, L. 2011. *Math. Prob. Eng.*, 2011, article 10, 538202–538210. Copyright 2009, Hindawi Publishing Corp. Reprinted with permission.)

a circuit with an externally driven damped *RLC* oscillator with all linear elements. A positive feedback loop consisting of a resistor and two diodes introduces the nonlinearity. Their electrical circuit resembles the Young–Silva oscillator [31], but is much simpler.

Chaos in Jerk Systems

Third-order explicit autonomous differential equations in one scalar variable, sometimes called jerky dynamics, constitute an interesting subclass of dynamical systems that can exhibit chaotic behavior. Consider an equation of the form $\dddot{x} = f(x, \dot{x}, \ddot{x})$, where x, $\dot{x}$, and $\ddot{x}$ are the displacement, velocity, and acceleration, respectively. The quantity $\dddot{x}$ is called jerk in a mechanical system [55]. Hans Gottlieb [19] posed the question "What is the simplest jerk equation that can give chaos?" In response, Linz [37] reported that the jerky dynamics for the Lorenz and Rössler models are not suitable for Gottlieb's simplest jerk function. One of the simplest jerk equations is $\dddot{x} = -a\ddot{x} + \dot{x}^2 - x$. Writing it as a system, we obtain $\dot{x} = t, \dot{t} = z$, and $\dot{z} = -az + t^2 - x$. Origin $(0, 0, 0)$ is the only equilibrium point of this system. Linearizing the system about the equilibrium point, we find that the eigenvalues of the Jacobian matrix are the roots of the equation $\lambda^3 + a\lambda^2 + 1 = 0$. For $a > 0$, this equation has no positive root and has only one negative root. The complex conjugate eigenvalues have positive real parts and the equilibrium is unstable. Sprott [58] had shown that this system exhibits chaos for $a = 2.02$. Differentiating each term of the equation $\dddot{x} = -a\ddot{x} + \dot{x}^2 - x$ with respect to time and defining $y \equiv 2\dot{x}$, we obtain the equation $\dddot{y} = -a\ddot{y} + y\dot{y} - y$. Malasoma [41] showed that these jerk forms can arise from six different 3D systems with only five terms and a single quadratic nonlinearity, which he called *"class P"* systems. The simplest cubic equation proposed by Malasoma [40] is $\dddot{x} = -a\ddot{x} + x\dot{x}^2 - x$ which is same as the above equation. For $a > 0$, origin $(0, 0, 0)$ which is the only equilibrium point is unstable. He had shown that the system exhibits chaos for $a = 2.03$, and conjectured that this is the simplest system that is invariant under the parity transformation $x \rightarrow -x$. Letellier and Valée [34] studied both the cases in detail. Malasoma [41] had also described three chaotic systems with five terms and a single quadratic nonlinearity whose jerk representation involves rational functions with four terms. One such system is $\dot{x} = z, \dot{y} = -\alpha y + z - 1, \dot{z} = xy$, which can be written in jerk form as $\dddot{x} = -x - \alpha\ddot{x} + x\dot{x} + (\dot{x}\ddot{x}/x)$. He had shown that this system exhibits chaos over the range $4.7293 < \alpha < 4.7558$. For example, for $\alpha = 4.73$, the system exhibits chaos with the initial condition $(1, 2, -15)$. Moore and Spiegel [45] described a model oscillator defined by $\dddot{x} = -5\dot{x} + 9\dot{x} - x^2\dot{x} - \ddot{x}$, in which the thermal dissipation causes instability due to the irregular variability in the luminosity of stars. They concluded that a great variety of oscillatory phenomena observed in variable stars can be generated from a single instability mechanism, provided the essential nonlinearities are retained and the law of dissipation is appropriately chosen [45]. Sun and Sprott [63] investigated a simple jerk system with a piecewise exponential nonlinearity as

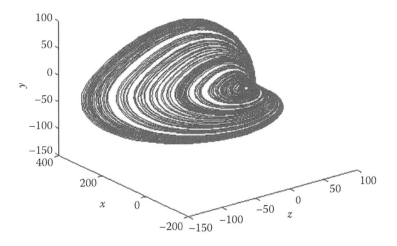

FIGURE 5.11
Chaotic attractor in a simple jerk system of Sun and Sprott ($a = 1.7$ and $b = 1.181$).

$\dddot{x} + a\ddot{x} - |\dot{x}|^b + x = 0$. The authors showed that this system has a period-doubling route to chaos and a narrow chaotic region in a parameter space. The model can be written as a system as $\dot{x} = y, \dot{y} = z, \dot{z} = -az + |y|^b - x$, where a and b are bifurcation parameters. The only equilibrium point is the origin (0, 0, 0), which is a saddle focus with instability index of 2 [63]. Linearizing about the equilibrium point, the characteristic equation of the Jacobian matrix is again obtained as $\lambda^3 + a\lambda^2 + 1 = 0$. For $a > 0$ origin (0, 0, 0) which is the only equilibrium point, is unstable. It is associated with a 2D unstable manifold in which trajectories are spiraling outward. A chaotic attractor of this system is plotted in Figure 5.11 for $a = 1.7$ and $b = 1.181$. Sun and Sprott [63] have plotted the chaotic attractor for the values $a = 2$ and $b = 1.5$. Chaos in this system was observed for all values of $b > 1$. For $b < 1$, the solution is unbounded and chaos does not exist for $b = 1$. In the limit as $b \rightarrow \infty$, the nonlinear term resembles a square well with sides at $\dot{x} = \pm 1$, and chaos arises from collisions with the side wall of the well at $\dot{x} = -1$, while the trajectories within the well are described by a linear system of equations [62].

Models in Electronic Circuits

All circuit elements are nonlinear and can only be approximated as linear elements over a certain range. The elements can be linear or nonlinear depending upon their characteristic curves, namely the *voltage–current curve (v–i)*, *voltage–charge* (v–Q), and *current–magnetic flux* (i–ϕ) curves. A two-terminal or one-port circuit elements whose current (i), voltage (v), and charge (Q) fall on some fixed characteristic curve in the different planes are represented by the following equations:

i. The constitutive relation of resistance R (measured in ohms Ω) is defined by $f(v, i) = 0$. Ohm's law states that the voltage drop across a resistor is proportional to the instantaneous current i, $E_R = Ri$, where $i = dQ/dt$, and Q is the instantaneous electric charge and measured in coulombs.

ii. The constitutive relation of capacitance C (measured in farads F) is defined by $f(v, Q) = 0$. The voltage drop across a capacitor is proportional to the instantaneous electric charge Q on the capacitor, $E_C = (Q/C) = (1/C)\int_{t_0}^{t} I(u)\, du$.

iii. The constitutive relation of inductance L (measured in Henry's H) is defined by $f(\phi, i) = 0$. The voltage drop across an inductor is proportional to the instantaneous time rate of change of current i, $E_L = L(di/dt)$. (Faraday's law).

Linear and nonlinear circuits: A linear circuit is the one which has no nonlinear electronic components in it. Some examples of linear circuits are small-signal amplifiers, differentiators, and integrators. Examples of nonlinear electronic components are diodes, transistors when they are operated in saturation, and so on. For example, the circuits of digital logic circuits, mixers, and modulators operate in a nonlinear way.

Kirchhoff's voltage law (KVL law) states that the algebraic sum of all voltage drops around any closed loop is zero or the voltage impressed on a closed loop is equal to the algebraic sum of the voltage drops in the rest of the loop. KVL law puts a physical restriction on v as $v_R + v_L - v_C = 0$.

Kirchhoff's current law (KCL law) states that at a point of a circuit, the sum of the inflowing currents is equal to the sum of the outflowing currents. That is, the algebraic sum of the currents meeting at a point in a circuit is zero.

Kirchhoff's laws govern the dynamics of a circuit.

RL circuit (resistance–inductance circuit) is given in Figure 5.12a. An Ohm's resistor of resistance R and an inductor of inductance L are connected in series to a source of electromotive force, $E(t)$. The voltage drops across the resistor and the inductor are RI and $L(di/dt)$, respectively. The equation governing the current $i(t)$ is given by

$$L\frac{di}{dt} + Ri = E(t).$$

RC circuit (resistance–capacitance circuit) is given in Figure 5.12b. The equation governing the current $i(t)$ is given by

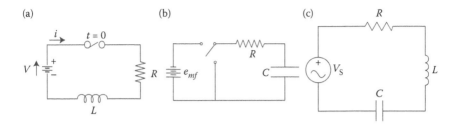

FIGURE 5.12
(a) *RL* circuit. (b) *RC* circuit. (c) *RLC* circuit.

$$Ri + \frac{1}{C}\int i(t)\,dt = E(t), \quad \text{or} \quad R\frac{di}{dt} + \frac{i}{C} = \frac{dE}{dt}.$$

RLC circuit is given in Figure 5.12c. The equation governing the current $i(t)$ is given by

$$L\frac{di}{dt} + Ri + \frac{1}{C}\int i(t)\,dt = E(t), \quad \text{or} \quad L\frac{d^2i}{dt^2} + R\frac{di}{dt} + \frac{i}{C} = \frac{dE}{dt}.$$

Since, $i = dQ/dt$, the first equation can also be written as

$$L\frac{d^2Q}{dt^2} + R\frac{dQ}{dt} + \frac{Q}{C} = E(t).$$

If there is no time-dependent charge, $E(t) = 0$. When $R > 2\sqrt{L/C}$, the solution is of the form

$$Q(t) = e^{-(R/2L)t}[Ae^{pt} + Be^{-pt}], \quad \text{where } p = \sqrt{(R/2L)^2 - \{1/(LC)\}}.$$

If R is positive, the resistor is said to be dissipative. The energy which was stored in the capacitor and inductor is dissipated as heat energy.

Junction law states that the sum of the currents entering any junction in a circuit is equal to the sum of the currents leaving that junction. The junction law is equivalent to stating that charge does not accumulate at a junction in a circuit. This is the conservation of charge law.

Loop law states the sum of the changes in electric potential across all elements in a complete circuit loop is zero. This is conservation of energy law.

In this section, we are interested in studying bifurcation and chaos in some nonlinear circuits.

Nonlinear Circuits

Chua's Diode and Chua's Circuit

A circuit is nonlinear if it contains at least one nonlinear circuit element. That is, it may contain a nonlinear resistor, a nonlinear capacitor, or a nonlinear inductor. Pioneering studies of nonlinear circuits which exhibited a strange attractor were made by Chua and coworkers [10,11,13]. The first announcement of a chaotic attractor in nonlinear circuits was made by Matsumoto [42] in his paper titled *A chaotic attractor from Chua's circuit*. The genesis of Chua's circuit was given by Chua himself in his excellent exposition [10]. The essence of the circuit is the use of a piecewise linear circuit element (the characteristic curve of the element is piecewise linear) along with other linear elements which are combined ingeniously. The nonlinear circuit was realized by the use of *Chua's diode*. Chua's diode is a nonlinear resistor with piecewise linear characteristics. Nonlinear resistors are easy to build. A good exposition of Chua's circuit was given by many authors (Lakshmanan and Rajasekar [33], Kennedy [27,28], Madan [39]). Chua's circuit is the only nonlinear circuit system for which the presence of chaos was proved experimentally, mathematically, and realized in computer simulations.

Chua's diode: As mentioned earlier, Chua's diode is a nonlinear resistor with piecewise linear characteristics. Its characteristic curve consists of odd-symmetric, three piecewise linear segments as given in Figure 5.13. The slope of the linear segment is negative between the two break points P and S. That is, $m_2 < 0$. Also, $m_1 < 0$. It is straightforward to write the equations of the three straight lines with the given slopes. The piecewise polynomial is given by

$$f(v_R) = \begin{cases} m_1 v_R + E_1(m_1 - m_2), & \text{for } v_R < -E_1, \\ m_2 v_R, & \text{for } -E_1 \leq v_R \leq E_1, \\ m_1 v_R + E_1(m_2 - m_1), & \text{for } v_R > E_1. \end{cases} \quad (5.14)$$

Chua's circuit: Consider a generalized *RLC* circuit in which the capacitor C is replaced by two linear capacitors C_1 and C_2 connected in parallel as in Figure 5.14. Chua's circuit is obtained by inserting in this *RLC* circuit, the Chua's diode as the nonlinear element. Chua's circuit is given in Figure 5.15.

Lakshmanan and Rajasekar [33] have discussed in detail the Chua's circuit. Applying Kirchhoff's laws to the branches of the circuit of Figure 5.14, the state equations of Chua's circuit are obtained as [26,33],

$$C_1 \frac{dv_1}{dt} = \frac{1}{R}(v_2 - v_1) - f(v_1), \quad C_2 \frac{dv_2}{dt} = \frac{1}{R}(v_1 - v_2) + i_L, \quad L \frac{di_L}{dt} = -v_2, \quad (5.15)$$

where $f(v_1)$ is the mathematical representation of Chua's diode.

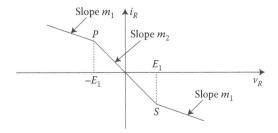

FIGURE 5.13
Characteristic curve of Chua's diode. (From Chua, L. O. 1992. *Arch. Elektron. Übertragungstech.* 46(4), 250–257. Copyright 1992, Hirzel-Verlag, Stuttgart. Reprinted with permission.)

The model given in Equation 5.15 constitutes a dynamical system. (There is no external signal injected into the system and allowed to evolve through its natural dynamics.) When the resistance R, inductance L, and capacitances C_1 and C_2 in the Chua's circuit are positive, then from an energy conservation point of view, the nonlinear resistor must be active for the circuit to oscillate. Chua's circuits can be constructed at low cost. The circuits exhibit a rich variety of bifurcation and chaos.

To nondimensionalize the equations in system (5.15), substitute $x = v_1/E_1$, $y = v_2/E_1$, $z = i_L R/E_1$, $t = C_2 \tau R$, $\alpha = C_2/C_1$, $\beta = C_2 R^2/L$. We obtain

$$\frac{dx}{d\tau} = \alpha[y - x - f(x)] = F(x,y,z), \quad \frac{dy}{d\tau} = x - y + z, \quad \frac{dz}{d\tau} = -\beta y, \quad (5.16)$$

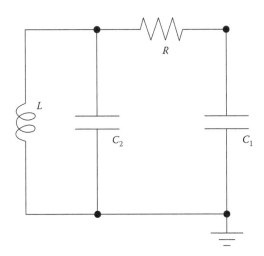

FIGURE 5.14
Generalized *RLC* circuit. (With kind permission from Springer Science+Business Media B.V.: *Nonlinear Dynamics: Integrability, Chaos and Pattern*, Lakshmanan, M., Rajasekar, S., Springer-Verlag, Berlin, Copyright 2003, Springer.)

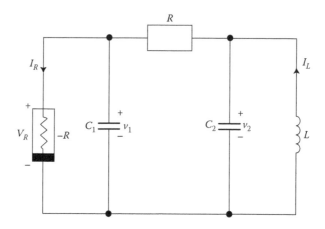

FIGURE 5.15
Chua's circuit. (From Chua, L. O. *Arch. Elektron. Übertragungstech.* 46(4), 250–257. Copyright 1992, Hirzel-Verlag, Stuttgart. Reprinted with permission.)

where

$$f(x) = \begin{cases} bx + (b - a), & \text{for } x < -1, \\ ax, & \text{for } -1 \le x \le 1, \\ bx + (a - b), & \text{for } x > 1. \end{cases} \tag{5.17}$$

System (5.16) is invariant under the transformation $(x, y, z) \to (-x, -y, -z)$. The fixed points are located in the x–z plane $(y = 0)$. In the region $\{(x, y) \mid -1 \le x \le 1\}$, origin $(0, 0, 0)$ is a fixed point. In the region $\{(x, y) \mid x < -1\}$, the fixed point is $(k, 0, -k)$; and in the region $\{(x, y) \mid x > 1\}$, the fixed point is $(-k, 0, k)$, where $k = (a - b)/(1 + b)$, $a \ne b$, $b \ne -1$. Let the parameter values be taken as $\alpha = 10$, $\beta = 14.87$, $a = -1.27$, and $b = -0.68$ as considered by Kapitaniak [26]. For these values, the equilibrium points are $(0, 0, 0)$, $(-59/32, 0, 59/32)$, and $(59/32, 0, -59/32)$. The Jacobian matrix of system (5.16) is given by

$$J = \begin{bmatrix} \partial F/\partial x & \partial F/\partial y & 0 \\ 1 & -1 & 1 \\ 0 & -\beta & 0 \end{bmatrix} = \begin{bmatrix} -\alpha[1 + f'(x)] & \alpha & 0 \\ 1 & -1 & 1 \\ 0 & -\beta & 0 \end{bmatrix}.$$

The characteristic equation is given by

$$\lambda^3 + \lambda^2 [1 + \alpha \{1 + f'(x)\}] + \lambda [\alpha f'(x) + \beta] + \alpha\beta \{1 + f'(x)\} = 0.$$

The equilibrium point is stable if the Routh–Hurwitz criterion is satisfied. That is,

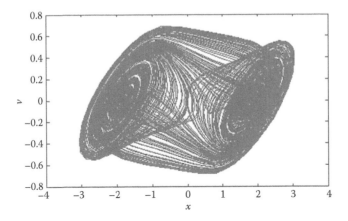

FIGURE 5.16
Double-scroll Chua's attractor.

$$[1 + \alpha \{1 + f'(x)\}] > 0, \; \alpha f'(x) + \beta > 0, \; \alpha\beta \{1 + f'(x)\} > 0,$$

and

$$[1 + \alpha \{1 + f'(x)\}][\alpha f'(x) + \beta] > \alpha\beta \{1 + f'(x)\}.$$

At the equilibrium point (0, 0, 0), the above conditions are not satisfied. The eigenvalues are 3.8478, −1.074 ± 3.0465i. The point (0, 0, 0) is unstable. For $x < −1$, the equilibrium point is $(k, 0, −k) = (−59/32, 0, 59/32)$. At this point, the fourth condition is not satisfied. The eigenvalues are −4.65967, 0.2298 ± 3.1873i. This equilibrium point is also unstable. For $x > 1$, the equilibrium point is $(−k, 0, k) = (59/32, 0, −59/32)$. We obtain the same characteristic equation and hence this equilibrium point is also unstable. Chua's circuit exhibits a typical chaotic attractor called the double-scroll Chua's attractor as given in Figure 5.16, for the parameter values $\alpha = 10$, $\beta = 14.87$, $a = −1.27$, and $b = −0.68$. Chua et al. [13] have plotted the figure for the parameter values $\alpha = 9.85$, $\beta = 14.3$, $a = −1/7$, and $b = 2/7$. Three unstable equilibrium states are visible in this attractor, which indicate that the double-scroll Chua's attractor is multistructural [33]. Kapitaniak [26] has also shown that (i) period-doubling bifurcations leading to chaos can be observed for $\beta = 16$, $a = −8/7$, $b = −5/7$, and for different values of control parameter α (see Figure 5.17), and (ii) intermittency can be observed for $\alpha = 4.295$, $\beta = 5$, $a = −1.27$, and $b = −0.68$ (see Figure 5.18).

Murali–Lakshmanan–Chua Circuit

The MLC circuit [33,46] is the simplest nonlinear dissipative second-order nonautonomous electronic circuit consisting of a forced *RLC* series circuit to

(a)   (b)

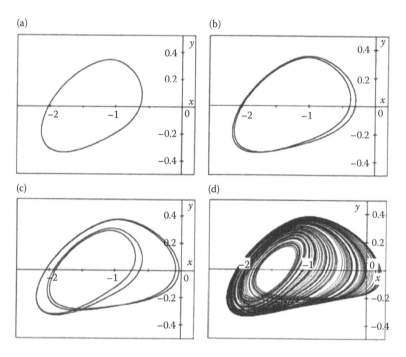

(c)   (d)

FIGURE 5.17

Period doubling bifurcation for (a) $\alpha = 8.8$, (b) $\alpha = 8.86$, (c) $\alpha = 9.12$, (d) $\alpha = 9.4$. (With kind permission from Springer Science+Business Media B.V.: *Chaos for Engineers: Theory, Applications, and Control*, Kapitaniak, T. Springer-Verlag, Germany, Copyright 2003, Springer.)

which the Chua's diode (N) is connected in parallel. The circuit was introduced by Murali, Lakshmanan, and Chua [46]. The circuit realization is given in Figure 5.19. The external periodic forcing is taken as $f(t) = F\sin(\omega t)$. A small current sensing resistor R_s was used. Applying the Kirchhoff's laws to the circuit, the governing equations for the voltage v across the capacitor C

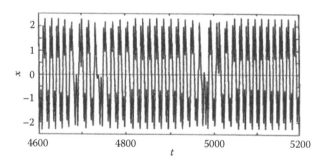

FIGURE 5.18

Intermittency in Chua's circuit. (With kind permission from Springer Science+Business Media B.V.: *Chaos for Engineers: Theory, Applications, and Control*, Kapitaniak, T. Springer-Verlag, Germany, Copyright 2003, Springer.)

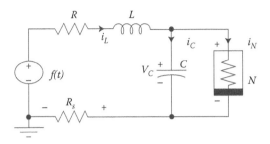

FIGURE 5.19
MLC circuit. (With kind permission from Springer Science+Business Media B.V.: *Nonlinear Dynamics: Integrability, Chaos and Pattern*, Lakshmanan, M., Rajasekar, S., Springer-Verlag, Berlin, Copyright 2003, Springer.)

and the current i_L through the inductor L are obtained as the two first-order coupled nonautonomous equations [47]:

$$C\frac{dv}{dt} = i_L - h(v), \quad L\frac{di_L}{dt} = -Ri_L - R_s i_L - v + F\sin(\omega t), \qquad (5.18)$$

where the characteristic function $h(v)$ is the piecewise linear function $f(x)$ defined in Equation 5.14. F and ω are the amplitude and angular frequency, respectively of the external periodic signal. First, consider the case when there is no external forcing, that is, $F = 0$. To nondimensionalize the equations in system (5.18), substitute $x = v/E_1, y = Ri_L/E_1, \omega^* = \omega RC, t = \tau CR$. We obtain

$$\frac{dx}{d\tau} = y - h(x), \quad \frac{dy}{d\tau} = -\beta[x + y(1 + v)] + f\sin(\omega^*\tau), \qquad (5.19)$$

where $v = R_s/R, \beta = CR^2/L, f = F\beta/E_1$, and $h(x)$ is $f(x)$ as given in Equation 5.17. When $f = 0$, the system has three equilibrium points. One equilibrium point is an unstable hyperbolic point and the remaining two points are stable focuses. In ref. [47], analytical solutions of the system were also derived in the three regions.

The detailed analysis of the nonlinear dynamics of the MLC circuit is given in Murali et al. [46,47] and Lakshmanan and Rajasekar [33]. For the values of the parameters $\beta = 1.0, v = 0.015, a = -1.02$, and $b = -0.55, \omega = 0.75$ with $f =$ (a) 0.065, (b) 0.08, (c) 0.091, (d) 0.1, and (e) 0.15 as the control parameter, the period-doubling route to chaos is given in Figure 5.20 (see Ref. [33]). Classification of bifurcations and routes to chaos in a variant of MLC circuit was given by Thamilmaran and Lakshmanan [68].

Chaos was observed in very simple circuits such as the nonideal operational amplifier with a single feedback resistor described by Yim et al. [78]). Chua et al. [12] numerically analyzed a circuit assuming a piecewise linear

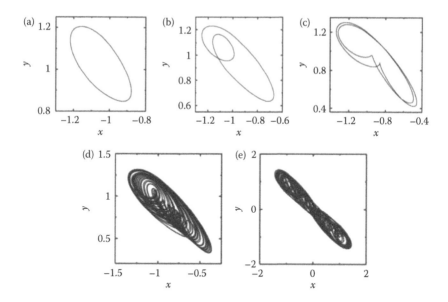

FIGURE 5.20
Period–doubling route to chaos. (With kind permission from Springer Science+Business Media B.V.: *Nonlinear Dynamics: Integrability, Chaos and Pattern*, Lakshmanan, M., Rajasekar, S., Springer-Verlag, Berlin, Copyright 2003, Springer.)

variation of inductance with current and showed that chaotic solution is expected. Chaotic electrical circuits are used for secure communications. Dean [14] experimentally observed chaos in a circuit when forced with a square wave rather than a sine wave. He assumed a piecewise linear variation of inductance with current and included hysteresis, which raises the dimension of the system by 1. A more realistic model without hysteresis was studied by Bartuccelli et al. [2].

A Third-Order Autonomous Chaotic Oscillator

Tamaševicius et al. [65] described a novel, simple third-order autonomous chaotic oscillator consisting of an operational amplifier, an *LCR* resonance loop, an extra capacitor C^*, a diode D as a nonlinear element, and three auxiliary resistors as given in Figure 5.21. Consider the three dynamical variables in the oscillator as the voltage across the capacitor C (V_C), the current through the inductor L (I_L), and the voltage across the capacitor C^* (V_{C^*}). I_L can be taken either as a voltage drop across the resistor R or as the output signal from the amplifier *OA*. Using the Kirchhoff's law, Tamaševicius et al. [65] derived the equations governing the dynamics of the oscillator as

$$C\frac{dV_C}{dt} = I_L, \quad L\frac{dI_L}{dt} = (k-1)RI_L - V_C - V_{C^*}, \quad C^*\frac{dV_{C^*}}{dt} = I_0 + I_L - I_D, \quad (5.20)$$

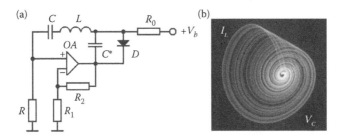

FIGURE 5.21
Circuit diagram of (a) chaotic oscillator, and (b) phase portrait (I_L vs. V_C). (From Tamaševicius, A. et al. A simple chaotic oscillator for educational purposes. *Eur. J. Phys.* 26, 61–63. Copyright 2005, IOP Publishing Ltd. Reprinted with permission.)

where $I_0 \approx V_b/R_0$, $R_0 \gg R\rho$, is a *dc* bias current. $I_D = f(V_{C^*})$ is a nonlinear current–voltage characteristic of the diode, $I_D = I_S[\exp(eV_{C^*}/k_B T) - 1]$, where I_S and I_D are the saturation current and the voltage across the diode, respectively, e is the electron charge, k_B is the Boltzmann constant, and T is the temperature.

To nondimensionalize Equations 5.20, define the variables and parameters as

$$x = \frac{V_C}{V_T}, y = \frac{\rho I_L}{V_T}, z = \frac{V_{C^*}}{V_T}, \theta = \frac{t}{\tau}, V_T = \frac{k_B T}{e}, \rho = \sqrt{L/C},$$

$$\tau = \sqrt{LC}, a = (k-1)\frac{R}{\rho}, b = \frac{\rho I_0}{V_T}, c = \frac{\rho I_S}{V_T}, \varepsilon = \frac{C^*}{C}.$$

Equation 5.20 simplify as

$$\frac{dx}{d\theta} = y, \quad \frac{dy}{d\theta} = ay - x - z, \quad \varepsilon\frac{dz}{d\theta} = b + y - c(e^z - 1). \tag{5.21}$$

The equilibrium point is in the x–z plane given by $[-\ln\{1 + (b/c)\}, 0, \ln\{1 + (b/c)\}]$. The characteristic equation of the Jacobian matrix is given by

$$\varepsilon\lambda^3 + \lambda^2 (e^z - a\varepsilon) + \lambda (\varepsilon + 1 - ae^z) + e^z = 0.$$

At the equilibrium point, we get the equation as

$$\varepsilon\lambda^3 + \lambda^2 (1 + s - a\varepsilon) + \lambda [\varepsilon + 1 - a(1 + s)] + 1 + s = 0, \text{ where } s = b/c > 0.$$

By the Routh–Hurwitz criterion, the equilibrium point is stable when

$$(1 + s - a\varepsilon) > 0, [\varepsilon + 1 - a(1 + s)] > 0, \text{ and } (1 + s - a\varepsilon)[\varepsilon + 1 - a(1 + s)] - \varepsilon(1 + s) > 0.$$

Simplifying the third condition, we get $(1 + s)[1 - a(1 + s) + a^2\varepsilon] - a\varepsilon(1 + \varepsilon) > 0$.

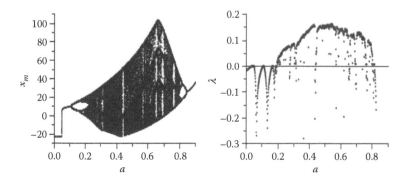

FIGURE 5.22

Bifurcation diagram and Lyapunov exponent. (From Tamaševicius, A. et al. 2005. A simple chaotic oscillator for educational purposes. *Eur. J. Phys.* 26, 61–63. Copyright 2005, IOP Publishing Ltd. Reprinted with permission.)

For the values of the parameters, $(a, b, c, \varepsilon) = (0.25, 30, 4 \times 10^{-9}, 0.13)$, we find that the equilibrium point is unstable. For the values of the control parameters $(b, c, \varepsilon) = (30, 4 \times 10^{-9}, 0.13)$, system (5.21) generates chaotic oscillations. The oscillator exhibits a period-doubling route to chaos. The phase portrait is given in Figure 5.21b. The bifurcation diagram and Lyapunov exponent are given in Figure 5.22. Positive values of the Lyapunov exponent confirm the chaotic behavior of the oscillator. The authors (Tamaševicius et al. [65]) had also developed an experimental prototype and exhibited the existence of a chaotic attractor.

Sprott's Chaotic Electrical Circuit

Sprott [59] described a new class of chaotic electrical circuits using only resistors, capacitors, diodes, and inverting operational amplifiers. This circuit solves the jerk equation of the form $\dddot{x} + A\ddot{x} + \dot{x} = G(x)$, where $G(x)$ belongs to a class of elementary piecewise linear functions. This class of chaotic electrical circuits is simple to construct, analyze, and scale to any desired frequency. Integration of each term of this equation reveals that it is a damped harmonic oscillator driven by a nonlinear memory term involving the integral of $G(x)$, [57]. Such an equation often arises in the feedback control of an oscillator in which the experimentally accessible variable is a transformed and integrated version of the fundamental dynamical variable [59]. Sprott used a general circuit as given in Figure 5.23, and considered various forms of $G(x)$ as $G(x) = |x| - 2.0$, $0.5 - 6 \max (x, 0)$, $1.2x - 4.5 \operatorname{sgn}(x)$, $-1.2x + 2 \operatorname{sgn}(x)$. Chaotic attractors produced by (i) $G(x) = |x| - 2.0$ and (ii) $G(x) = -1.2x + 2 \operatorname{sgn}(x)$ are presented in Figures 5.24a,b.

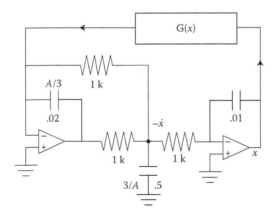

FIGURE 5.23
A general circuit for solving $\ddot{x} + A\ddot{x} + \dot{x} = G(x)$. (Reprinted from *Phys. Lett. A*, 266, Sprott, J. C., A new class of chaotic circuit, 19–23, Copyright 2000, with permission from Elsevier.)

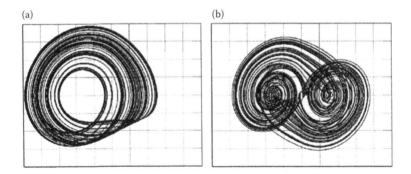

FIGURE 5.24
(a, b) Chaotic attractors obtained in cases (i), (ii). (Reprinted from *Phys. Lett. A*, 266, Sprott, J. C., A new class of chaotic circuit, 19–23, Copyright 2000, with permission from Elsevier.)

References

1. Alligood, K. T., Sauer, T. D., Yorke, J. A. 1996. *Chaos: An Introduction to Dynamical Systems*. Springer-Verlag, New York.
2. Bartuccelli, M. V., Jonathan, H. B. D., Gentile, G. 2007. Bifurcation phenomena and attractive periodic solution in the saturating inductor circuit. *Proc. R. Soc. A* 463, 2351–2369.
3. Bazoune, A. A. *ME 413*, Chapter 3. *Systems Dynam. Control* 1–34. www.open courseware.kfupm.edu.sa
4. Blackburn, J. A., Baker, G. L. 1998. A comparison of commercial chaotic pendulums. *Am. J. Phys.* 66(9), 821–829.

5. Bonatto, C., Gallas, J. A. C., Ueda, Y. 2008. Chaotic phase similarities and recurrences in a damped-driven Duffing oscillator. *Phys. Rev. E* 77, 026217-1-5.

6. Cartwright, J. H. E., Hernandez-Garcia, E., Piro, O. 1997. Burridge–Knopoff models as elastic excitable media. *Phys. Rev. Lett.* 79, 527–530.

7. Cartwright, M. L., Littlewood, J. E. 1945. On nonlinear differential equation of the second order. I. The equation $\ddot{x} - k(1 - y^2)\dot{x} + y = b\lambda k \cos(\lambda t + \alpha)$, k large. *J. Lond. Math. Soc.* 20, 180–189.

8. Chen, G. 1999. *Controlling Chaos and Bifurcations in Engineering System*. CRC Press, Boca Raton.

9. Chen, G., Ueta, T. (eds.) 2002. *Chaos in Circuits and Systems*. World Scientific, Singapore.

10. Chua, L. O. 1992. The genesis of Chua's circuit. *Arch. Elektron. Übertragungstech.* 46(4), 250–257.

11. Chua, L. O. 1994. Chua's circuit: Ten years later. *IEICE Trans. Fund. Electron. Comm. Comput. Sci.* E77-A, 1811–1822.

12. Chua, L. O., Hasler, M., Neirynck, I., Yerburgh, P. 1982. Dynamics of a piecewise-linear resonant circuit. *IEEE Trans.* CAS-29, 535–547.

13. Chua, L. O., Komuro, M., Matsumoto, T. 1986. The double scroll family. Parts I & II. *IEEE Trans.* CAS-33(11), 1073–1118.

14. Dean, J. H. B. 1994. Modelling the dynamics of nonlinear inductor circuits. *IEEE Trans. Magnet.* 30, 2795–2801.

15. DeSerio, R. 2003. Chaotic pendulum: The complete attractor. *Am. J. Phys.* 71(3), 250–256.

16. Dressier, U., Ritz, T., Schenck zu Schweinsberg, A. 1995. Tracking unstable periodic orbits in a bronze ribbon experiment. *Phys. Rev. E* 51(3), 1845–1848.

17. Duffing, G. 1918. *Erzwungene Schwingungen bei Veranderlicher Eigenfrequenz und ihre Technische bedeutung, sannlung Vieweg*, Heft 41/42. Vieweg, Braunschweig.

18. Franca, L. F. P., Savi, M. A. 2001. Distinguishing periodic and chaotic time series obtained from an experimental nonlinear pendulum. *Nonl. Dyn.* 26(3), 253–271.

19. Gottlieb, H. P. W. 1996. Question 38. What is the simplest jerk function that gives chaos? *Am. J. Phys.* 64, 525.

20. Hayashi, C., Ueda, Y., Akamatsu, N., Itakura, H. 1970. On the behavior of self-oscillatory systems with external force (in Japanese). *Trans. Inst. Electron. Comm. Eng. Jpn.* 53-A, 150–158.

21. Hegedorn, P. 1988. *Nonlinear Oscillations*. Oxford Sci. Publ., Oxford.

22. Holmes, P. 1979. A nonlinear oscillator with a strange attractor. *Philos. Trans. Roy. Soc. London A* 292, 419–448.

23. Huang, J., Jing, Z. 2009. Bifurcations and chaos in three-well Duffing system with one external forcing. *Chaos, Solitons Fractals* 40, 1449–1466.

24. Hübinger, B., Doerner, R., Martienssen, W., Herdering, M., Pitka, R., Dressier, U. 1994. Controlling chaos experimentally in systems exhibiting large effective Lyapunov exponents. *Phys. Rev. E* 50(2), 932–948.

25. Kandangath, A. K. 2004. *Inducing Chaos in Electronic Circuits by Resonant Perturbations*. Master thesis. Arizona State University, Tempe, AZ.

26. Kapitaniak, T. 2000. *Chaos for Engineers: Theory, Applications, and Control*. Springer-Verlag, Germany.

27. Kennedy, M. P. 1992. Robust op amp realization of Chua's circuit. *Frequenz* 46(3–4), 66–80.

28. Kennedy, M. P. 1993. Three steps to chaos-Part I: Evolution, Part II: A Chua's circuit primer. *IEEE Trans. Circuits Systems I: Fundam. Theory Appl.* 40(10), 640–674.

29. Knop, W., Lauterborn, W. 1990. Bifurcation structure of the classical Morse oscillator. *J. Chem. Phys.* 93, 3950–3958.

30. Kunick, A., Steeb, W.-H. 1987. Chaos in systems with limit cycle. *Int. J. Nonlinear Mech.* 22, 349–362.

31. Lai, Y.-Ch., Kandangath, A., Krishnamoorthy, S., Gaudet, J. A., Moura, P. S. de, 2005. Inducing chaos by resonant perturbations: Theory and experiment. *Phys. Rev. Lett.* 94, 214101-1-4.

32. Laje, R., Gardner, T., Mindlin, G. B. 2001. Continuous model for vocal fold oscillations to study the effect of feedback. *Phys. Rev. E* 64, 056201-1-7.

33. Lakshmanan, M., Rajasekar, S. 2003. *Nonlinear Dynamics: Integrability, Chaos and Pattern.* Springer-Verlag, Berlin.

34. Letellier, C., Valée, O. 2003. Analytical results and feedback circuit analysis for simple chaotic flows. *J. Phys. Math. Gen.* 36, 11229–11245.

35. Levinson, N. 1949. A second order differential equation with singular solution. *Ann. Math.* 50, 127–153.

36. Li, G. X., Moon, F. C. 1990. Criteria for chaos of a three-well potential oscillator with homoclinic and heteroclinic orbits. *J. Sound Vib.* 136(1), 17–34.

37. Linz, S. J. 1997. Nonlinear dynamical models and jerky motion. *Am. J. Phys.* 65, 523–526.

38. Longsuo, L. 2011. Suppressing chaos of Duffing–Holmes system using random phase. *Math. Prob. Eng.*, 2011, article 10, 538202–538210.

39. Madan, R. 1993. *Chua's Circuit: A Paradigm for Chaos.* World Scientific, Singapore.

40. Malasoma, J.-M. 2000. What is the simplest dissipative chaotic jerk equation which is parity invariant? *Phys. Lett. A* 264, 383–389.

41. Malasoma, J.-M. 2002. A new class of minimal chaotic flow. *Phys. Lett. A* 305, 52–58.

42. Matsumoto, T. 1984. A chaotic attractor from Chua's circuit. *IEEE Trans. CAS-*31(12), 1055–1058.

43. Moon, F. C. 1987. *Chaotic Vibrations.* Vol. 1. Wiley, New York.

44. Moon, F. C., Holmes, W. T. 1979. A magnetoelastic strange attractor. *J. Sound Vib.* 65, 275–296.

45. Moore, D. W., Spiegel, E. A. 1966. A thermally excited nonlinear oscillator. *J. Astrophys.* 143, 871–887.

46. Murali, K., Lakshmanan, M., Chua, L. O. 1994a. The simplest dissipative non-autonomous chaotic circuit. *IEEE Trans. Circuits Systems* I41, 462–463.

47. Murali, K., Lakshmanan, M., Chua, L. O. 1994b. Bifurcation and chaos in the simplest dissipative non-autonomous circuit. *Int. J. Bif. Chaos* 4, 1511–1524.

48. Ott, E., Grebogi, C., Yorke, J. A. 1990. Controlling chaos. *Phys. Rev. Lett.* 64(11), 1196–1199.

49. Passos, D., Lopez, I. 2008. Phase space analysis: The equilibrium of the solar magnetic cycle. *Solar Phys.* 250, 403–410.

50. Paula, A. S. de, Savi, M. A., Pereira-Pinto, F. H. I. 2006. Chaos and transient chaos in an experimental nonlinear pendulum. *J. Sound Vib.* 294, 585–595.

51. Pereira-Pinto, F. H. I., Ferreira, A. M. 2005. State space reconstruction using extended state observers to control chaos in a nonlinear pendulum. *Int. J. Bif. Chaos* 15(12), 4051–4063.

52. Richards, R. 1999. The subtle attraction: Beauty as a force in awareness, creativity, and survival. In: *Affect, Creative Experience, and Psychological Adjustment*, S. W. Russ, ed., pp. 195–219, Brunner/Mazel, Philadelphia.

53. Sabatini, M. 2004. On the periodic function of $\ddot{x} + f(x)\dot{x}^2 + g(x) = 0$. *J. Diff. Eq.* 196, 151–168.

54. Savi, M. A., Pacheco, P. M. C. L. 2002. Chaos in a two-degree of freedom Duffing oscillator. *J. Braz. Soc. Mech. Sci.* 24(2), 115–121.

55. Schot, S. H. 1978. The time rate of change of acceleration. *Am. J. Phys.* 46, 1090–1094.

56. Silva, C. P., Young, A. M. 1998. Implementing RF broadband chaotic oscillators: Design issues and results. *Proc. IEEE Int. Symp. Circuits and Systems* 4, 489 – 493, IEEE, Piscataway, NJ.

57. Silva, C. P., Young, A. M. 2000. *High frequency anharmonic oscillator for the generation of broadband deterministic noise.* U.S. Patent No. 6, 127, 899.

58. Sprott, J. C. 1997. Some simple chaotic jerk functions. *Am. J. Phys.* 65, 537–543.

59. Sprott, J. C. 2000. A new class of chaotic circuit. *Phys. Lett.* A 266, 19–23.

60. Sprott, J. C. 2003. *Chaos and Time Series Analysis.* Oxford University Press, Oxford.

61. Sprott, J. C. 2005. Dynamical models of happiness. *Nonl. Dyn. Psychol. Life Sci.* 9, 23–36.

62. Sprott, J. C. 2010. *Elegant Chaos.* World Scientific, USA.

63. Sun, K., Sprott, J. C. 2009. A simple jerk system with piecewise exponential nonlinearity. *Intl. J. Nonl. Sci. Num. Simul.* 10(11–12), 1443–1450.

64. Sun, K., Sprott, J. C. 2010. Periodically forced chaotic system with signum nonlinearity. *Intl. J. Bif. Chaos* 20(5), 1499–1507.

65. Tamaševicius, A., Mykolaitis, G., Pyragas, V., Pyragas, K. 2005. A simple chaotic oscillator for educational purposes. *Eur. J. Phys.* 26, 61–63.

66. Tamaševicius, A., Tamaseviciute, E., Mykolaitis, G., Bumelien'e, S. 2007. Stabilization of unstable periodic orbit in chaotic Duffing–Holmes oscillator by second order resonant negative feedback. *Lith. J. Phys.* 47, 235–239.

67. Tamasevieiute, E., Tamasevicius, A., Mykolaitis, G., Bumeliene, S., Lindberg, E. 2008. Analogue, electrical circuit for simulation of Duffing–Holmes equation. *Nonlinear Anal: Model. Control* 13(2), 241–252.

68. Thamilmaran, K., Lakshmanan, M. 2002. Classification of bifurcations and routes to chaos in a variant of MLC circuit. *Int. J. Bif. Chaos* 12(4), 783–813.

69. Ueda, Y. 1979. Randomly transitional phenomena in the system governed by Duffing's equation. *J. Stat. Phys.* 20, 181–196.

70. Ueda, Y. 2001. *The Road to Chaos* (2nd edn). Aerial Press, Santa Cruz, CA.

71. van der Pol, B. 1920. A theory of the amplitude of free and forced triode vibrations. *Radio Rev.* 1, 701–710, 754–762.

72. van der Pol, B. 1926. On relaxation oscillations. *Philos. Mag. Ser.* 7(2), 978–992.

73. van der Pol, B., van der Mark, J. 1928. The heartbeat considered as a relaxation oscillation, and the electrical model of the heart. *Philos. Mag. Ser.* 76, 763–765.

74. Vasegh, N., Khellat, F. 2009. Periodic, quasi-periodic and chaotic motion in mass–spring model. *PhysCon-2009*, Catania, Italy, pp. 1–4.

75. Virgin, L. N. 2000. *Introduction to Experimental Nonlinear Dynamics: A Case Study in Mechanical Vibration.* Cambridge University Press, Cambridge.

76. Weinert, K., Webber, O., Husken, M., Theis, W. 2002. Analysis and prediction of dynamic disturbances of the BTA deep hole drilling process. In: *Proc. of the Third CIRP International Seminar on Intelligent Computation in Manufacturing Engineering*, R. Teli, ed., ICME 2002, Ischia, Italy, pp. 297–302.

77. Wolf, A., Swift, J. B., Swinney, L., Vastano, J. A. 1985. Determining Lyapunov exponents from a time series. *Physica D* 16, 285–317.

78. Yim, G., Ryu, J., Park, Y. 2004. Chaotic behaviors of operational amplifiers. *Phys. Rev. E* 69, 045201-1-4.

Solutions to Odd-Numbered Problems

Chapter 1

Exercise 1.1

1. Cost function = $\{30(l^3 + 10)/l\}$, l = width of the box. Cost function is minimum when $l^3 = 5$. Minimum cost = $90(5^{2/3})\$$.

3. 17.5 min (total of 27.5 min).

5. Asymptotically stable.

7. Asymptotically stable.

9. Origin $(0, 0)$ is asymptotically stable.

11. Asymptotically stable.

13. The roots of $P(\lambda) = 0$ are negative or have negative real parts.

15. For $0 < p < (2/3)$, the roots of $P(\lambda) = 0$ are negative or have negative real parts.

17. Zero solution is asymptotically stable.

19. Zero solution is asymptotically stable.

21. Unstable.

23. Study of the nonlinear terms is required to decide whether C is semistable.

25. No closed orbits.

27. Unique limit cycle.

29. Origin $(0, 0)$ is an unstable focus.

31. Origin is a stable focus.

33. Dissipative for $|x| > 1$. Area expanding for $|x| < 1$.

35. Dissipative for $0 < x < 5.5$. Area expanding for $x > 5.5$.

37. Dissipative when $8ab < 1$, and area expanding when $8ab > 1$.

39. Hamiltonian system. $(0, 0)$ is a saddle point.

41. Hamiltonian system. $(0, 0)$, $(-4, \pm 2/\sqrt{3})$ are saddle points.

43. % MATLAB® code for Lotka–Volterra model in Problem 43

```
g=inline('[1.*x(1)-(0.2.*x(1).*x(2));0.5.*x(2)+(0.04.*x(1).*x(2))]','t','x');
[t xa]=ode45(g,[0 150],[10 10])
plot(xa(:,1),xa(:,2))
%plot3(xa(:,1),xa(:,2),xa(:,3));
```

```
%comet3(xa(:,1),xa(:,2),xa(:,3));
hold on;
figure;
plot(t,xa(:,1),'r');
hold on;
plot(t,xa(:,2),'g');
```

The phase plot and time series are given in Figures 1.1 and 1.2.

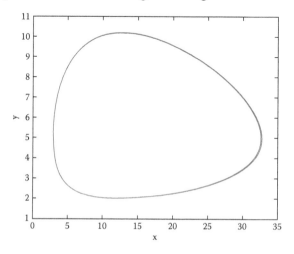

FIGURE 1.1
Phase plot for Problem 43.

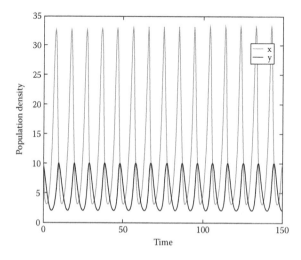

FIGURE 1.2
Time series for Problem 43.

45. % MATLAB code

```
g=inline('[2.5.*u(1)−0.05.*u(1).*u(1)−(0.85.*u(1).*u(2))./(0.45 + 0.2.*
u(2) + 0.6.*u(1));
−0.95.*u(2)+(1.65.*u(1).*u(2))/(0.45+0.2.*u(2) + 0.6.*u(1))]','t','u');
[t ua]=ode45(g,[400 600],[5.9157 35.6279])
figure;
plot(t,ua(:,1),'b');
hold on;
plot(t,ua(:,2),'r');
xlabel('time');
ylabel('state variable');
legend('x','y');
figure;
plot(ua(:,1),ua(:,2));
xlabel('x');
ylabel('y');
```

The phase plot is given in Figure 1.3 and the times series is plotted in Figure 1.4.

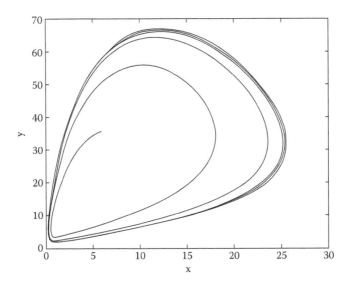

FIGURE 1.3
Phase plot for Problem 45.

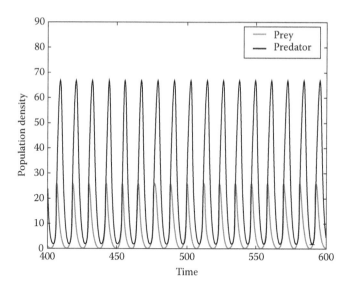

FIGURE 1.4
Time series for Problem 45.

Chapter 2

Exercise 2.1

1. (i) $P_1 = 0$ is unstable and $P_2 = 500$ is stable.
   (ii) $P(t) = 500/[1 + 9e^{-0.5t}]$.
3. 3.36 P.M (approximately).
5. Asymptotically stable for all r.
7. Asymptotically stable for $0 < r < 2$. Maximum value is $\exp{(r-1)}/r$.
9. 2454 students.
11. $P_1 = 3 - \sqrt{9-h}$, $P_2 = 3 + \sqrt{9-h}$, $h < 9$. P_1 is unstable and P_2 is asymptotically stable. $P_0 > 3$.

Exercise 2.2

1. $X^* = 212/3$, $Y^* = 2332/15$. (X^*, Y^*) is asymptotically stable.
3. $w_1 > a_2$, $K(w_1 - a_2) > D_1 a_2$.
5. $(0, 0)$ is a hyperbolic saddle point. $(1, 0)$ is a saddle point. Sufficient conditions for the equilibrium point (u^*, v^*) to be asymptotically stable are $v^* < (h_p + u^*)^2$, and $(v^*)^2 < h_z^2$. If the above conditions are

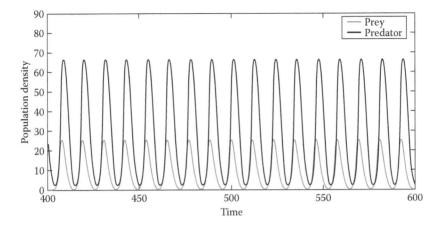

FIGURE 2.1
Time-series displaying oscillatory dynamics in the model (2.71) and (2.72) for Problem 7.

satisfied, then both predator and prey species coexist, and they settle down at its equilibrium point.

7. $w > a_1\beta$, $w_1 > a_2\gamma$ and $(a_1/b_1) > (a_2\alpha/(w_1 - a_2\gamma))$. An oscillatory predator–prey dynamics (time series) exhibited by the model system for the given set of parameter values, is presented in Figure 2.1.

9. The system has <u>four</u> equilibrium points. They are $(0, 0)$, $(K, 0)$ and
$u^* = [-S \pm \sqrt{S^2 - 4\alpha}]/2$, $S = \alpha [1 - (\beta/\gamma)]$, $v^* = (\beta/\gamma K)[K - u^*]u^*$.
The nontrivial solutions exist if $S^2 > 4\alpha$, and $u^* < K$, that is if $[1 - (\beta/\gamma)]^2 > (4/\alpha)$, and $u^* < K$. If $S < 0$, that is $\beta > \gamma$, we obtain two positive equilibrium points.

11. Equilibrium solution is $(u^*, v^*) = (0.1511, 0.9836)$. Phase plot and time series is given in Figure 2.2.

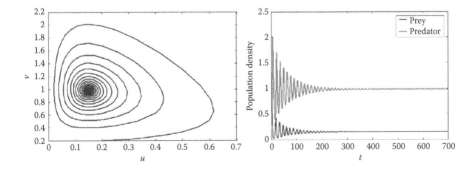

FIGURE 2.2
Phase plot and time series for $K = 1$, $\alpha = 3$, $\beta = 2.3$, and $\gamma = 0.3$.

13. Condition (viii) is violated. We obtain $G(C^*, 0) = c \neq 0$.

15. $(K, 0)$ and (Z^*, U^*), where $2Z^* = [-(p - K) + \sqrt{(p - K)^2 + 4q}]$, $U^* = cZ^*/w_4$.

17. The equilibrium point is $X_1^* = (K_1 - K_2 b_1)/(1 - b_1 b_2)$, $X_2^* = (K_2 - K_1 b_2)/(1 - b_1 b_2)$. The positive equilibrium point is asymptotically stable when $b_1 < (K_1/K_2) < (1/b_2)$.

19. For $r = 1.5$, $\alpha = 3$, we get $(1/3) < u^* < 1$. $u^* = 0.48005$, $v^* = r(1 - u^*) = 0.77993$.

The equilibrium point is asymptotically stable.

Chapter 3

Exercise 3.1

1. MATLAB 7.0 is used to compute the phase plane diagram to generate the chaotic attractor and time series. Chaotic attractor and the temporal evolution for (i) t versus x, (ii) t versus y, (iii) t versus z are plotted in Figure 3.1a–d.

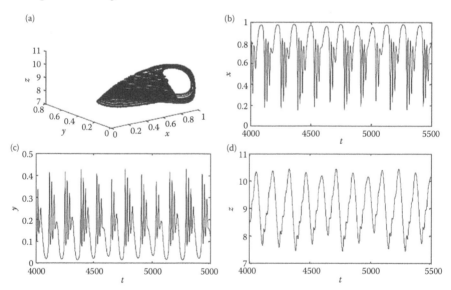

FIGURE 3.1
(a) Chaotic attractor, (b) Temporal evolution (t vs. x), (c) Temporal evolution (t vs. y), and (d) Temporal evolution (t vs. z), in Problem 1. (Reprinted from *Chaos, Solitons Fractals*, 40, Upadhyay, R. K., Rai, V., Complex dynamics and synchronization in two non-identical chaotic ecological systems, 2233–2241, Copyright 2009, with permission from Elsevier.)

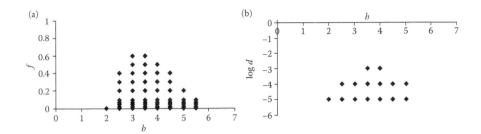

FIGURE 3.2
Points in the 2D parameter spaces (*a*) (*b*, *f*), (b) (*b*, log *d*) for Problem 3. (Reprinted from *Chaos, Solitons Fractals*, 42(1), Upadhyay, R. K., Kumari, N., Rai, V., Exploring dynamical complexity in diffusion driven predator–prey systems: Effect of toxin production by phytoplankton and spatial heterogeneities, 584–594, Copyright 2009, with permission from Elsevier.)

3. MATLAB 7.0 is used to compute the discrete points at which chaos was observed and Microsoft Excel 2007 to draw the 2D scan diagram. All the points in the two-dimensional parameter spaces where the model system exhibits chaotic dynamics are shown in Figure 3.2a and b.

5. *BAS* routine from *Dynamics: Numerical Explorations* software is used to compute the basin boundary structure. Basin boundary structure is plotted in Figure 3.3.

7. MATLAB 7.0 is used to generate the bifurcation diagrams. The successive maxima of *y* as a function of w_5 for the given parameter

FIGURE 3.3
Basin boundary structure for the model given in Problem 5. (Reprinted from *Chaos, Solitons Fractals*, 39, Upadhyay, R. K., Rao, V. S. H., Short term recurrent chaos and role of toxin producing phytoplankton on chaotic dynamics in aquatic systems, 1550–1564, Copyright 2009, with permission from Elsevier.)

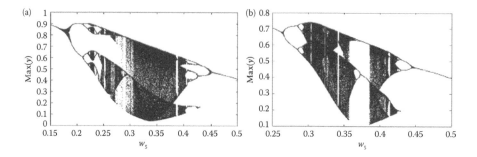

FIGURE 3.4
Bifurcation diagram of the model system for Problem 7. (Upadhyay, R. K., Raw, S. N., Rai, V. 2010. Dynamical complexities in a tri-trophic hybrid food chain model with Holling type II and Crawley–Martin functional responses. *Nonlinear Anal.: Model. Control* 15(3), 361–375. Copyright © 2010, Lithuanian Association of Nonlinear Analysts, (LANA). Reprinted with permission.)

values with $w_{11} = 0.03$ and $0.15 < w_5 < 0.5$, are plotted in Figure 3.4a; and with $w_{11} = 0.06$ and $0.25 < w_5 < 0.5$, are plotted in Figure 3.4b.

9. The fixed points are $(0, 0)$, and $x^* = x_{1,2} = [1 \pm \sqrt{1 - 36a^2\mu^2}]/(2a\mu)$. The critical value is $\mu_c = 1/6a$. At the bifurcation, the value of the fixed point is $x^* = 3$. $(0, 0)$ is always a stable fixed point. For $0 < x^* < 3$, the fixed point is a saddle point. For $x^* > 3$, the fixed point is a stable node.

11. i. The fixed points are $x = 0$ and $x = \pm\sqrt{\mu}$. $x = 0$ is unstable and the other two fixed points are stable giving rise to supercritical pitchfork bifurcation.

    ii. The fixed points are $x = 0$ and $x = \pm\sqrt{-\mu}$. For $\mu < 0$, the fixed point $x = 0$ is stable and $x = \pm\sqrt{-\mu}$ are unstable. The stationary solution (node) becomes an unstable saddle (a saddle-node, a saddle-focus), and together with it, the other two unstable stationary solutions disappear. The bifurcation is a crisis giving rise to subcritical pitchfork bifurcation.

13. The fixed points are $x = 0$ and $x = \mu/2$. An exchange of stability occurs at $\mu = 0$ with respect to both the fixed points. Transcritical bifurcation occurs at $(x, \mu) = (0, 0)$.

15. The fixed point $x_0 = [-1 + \sqrt{1 + 4\mu}]/2 > 0$ is called a period-1 cycle. It is stable for $\mu \in (0, 3/4]$. At $\mu = \mu_1 = 3/4$, the first period-doubling bifurcation takes place giving rise to a stable period-2 cycle. The 2-cycle is stable in the interval $\mu \in [3/4, 5/4]$. At $\mu = \mu_2 = 5/4$, the second period-doubling bifurcation takes place giving rise to a stable period-2^2 cycle.

17. The conditions to be satisfied are

$$\frac{w_2 \gamma_2 \, Z^*}{(\alpha_2 + \beta_2 Z^* + \gamma_2 Y^*)^2} > \frac{w_1 \beta \, X^*}{(\alpha + \beta Y^* + \gamma X^*)^2}.$$

$$\frac{Y^*}{X^*}\left[\frac{(w\gamma - w_1\beta)X^*}{(\alpha + \beta Y^* + \gamma X^*)^2} + \frac{w\gamma_2 Z^*}{(\alpha_2 + \beta_2 Z^* + \gamma_2 Y^*)^2}\right] < b_1, \quad \text{and}$$

$$\left(b_1 - \frac{\gamma \, wY^*}{(\alpha + \beta Y^* + \gamma X^*)^2}\right)\left(\frac{w_2 \gamma_2 Z^*}{(\alpha_2 + \beta_2 Z^* + \gamma_2 Y^*)^2} - \frac{w_1 \beta \, X^*}{(\alpha + \beta Y^* + \gamma X^*)^2}\right)$$

$$< \frac{ww_1(\alpha + \gamma \, X^*)(\alpha + \beta \, Y^*)}{(\alpha + \beta Y^* + \gamma X^*)^4}.$$

19. $f'(x) = 2$, and $\lambda = \log 2$.

Chapter 4

Exercise 4.1

1. The positive equilibrium point E^* is locally asymptotically stable in the presence of diffusion if the sufficient conditions, $abz^* < r_x(1 + bp^*)^2$, and $g^2 z^{*2} < 1$, are satisfied. There exist parameter values and a range for d, for which the system may be unstable.

3. Linearize the system about the steady state (P^*, H^*). Assume the solution of this system in the form $U(x, t) = se^{\lambda t} \cos (kx)$, $V(x, t) = we^{\lambda t} \sin (kx)$, where k takes the discrete values, $k = (n\pi/L)$, n is an integer, and λ is the eigenvalue determining the temporal growth (frequency). Equilibrium point for the given set of parameter values is $P^* = 1.184922937$, and $H^* = 1.205819205$. In the absence of diffusion, the system is unstable. Let, $d_1 = 1$, $d_2 = 8$. For all wave numbers $k^2 > 0.134231$, the system is stable under the action of diffusion. For, $0.09973 < k^2 < 0.13423$, diffusive instability sets in.

5. The system has four equilibrium points $E_0(0, 0, 0)$, $E_1(a_1/b_1, 0, 0)$, $E_2(\hat{x}, \hat{y}, 0)$, and $E_3(x^*, y^*, z^*)$. For E_3 to exist, we require $w_3 > cD_3$, $w_1 x^* > (x^* + D_1)(a_2 + \theta f(x^*))$. $E_0(0, 0, 0)$, $E_1(a_1/b_1, 0, 0)$ are nonhyperbolic fixed points. For the given set of parameter values, the characteristic equation of the Jacobian matrix for $E_3(x^*, y^*, z^*)$ has one negative root -1.197, and a complex pair with positive real parts. The equilibrium point is unstable.

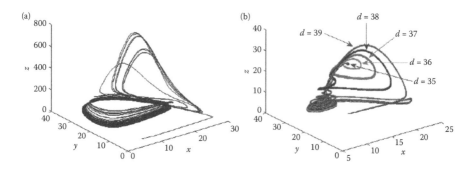

FIGURE 4.1
(a) Chaotic attractor for the first set of parameter values. (b) Stable limit cycles for the second set of parameter values. (Reprinted from *Appl. Math. Comp.*, 217, Upadhyay, R. K., Rai, V., Raw, S. N., Challenges of living in the harsh environments: A mathematical modeling study, 10105–10117, Copyright 2011, with permission from Elsevier.)

7. The system has four equilibrium points $E_0(0, 0, 0)$, $E_1(a_1/b_1, 0, 0)$, $E_2(\hat{x}, \hat{y}, 0)$, and $E_3(x^*, y^*, z^*)$. For E_3 to exist, we require the conditions $cd < w_3$, $y^* < a_1 a/w$, and $w_1 x^* > a_2 T_1(x^*)$, where $T_1(x^*) = (x^{*2}/i_1) + x^* + a$. $E_0(0, 0, 0)$, $E_1(a_1/b_1, 0, 0)$ are nonhyperbolic fixed points. (i) We get one positive equilibrium point (22.718741, 23.148148, 70.790948). The characteristic equation has one negative root −1.14881, and the other roots are 0.104077, 0.00649. The equilibrium point is unstable. (ii) We get one positive equilibrium point (15.812703, 25.384615, 22.974668). The characteristic equation has one negative root −0.47692, and the complex pair −0.027594 ± 0.01064i. The equilibrium point is asymptotically stable. The chaotic attractor for the first set of parameter values is given in Figure 4.1a. Stable limit cycles for the second set of parameter values are given in Figure 4.1b.

Index